Ian S. Butler (Ed.)
High-pressure Molecular Spectroscopy

Also of interest

Micro-Raman Spectroscopy.
Theory and Application
Popp, Mayerhöfer (Eds.), *2020*
ISBN 978-3-11-051479-7, e-ISBN (PDF) 978-3-11-051531-2

Quantum Crystallography
Massa, Matta, Huang, 2022
ISBN 978-3-11-056566-9, e-ISBN (PDF) 978-3-11-056667-3

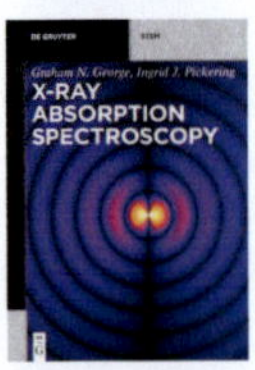

X-ray Absorption Spectroscopy
George, Pickering, 2023
ISBN 978-3-11-057037-3, e-ISBN (PDF) 978-3-11-057044-1

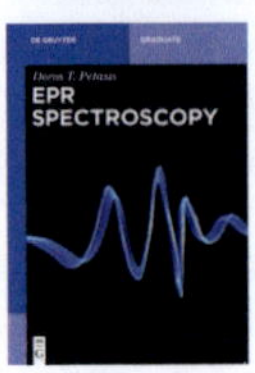

EPR Spectroscopy
Petasis, Hendrich, 2022
ISBN 978-3-11-041753-1, e-ISBN (PDF) 978-3-11-041756-2

Instrumental Analysis.
Chemical IT
Schlemmer G, Schlemmer J, 2022
ISBN 978-3-11-068964-8, e-ISBN (PDF) 978-3-11-068966-2

High-pressure Molecular Spectroscopy

Edited by
Ian S. Butler

DE GRUYTER

Editor
Prof. em. Ian S. Butler
Department of Chemistry
McGill University
801 Sherbrooke St. W.
Montreal
QC
Canada H3A 0B8

ISBN 978-3-11-066528-4
e-ISBN (PDF) 978-3-11-066861-2
e-ISBN (EPUB) 978-3-11-066544-4

Library of Congress Control Number: 2022940992

Bibliographic information published by the Deutsche Nationalbibliothek
The Deutsche Nationalbibliothek lists this publication in the Deutsche Nationalbibliografie; detailed bibliographic data are available on the Internet at http://dnb.dnb.de.

Cover image: Science Photo Library/Alexander, Max
Typesetting: Integra Software Services Pvt. Ltd.
Printing and binding: CPI books GmbH, Leck

www.degruyter.com

Preface

Pressure is a variable that encompasses approximately 12 orders of magnitude (10^{-6} to 10^{6} atm), and a variety of specialized equipment has been designed over the years to investigate all kinds of materials throughout most of this pressure range. The purpose of this monograph is to expose the reader to a selection of high-pressure studies that have been undertaken recently using various molecular spectroscopy techniques. There are eight chapters in all, the first three of which are focused on applications involving infrared and Raman spectroscopy. Initially, Dr. Yang Song from Western University in Ontario, Canada, describes some of his research on pressure-mediated hydrogen bonding, pressure-induced photochemical reactions, and porous materials and guest–host interactions. Then, Dr. Ian Butler, at McGill University in Quebec, Canada, describes his diamond-anvil vibrational research on solid materials, especially organometallic complexes. Next, Dr. Janice Musfeldt at the University of Tennessee in the USA presents some of her research on molecule-based multiferroics. This chapter is followed by one from Dr. Alexander Goncharov at the Carnegie Institution for Science in Washington, DC, USA, on recent experimental and theoretical investigations of simple molecules at very high pressures approaching and reaching molecular dissociation, and transformations to polymeric and monoatomic structures. In the next chapter, Dr. Helen Walker at the Rutherford Appleton Laboratory in Harwell, UK, describes her research on inelastic neutron scattering under pressure. Another Canadian scientist Dr. John Tse at the University of Saskatchewan in Canada and Dr. Ercan Alp at the Argonne National Laboratory in Illinois, USA, present a report on the application of inelastic scattering of X-rays at high pressure. Moving on to Brazil, Dr. Naira Balzaretti at the Universidade Federal do Rio Grande do Sul in Porto Alegre describes her luminescence work on lanthanide-doped glasses under high pressure. In the final chapter, Dr. Christian Reber at the University of Montreal in Quebec, Canada, introduces us to the application of variable-pressure fluorescence measurements on six-coordinate d-block compounds. One of the rationales for this particular monograph was that the field is steadily growing. A brief survey, using SciFinder, for the two concepts *high-pressure spectroscopy* and *high-pressure molecular spectroscopy* revealed that there are over 19,000 and 1,100 citations, respectively. We trust that the reader will enjoy our collection of different high-pressure studies and consequently will be stimulated to undertake further research in this fascinating area.

Ian S. Butler
Professor Emeritus of Chemistry
McGill University, Montreal
QC, Canada

https://doi.org/10.1515/9783110668612-202

Contents

List of Contributors

Christian Reber
Département de chimie
Université de Montréal
Canada
christian.reber@umontreal.ca

Chahinez Dab
Département de chimie
Université de Montréal
Canada

Pierre-François Blanc
Département de chimie
Université de Montréal
Canada

Janice L. Musfeldt
Departments of Chemistry and Physics
University of Tennessee
Knoxville, TN 37996, USA

Dr Helen Walker
MAPS Instrument Scientist
ISIS Neutron and Muon Source
Rutherford Appleton Laboratory
Harwell, Oxfordshire OX11 0QX
UK

Ian S. Butler
Department of Chemistry
Otto Maass Building, Room 002
McGill University
801 Sherbrooke Street West
Montreal, QC
Canada H3A 0B8

John S. Tse
Fellow of Royal Society, Canada
Distinguished Professor
Centennial Enhancement Chair
Department of Physics and Engineering Physics
University of Saskatchewan Saskatoon
Saskatchewan, SK
Canada S7N 5E2

E. Ercan Alp
Advanced Photon Source
Argonne National Laboratory
Lemont, IL 60439
USA

Alexander Goncharov
Earth and Planets Laboratory
Carnegie Institution for Science
5241 Broad Branch Rd. NW
Washington, DC 20015
USA

Yang Song
Department of Chemistry
Western University
1151 Richmond Street
London
ON
Canada N6A 5B7

Israel Roger Montoya Matos
Department of Civil Engineering
University of Lima
Lima, Peru

Alvaro Carrillo Herrera
Department of Physics
Universidad de Antioquia
Medellin, Colombia

Silvio Buchner
Department of Physics
Universidade Federal do Rio Grande do Sul
Porto Alegre, Brazil

Naira Maria Balzaretti
Department of Physics
Universidade Federal do Rio Grande do Sul
Porto Alegre, Brazil

https://doi.org/10.1515/9783110668612-204

Yang Song

Characterization of pressure-mediated molecular evolutions in confined space probed by *in situ* vibrational spectroscopy

1 Introduction

As a fundamental thermodynamic parameter, pressure can profoundly change the interatomic distances, the nature of intermolecular interaction, chemical bonding, molecular configuration, and crystal structure, resulting in the formation of exotic structures and characteristic properties of new materials. Extreme pressure can even induce transformations involving the strongest chemical interactions that exceed 10 eV (965 kJ mol^{-1}) such that the chemical bonds and even the well-known properties of atoms and molecules can be completely changed. Correspondingly, understanding of pressure-induced structural transformations and formation of novel functional materials by using a wide range of experimental and theoretical techniques is of fundamental importance. Especially on the experimental side, major advances in high-pressure characterization techniques accommodating diamond anvil cell (DAC) as the centerpiece of high-pressure apparatus have allowed the study of molecules and materials in an unprecedented pressure-temperature range. For instance, the compatible micro-spectroscopic probes have made possible the characterization of structures and transformational processes *in situ* with great spectral and spatial resolution. The recent advances in high-pressure science and technology, and their applications in materials research have been provided in several excellent review articles or book chapters [1–12].

The prevailing structural characterization techniques include predominantly synchrotron radiation enabled as well as local laboratory-based tools involving spectroscopy, microscopy, scattering, and imaging as well as other advanced techniques such as time-resolved *in situ* and *in operando*. Among all the available *in situ* structural characterization probes for materials under extreme conditions, in particular,

Acknowledgments: This work was supported by a Discovery Grant, a Research Tools and Instruments Grant from the Natural Science and Engineering Research Council of Canada, a Leaders Opportunity Fund from the Canadian Foundation for Innovation, an Early Researcher Award from the Ontario Ministry of Research and Innovation, a Petro-Canada Young Innovator Award, and by Defense Research and Development Canada under contract No. W7702-135601. The synchrotron IR measurements presented were performed at U2A beamline of National Synchrotron Light Source of US Brookhaven National Lab.

Yang Song, Department of Chemistry, Department of Physics and Astronomy, Western University, London, Ontario N6A 5B7, Canada, e-mail: yang.song@uwo.ca

https://doi.org/10.1515/9783110668612-001

Raman and Infrared (IR) spectroscopy has demonstrated strong sensitivity and accuracy as well as efficiency in monitoring the pressure-induced transformations at molecular level in a wide range of materials. Raman and IR spectroscopy, complementary to each other, provide valuable information on molecular structures, nature of bonding, lattice dynamics, and intermolecular interactions. In this chapter, a comprehensive and critical review of examples of pressure-induced molecular transformations in a wide variety of molecules and materials probed by vibrational spectroscopy will be provided. The examples include 1) pressure-mediated hydrogen bonding; 2) pressure-induced photochemical reactions; and 3) porous materials and guest–host interactions. Using author-selected case studies in these three themes, this chapter will provide readers with the most recent advances in high-pressure chemistry and materials research by demonstrating that vibrational spectroscopy is an indispensable and sensitive tool in these studies.

2 Experimental methods

2.1 The diamond-anvil cell

Invented and developed by Percy Bridgman who was awarded the 1946 Nobel Prize in Physics, diamond-anvil cell has been used as a prevailing and fundamental apparatus to achieve static high pressures in the gigapascal (GPa) range. Due to the supervisor hardness, diamonds are the most suitable as anvil materials to generate very high pressure up to several hundreds of GPa. In addition, diamonds are transparent to a wide spectral range of electromagnetic radiation from far-IR to hard X-ray. Therefore, structural characterization of material loaded in DACs can be achieved using various analytical probes, including optical spectroscopy, synchrotron, and neutron sources with unprecedented spatial, temporal, and spectral resolutions.

A typical DAC apparatus (Figure 1) consists of two brilliant cut diamonds as anvils to exert static pressure up to several million atmospheres (or several hundred GPa) using only moderate force delivered by spring washer screws. A metal gasket with a hole drilled at the center serves as the sample chamber. Ideally, the sample to be studied is loaded together with pressure-transmitting medium (PTM) to enhance the hydrostaticity, and a ruby chip which is used for pressure calibration. The extreme pressures can be accurately determined by monitoring using the well-established ruby fluorescence scale [13]:

$$P = \frac{1904}{B}\left[\left(1 + \frac{\Delta\lambda}{694.24}\right)^{B} - 1\right] \tag{1}$$

where P is the pressure in GPa, $\Delta\lambda$ is the ruby R_1 line shift in nm, and parameter B is 7.665 for quasi-hydrostatic conditions and 5 is for non-hydrostatic conditions. The ruby fluorescence can be conveniently collected using a Raman system or a separate portable and compact ruby fluorescence system available from commercial sources or custom house-build.

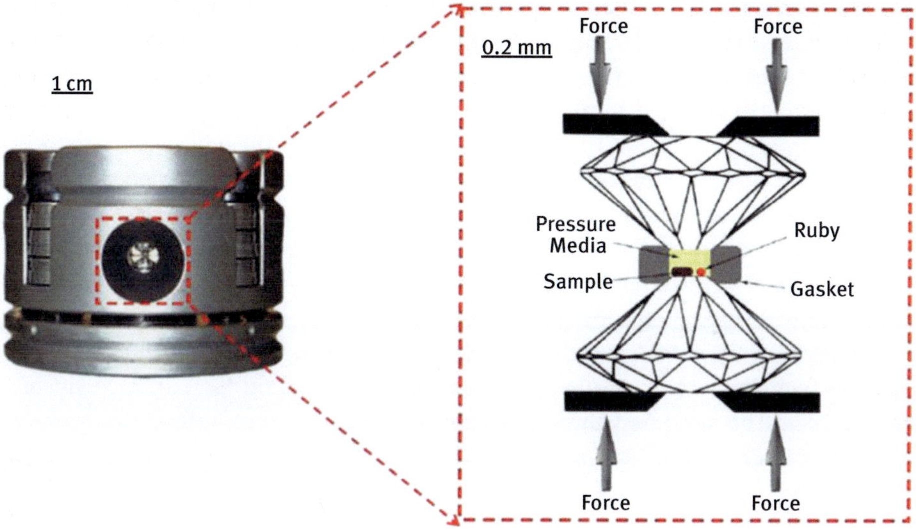

Figure 1: Photo and schematic diagram of a diamond-anvil cell.

For vibrational spectroscopic measurements on materials loaded in DAC, optical property and quality is a prime factor in selecting diamond anvils. Typically, two types of diamonds, i.e., type I and type II are used for different spectroscopic probes. Both types have the intense first-order Raman line at 1,332 cm^{-1} (F_{2g} mode of the diamond). The difference between the two types is in the infrared absorption spectrum. Type I anvils that have more nitrogen impurities show two strong IR absorption regions around 2,000 cm^{-1} and 1,000–1,350 cm^{-1}, respectively. In contrast, type II anvils that are mostly nitrogen free offer a relatively clean IR window bellow 2,000 cm^{-1} ideal for IR absorption measurements. Therefore, the low-fluorescent type I diamonds are only suitable for Raman spectroscopy, while the type II diamonds are mainly used in IR spectroscopy. More recently, new development in high-P and high-T technique to produce synthetic diamonds offer high-quality type II diamonds with low N-impurity and low fluorescence, so samples can be measured by Raman and IR spectroscopy in the same DAC.

2.2 Raman spectroscopy

Raman spectroscopy, also known as Raman scattering is a vibrational spectroscopy based on the inelastic scattering of light (typically a laser source) by materials. The Raman process has a much smaller cross-section than other spectroscopic processes (e.g., absorption, fluorescence, etc.). Although many commercial Raman microscopy systems (e.g., Renishaw) are available, they generally have a rigid design that does not allow *in situ* measurements with different DAC configurations and temperature requirements. Therefore, a state-of-the-art customized Raman system [Figure 2] was constructed to allow the DAC based measurements in a broad *P*–*T* range with multiple excitation laser sources, such as 488–514 nm lines from an Innova Ar^+ laser (Coherent Inc.), 532 and 782 mn lines from diode-pumped solid-state lasers. The laser was focused to less than 5 µm on the sample by an Olympus 20 × objective, while a 15 × eyepiece and a digital camera allowed precise alignment of the focused laser beam. The Raman scattering signal was collected by the same objective lens with backscattering geometry. The Rayleigh and anti-Stokes scattering were removed by an edge filter that enabled a spectral range above 100 cm^{-1} to be measured before the total scattered photons were focused on the entrance slit of a spectrometer. The scattered light was then dispersed using an imaging spectrograph (SpectroPro-2500i, Acton Research Corporation) that housed a 0.5 m focal

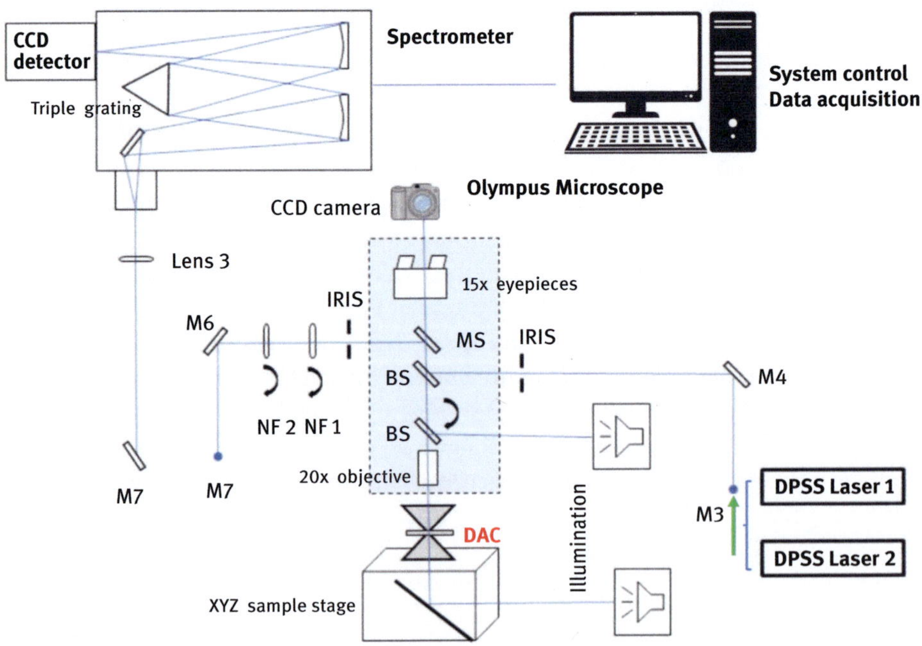

Figure 2: Schematic diagram of customized Raman microspectroscopy system [14, 15].

distance monochromator equipped with multiple gratings, such as a 1,200 lines/mm grating offering ± 0.2 cm^{-1} resolution. The Raman signal was recorded using an ultrasensitive liquid nitrogen cooled, back-illuminated, charge-coupled device (CCD) detector from Acton. The system was calibrated using Neon lines with an uncertainty of ± 1 cm^{-1} for each experiment.

2.3 FTIR spectroscopy

IR absorption spectroscopy provides sensitive and fingerprints information on materials loaded in DAC. This measurement is essential for samples with high fluorescence that prohibit effective Raman measurements. The IR measurements for the examples demonstrated in this chapter were mostly carried out using a customized IR micro-spectroscopy system constructed in-house [Figure 3], that consists of a commercial Fourier Transform Infrared (FTIR) spectrometer from Bruker Optics Inc. (model Vertex 80v) equipped with a Globar IR light source. The system was operated under a vacuum of <5 mbar such that absorption by H_2O and CO_2 was efficiently removed. A collimated IR beam of varying diameters set with apertures ranging from 0.25 to 8 mm was directed into a relay box through a KBr window on

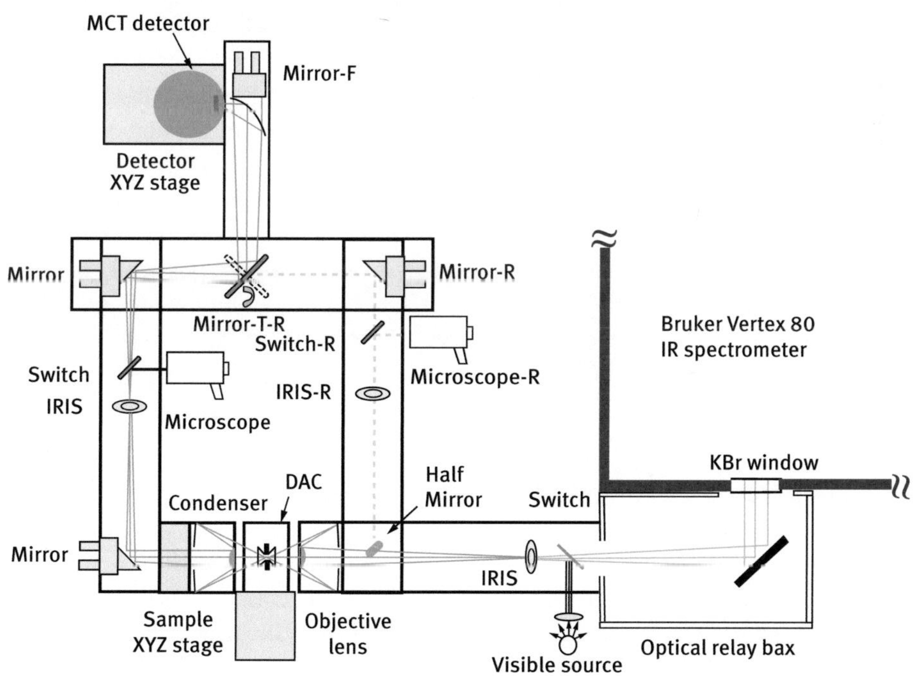

Figure 3: Schematic diagram of customized FTIR microspectroscopy system. Reproduced with permission from ref. [16].

the spectrometer. The beam was then focused onto the sample in the DAC by iris optics and 15 × reflective objective lens with a numerical aperture of 0.4. The DAC on the sample XYZ precision stage was aligned with the aid of an optical microscope equipped with an objective lens of variable magnifications (0.75 × to 3 ×) and a 20 × eyepiece from Edmond Optics. The size of the IR beam was set to be identical to the entire sample size set by a series of iris apertures. The transmitted IR beam was collected using another identical reflective objective as the condenser and was directed to a mid-band mercury cadmium telluride (MCT) detector equipped with a ZnSe window that allows measurements in the spectral range of 600 to 12,000 cm^{-1}. A resolution of 4 cm^{-1} and 512 scans were typically applied for each spectrum measurement, achieving an excellent signal-to-noise ratio. The reference spectrum, which is the absorption of diamond anvils loaded with KBr but without any sample, was later divided as background from each sample spectrum to obtain the absorbance.

2.4 Synchrotron-based FTIR spectroscopy

Synchrotron light is a source of electromagnetic radiation produced by a storage ring housing traveling electrons with a near speed of light. Although synchrotron source provides enormous advantages typically in the X-ray region, the infrared synchrotron light has unique applications for DAC-based measurements due to the very intense, very broad and highly focused source that allows very high spatial resolution and far-IR measurements. Some examples in this chapter were based on the experiments performed at the U2A beamline at the National Synchrotron Light Source I (NSLS-I) at Brookhaven National Laboratory (BNL). This facility was decommissioned in 2014 and replaced by a new generation called National Synchrotron Light Source II (NSLS-II). The corresponding IR beamline, now named Frontier Synchrotron Infrared Spectroscopy, underwent a major upgrade as well with the most advanced in-line IR spectroscopic capacities supported by extensive off-line optical tools dedicated to studies of structures and behavior of materials under conditions of extreme pressures and temperatures.

3 Pressure-mediated hydrogen bonding

The most dominant pressure effect in molecular systems is the significant reduction of the intermolecular distance. During such a process, systems with hydrogen bonding exhibit more prominent pressure responses. Hydrogen bonding plays an important role in stabilizing a wide range of molecular structures and influences the chemical and physical properties of molecular systems upon compression. Typically, the characterizations of hydrogen bonding are inferred from crystal structures or by theoretical modeling. Here four examples are shown to demonstrate that vibrational spectroscopy on materials loaded in DAC can reveal interesting pressure-mediated hydrogen bonding interactions. Each of the examples represents an interesting class of materials such as molecular crystals, energetic materials, photovoltaic materials, and hydrogen storage materials.

3.1 Ethylene glycol

Ethylene glycol (EG) represents a prototype multi-hydroxyl-containing biological compound for the study of interactions among more complicated systems such as sugars and polysaccharides. Using *in situ* high-pressure Raman and FTIR absorption spectroscopy, the structural and conformational transformations of EG were found to be substantially influenced by hydrogen bonding interactions under pressure up to 10 GPa [17]. The characteristic Raman modes reveal that EG exists as a liquid with a mixture of *trans* and *gauche* conformations up to 3.1 GPa. In the pressure range 4–7 GPa, the solid phase shows a pressure-dependent proportion of *trans* and *gauche* conformations. At pressures above 7 GPa, the structure is stabilized to *gauche* conformation and remains stable up to 10 GPa. The increase in the intensity and the large pressure-induced red shift of the infrared active OH mode $\boldsymbol{\nu_\varepsilon}$ suggests that intra-molecular hydrogen bond is formed and strengthened during the stabilization of *gauche* conformation (Figure 4). The pressure-induced changes were found to be completely reversible on release to ambient conditions.

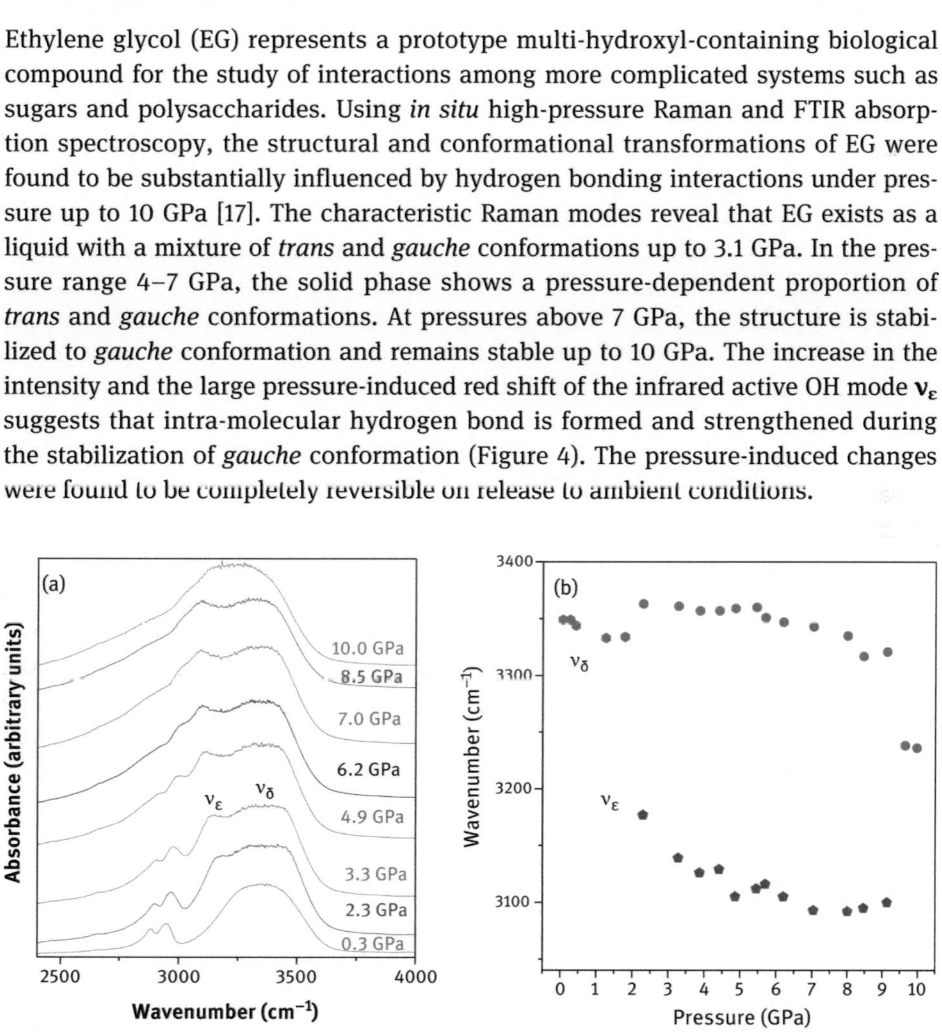

Figure 4: Infrared absorption spectra of ethylene glycol at different pressures under compression in the spectral region of 2,500–4,000 cm $^{-1}$ (a) and variation of OH stretching infrared active modes of ethylene glycol with pressure (b). Reproduced with permission from ref. [17].

3.2 Bis(1H-tetrazol-5-yl)amine monohydrate

Bis(1H-tetrazol-5-yl)amine (BTA) has been considered a promising high-energy density material due to the high nitrogen content (i.e., 82.5 wt%). Moreover, examining the possibility of converting this high nitrogen content precursor to other polymorphs with higher energy density using high pressure is of great interest. Using Raman and IR spectroscopy, X-ray diffraction, and ab initio simulations, *in situ* high-pressure study of BTA · H_2O was carried out up to 25 GPa at room temperature [18]. Upon compression, both the Raman and IR vibrational bands were found to undergo continuous and gradual broadening without significant change of the profile, indicating pressure-induced structural disordering rather than phase transition. Interestingly, in contrast to all other Raman and IR modes of BTA · H_2O which exhibit blue shifts, the N–H stretching mode shows a prominent red shift upon compression to ~8 GPa, strongly suggesting pressure-enhanced hydrogen bonding between BTA and H_2O (Figure 5). The analysis of X-ray diffraction patterns of BTA · H_2O suggests that the

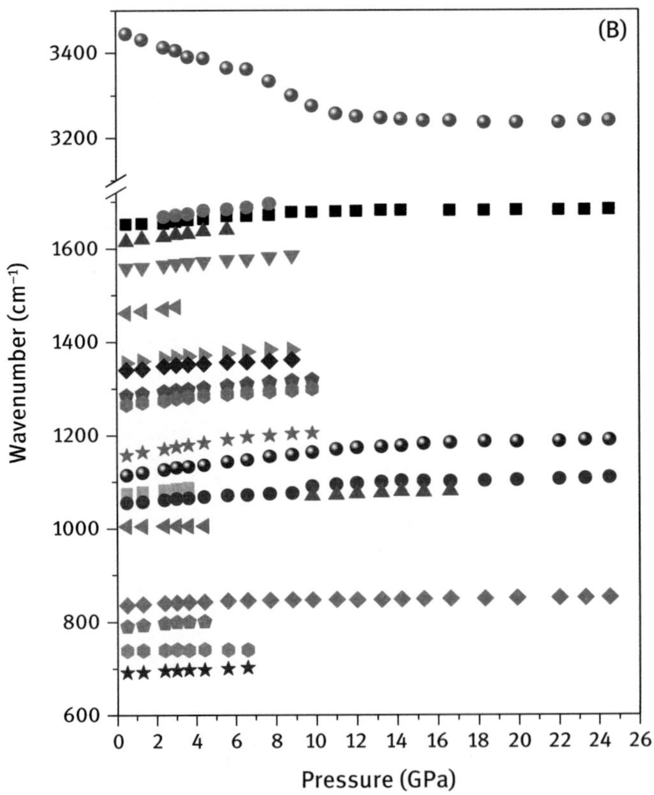

Figure 5: Pressure dependence of IR modes of BTA · H_2O on compression. Reproduced with permission from ref. [18].

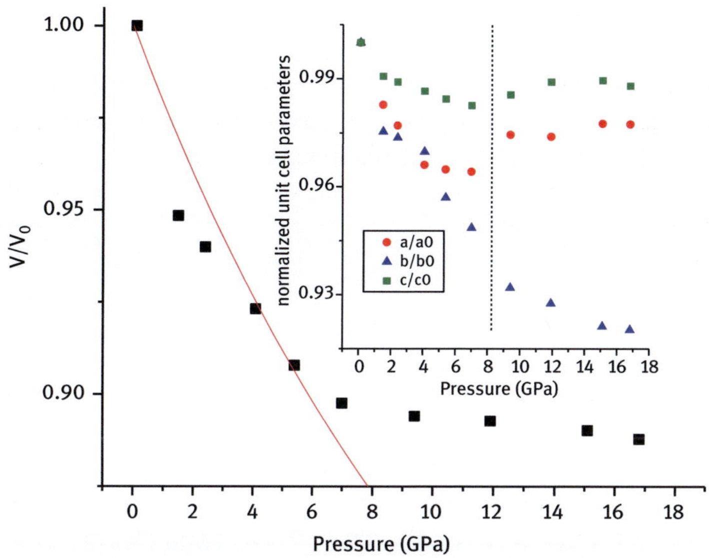

Figure 6: Normalized unit cell volume vs pressure (black squares) BTA · H_2O on compression and fitted equation of state (red curve) using second-order Birch–Murnaghan equation. The inset shows normalized monoclinic unit cell parameters for *a*, *b*, and *c* of BTA · H_2O on compression. The vertical dashed line denotes the pressure at which the monotonic contraction of *a* and *c* axes changed. Reproduced with permission from ref. [18].

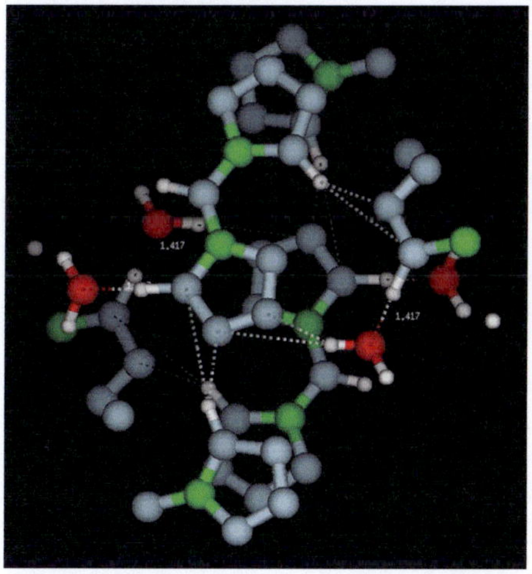

Figure 7: Proton hopping and molecular interactions of BTA · H_2O system based on first-principles simulations. Reproduced with permission from ref. [18].

unit cell parameters undergo anisotropic compression rate, in excellent correlation with the behavior of the hydrogen bonding enhancement (Figure 6). Aided by first-principles simulations (Figure 7), these pressure-mediated structural modifications consistently suggest that hydrogen bonding played an important role in the compression behavior and structural stability of BTA · H_2O under high pressures.

3.3 $FAPbI_3$

Organometal halide perovskite-based solar cells have drawn enormous attention due to their superior properties and high power conversion efficiency (PCE). Substantial efforts have been made over the past several years to improve the PCE for which formamidinium lead triiodide ($FAPbI_3$) has shown great performance among all organometal halide perovskite due to the smaller bandgap (1.45 eV) and broader spectral absorption. Two polymorphs of $FAPbI_3$ were investigated at high pressures using *in situ* synchrotron X-ray diffraction, FTIR spectroscopy, photoluminescence (PL) spectroscopy, electrical conductivity (EC) measurements, and ab initio calculations [19]. Starting from ambient pressure, two new high-pressure polymorphs (i.e., *I*mm2 and *I*mmm shown in Figure 8) were identified for α-$FAPbI_3$.

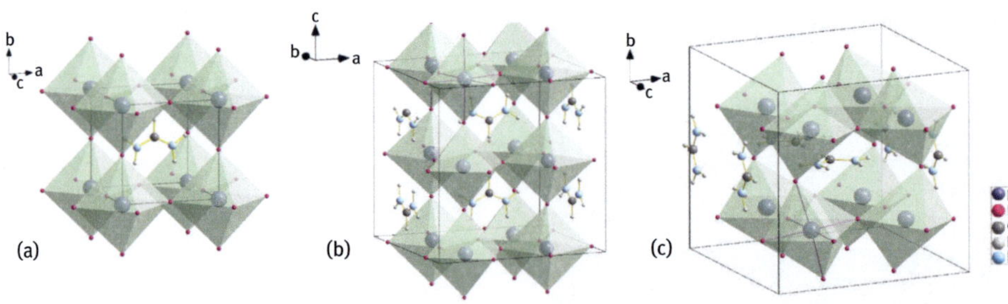

Figure 8: Crystal structures of α-$FAPbI_3$ at ambient pressure (a), high-pressure phase II (b), and phase III (c). Reproduced with permission from ref. [19].

To explore the pressure effect on the FA moiety and cation–anion interactions, *in situ* FTIR spectroscopy (Figure 9a) was carried out. In the mid-IR region, the optical absorption is exclusively attributed to the vibrations from organic FA ($H_2N^+ = CHNH_2$). Upon compression, the pressure dependence collectively exhibits discontinuities at 0.5 GPa and 1.7 GPa for α-$FAPbI_3$, in agreement with the identified transition boundaries (Figure 9b). Most of the observed IR modes especially the N–H and C–H stretching region (3,250–3,400 cm^{-1}) exhibit a prominent red shift in the entire compression region for both α and δ-$FAPbI_3$, indicating a pressure-induced enhancement of hydrogen bond interactions. MD simulation reveals that the rotations of FA molecules are reduced at high pressure to ordered oscillations, which engages hydrogen

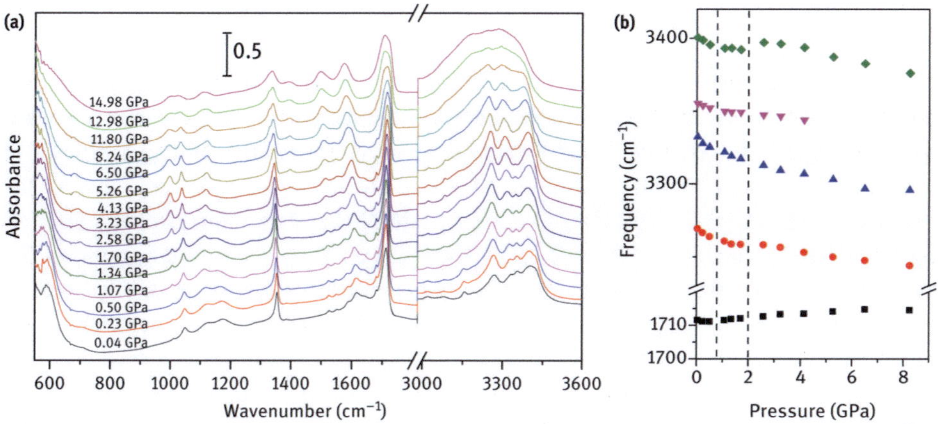

Figure 9: FTIR spectra of α-$FAPbI_3$ upon compression (a) and pressure dependence of selected IR modes (b). Reproduced with permission from ref. [19].

bonding between the FA and PbI_6 via highly effective H· · ·I interactions, with the shortest H· · ·I distance of 2.4 Å (Figure 10). In contrast to the MA-based counterparts, the FA perovskite is clearly influenced by hydrogen bonding, which in turn causes the polymorphic transitions to be significantly different from the former.

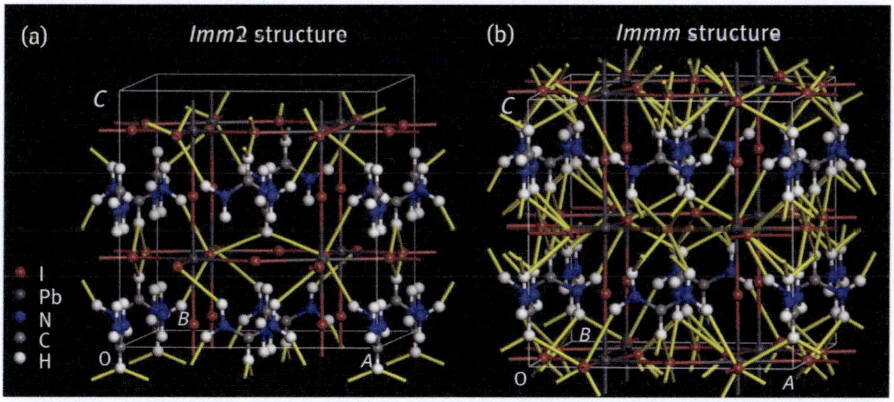

Figure 10: Schematic hydrogen bonding interactions between FA and PbI_6 polyhedra in Imm2 structure at 0.34 GPa (a) and in *Immm* structure at 1.67GPa (b). The distances of H – I ranging from 2.4 Å to 3.2 Å, i.e., those in strong to moderate hydrogen bonding region are highlighted by yellow lines. Hydrogen bonds are clearly more populated and symmetric, in the *I*mmm structure. Reproduced with permission from ref. [19].

3.4 EDAB

Ammonia borane (NH_3BH_3) has been regarded as a prime candidate for solid hydrogen storage owing to its high hydrogen content (19.6% by weight). However, on-board use of ammonia borane has faced several challenges including slow decomposition kinetics and damage to fuels cells caused by NH_3BH_3 decomposition by-products. One way to address these difficulties is to employ derivatives of ammonia borane, such as ethylenediamine bisborane (EDAB), $BH_3NH_2CH_2CH_2NH_2BH_3$. Guan et al. reported the first detailed investigation of structural transformations of EDAB at high pressures using a combination of *in situ* Fourier-transform IR spectroscopy, Raman spectroscopy, XRD, and density functional theory (DFT) [20]. The detailed analysis shows that EDAB exists as a *Pbca* structure until around 1 GPa when it transforms into the $P2_1/c$ (A) structure. The second phase transition starts taking place around 7 GPa and gives the $P2_1/c$ (B) structure which coexists with $P2_1/c$ (A) structure to at least 12 GPa. These transformation sequences are illustrated in Figure 11.

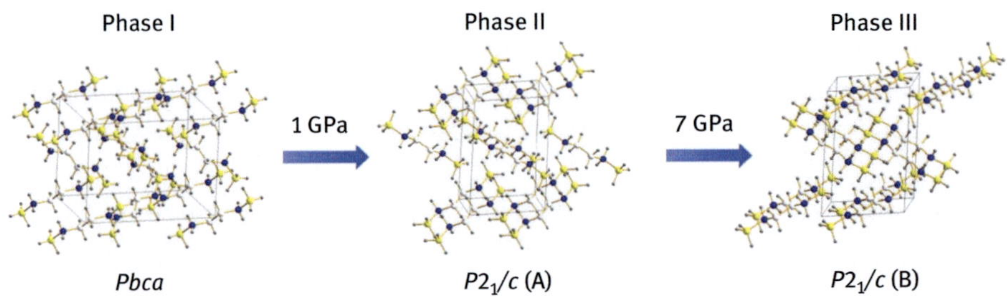

Figure 11: Identified pressure-induced polymorphic transformations of EDAB in the 0–17 GPa pressure range. Reproduced with permission from ref. [20].

In situ IR and Raman spectra of EDAB sample in the 0–17 GPa pressure range (Figure 12) suggest that EDAB undergoes two possible phase transitions: the first near 1 GPa-from phase I to phase II, and the second near 7 GPa, from phase II to another polymorph labeled phase III. As the pressure increases, the distance between EDAB molecules decreases, which favors the formation of N–H ···H–B dihydrogen bonds. Spectroscopic evidence for such bonds includes the progressive blue shifts of the B–H stretching modes as well as red shifts of the N–H stretching modes (Figure 12c).

Dihydrogen bonds are found in amino borane and its derivatives are relevant to the ability of these compounds to store and release molecular hydrogen. A useful test for the presence of dihydrogen bonds in EDAB is based on the empirical observation that the typical H–H distance in intermolecular $N–H^{\delta+} \cdots {}^{\delta-}H–B$ bonds is between 1.7 and 2.2 Å. At ambient pressure, all of the intermolecular H–H distances in phase I are greater than 2.2 Å, indicating that there are no dihydrogen bonds whereas the number of dihydrogen bonds increases upon compression. Dihydrogen bonds form around 2

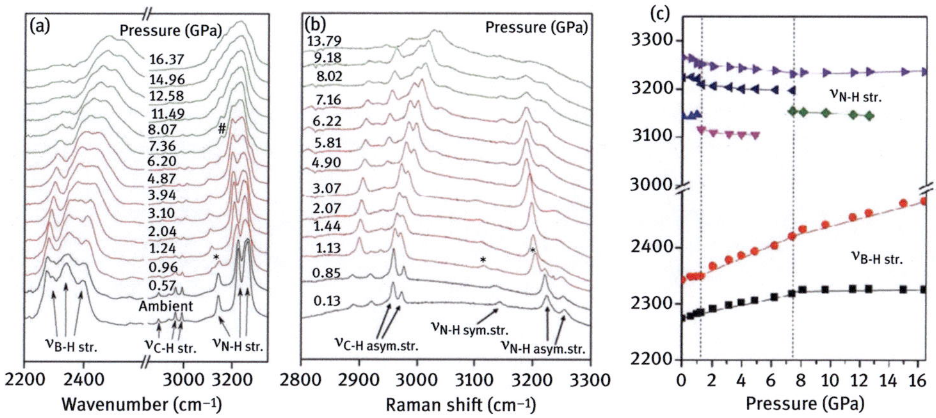

Figure 12: IR (a) and Raman (b) spectra of EDAB collected at room temperature at various pressures in a selected spectral region. The astersisks (*) and pound signs (#) mark the appearance of new vibrational modes associated with phases II and III, respectively. Pressure dependence of selected IR modes of EDAB (c). Reproduced with permission from ref. [20].

GPa and again near 12 GPa, at which point their total number reaches 20 per molecule (Figure 13). In particular, the structural fragment of phase II at 2 GPa has 8 dihydrogen bonds per molecule and 16 in phase III at 8 GPa (Figure 14). These results suggest that the origin of the observed phase transitions in EDAB may be significantly facilitated by the evolution of dihydrogen bond frameworks.

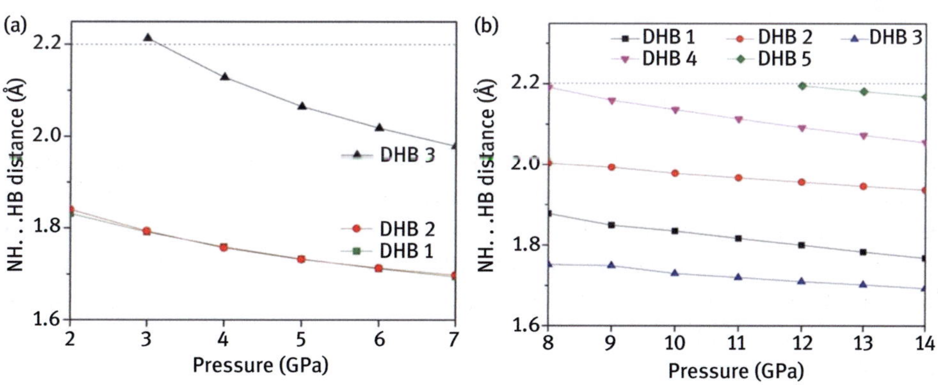

Figure 13: Calculated (vdW-DF) N–H ···H–B distances vs pressure: (a) in phase II (there are four occurrences of DHB 1, four of DHB 2, and two of DHB 3. The different DHBs are distinguished by the N–H ···H–B distance); (b) in phase III. The number of DHBs is four of each type. Reproduced with permission from ref. [20].

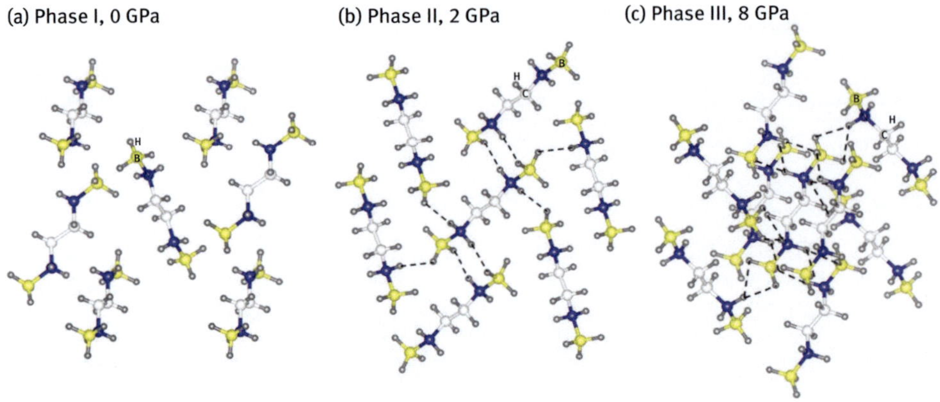

Figure 14: Dihydrogen bonds (represented by dashed lines) in phases I, II, and III at selected pressures. Each panel shows all of such bonds formed by the central molecule only. Reproduced with permission from ref. [20].

4 Elucidation of pressure and photon-mediated reaction mechanisms using *in situ* vibrational spectroscopy

4.1 Ethylene glycol

Using *in situ* vibrational spectroscopy, the structures and hydrogen bonding interactions at high pressures of ethylene glycol (EG) were reported previously as discussed in Section 3.1. Here, the photon-induced reactivity of liquid ethylene glycol (EG) was investigated at room temperature using multi-line UV radiation at ~350 nm and monitored by FTIR spectroscopy [21]. Representative IR spectra of EG show two sets of profiles after a specific reaction time, suggesting multiple photon-induced chemical reactions designated as primary and secondary processes (Figure 15). Spectral analysis results in the identification of glycolaldehyde, acetaldehyde, and methanol as the primary products. Subsequent photoreactions of these primary products gave rise to secondary products including methane, formaldehyde, methoxymethanol, methylformate, and carbon dioxide. It was further found that the initial loading pressure of EG not only can substantially influence the reaction kinetics but controls the accessibilities for some reaction channels such as for CH_4 (Figure 16a). Quantitative analysis of the antisymmetric stretching mode of CO_2 formed at different loading pressures suggests the formation of CO_2 clathratehydrate and CO_2 clusters. The

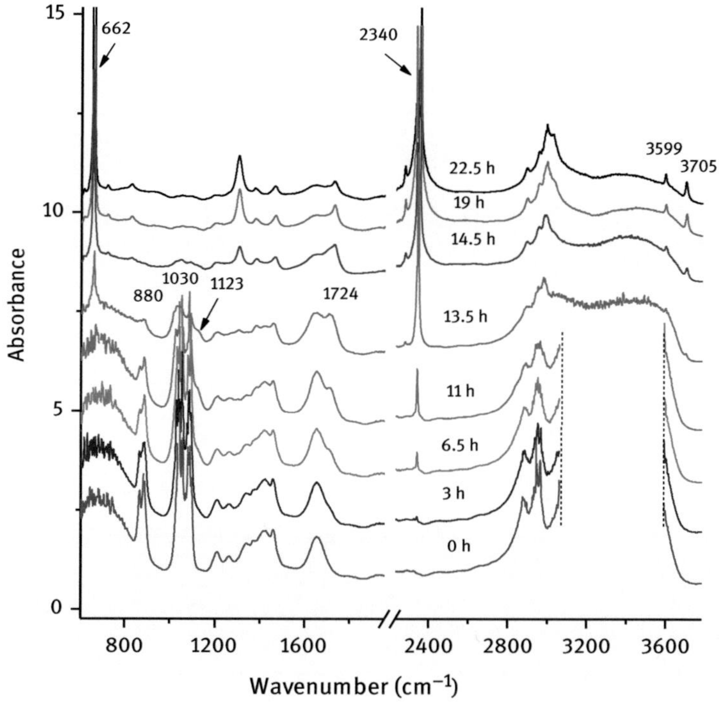

Figure 15: Selected FTIR spectra of EG with an initial loading pressure of 0.1 GPa upon UV irradiation (with λ of ~350 nm and power of ~700 mW) collected at different radiation time. Selected characteristic IR bands emerged at 13.5 h and observed at 22.5 h indicating sequential photochemical reactions are labeled. The spectral region in 3,000–3,500 cm^{-1} before 13.5 h is truncated due to the saturated IR absorption intensity in this region. Reproduced with permission from ref. [21].

stabilities as well as relative abundance of these CO_2 species are found to be dependent on both pressure and radiation time (Figure 16b). These observations revealed interesting pressure-induced CO_2 sequestration behaviors as a result of photochemical reactions of ethylene glycol.

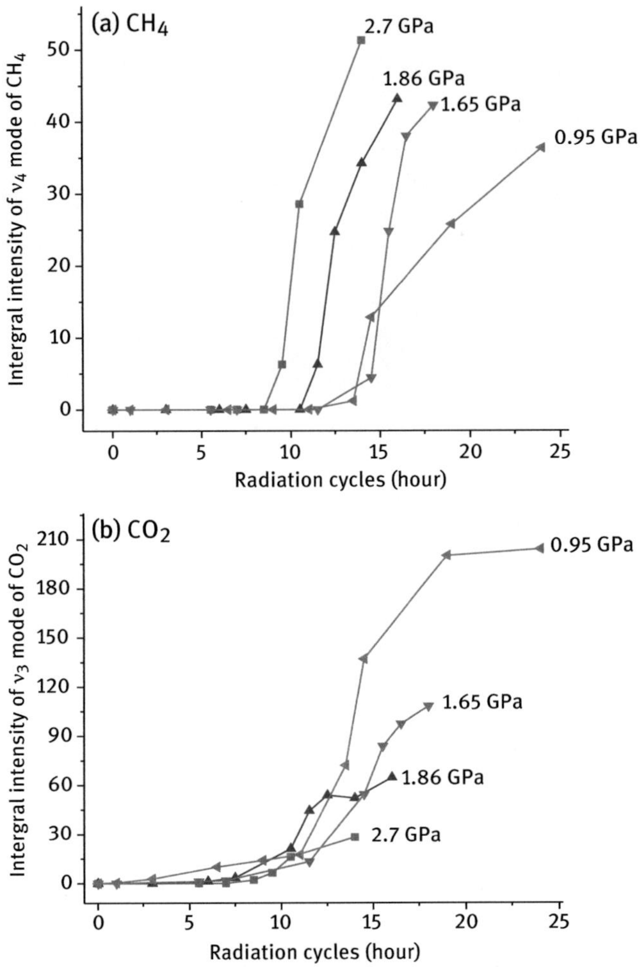

Figure 16: Relative photochemical reaction yields of CH_4 (a) and CO_2 (b) estimated by integrating the IR absorbance of the respective characteristic IR modes (ν_4 of CH_4 and ν_3 of CO_2) as a function of radiation time for EG samples with an initial loading pressure of 0.1 (◀), 0.5 (▼), 1.0 (▲), and 2.0 (■) GPa. The pressures labeled for each sample denote the final pressure of the system. Reproduced with permission from ref. [21].

4.2 Styrene

The polymerization of styrene to produce polystyrene has been one of the most important polymeric reactions in the industry. The known tacticity of polystyrene includes crystalline syndiotactic polystyrene (SPS), isotactic polystyrene (IPS), and amorphous atactic polystyrene (APS). Guan et al. reported pressure-assisted photopolymerization of monomeric styrene under UV radiation conditions using FTIR [22]. As shown in

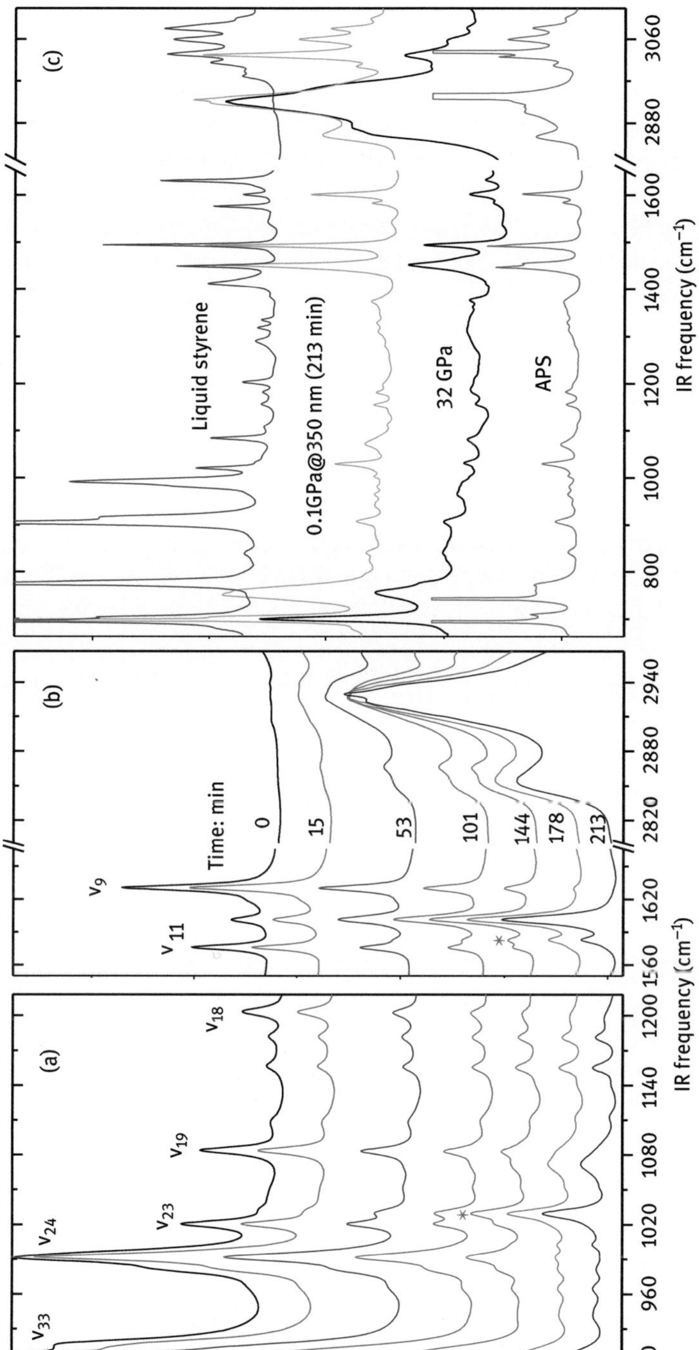

Figure 17: (a)–(b) Selected FTIR spectra of styrene with an initial loading pressure of 0.1 GPa upon UV irradiation (with UVML 350 nm and power of 35 mW) collected at different radiation times under the same condition as a function of time. (c) The collected IR spectra for liquid styrene, photoproducts at 213 min, pressure-induced products at 32 GPa and commercial APS. Reproduced with permission from ref. [22].

selected IR spectra (Figure 17a and 17b), when the liquid styrene is irradiated upon 350 nm UV radiation (35 mW) with 15 min, a new IR peak at 2,900 cm^{-1} appears indicating the formation of APS. Upon further radiation, the completion of the photoreaction was found after 213 min radiation. Figure 17c shows by a comparison the IR spectrum of the final product is consistent with the one collected from the commercial APS.

In order to examine the pressure selectivity in producing different tacticities, the sample was loaded with different initial pressures. As shown in Figure 18a, the IR spectra of final photoproducts, where no IR spectral changes are observed upon further radiation, are collected from the initial loading pressure at 0.02, 0.13, 0.34, 0.85 GPa, respectively. The top two present the IR spectra of the crystalline SPS and amorphous APS. Apparently, the formation of SPS can be excluded due to the missing of the IR peak at 1,222 cm^{-1}. At loading pressure 0.02 GPa, IR spectrum of the photoproducts shows a peak at 1,052 cm^{-1} along with the blue shift of the peak at 1,075 cm^{-1} compared with the one in APS. In the case of the 0.13 and 0.34 GPa, there is a shoulder at 1,052 cm^{-1}. Figure 18b shows that the broad peak at 1,075 cm^{-1} is the superposition of two peaks at 1,070 and 1,083 cm^{-1}. The peak at 1,052 and 1,083 cm^{-1}, which are very characteristic peaks to identify the IPS, suggests the formation of small amount of IPS at low loading pressures. In the case of 0.85 GPa, no IR signatures of IPS can be observed.

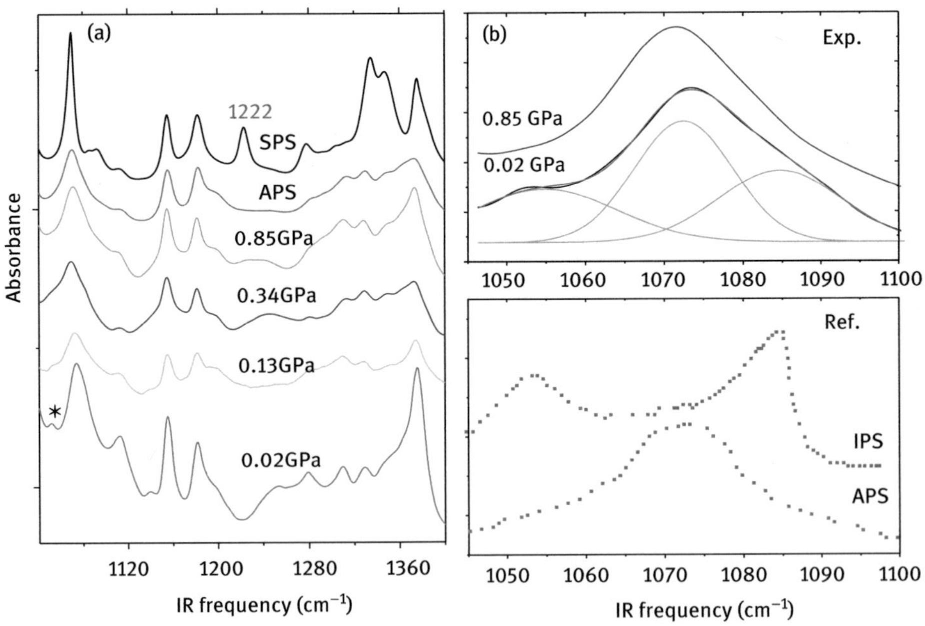

Figure 18: (a) Selected FTIR spectra of the final photoproducts recovered from different loading pressures and radiation time. The asterisk labels the peak at 1,052 cm^{-1}. (b) The deconvoluted IR spectrum in 1,040–1,100 cm^{-1} for the one at 0.02 GPa and IR spectrum from 0.85 GPa is also shown at the top as a comparison. The reported IR spectra of the IPS and APS in the same region are shown in lower right panel. Reproduced with permission from ref. [22].

The kinetics of the polymerization reaction can be monitored *in situ* by the decrease in intensity of vibrational bands associated with C = C double bonds. In the present experiments, the polymerization yield (PY) at the time *t*, or fraction of polymerized component, was determined from relative intensities of characteristic vibrational bands of styrene observed at 1,631 cm^{-1} associated with the $C_e = C_e$ vinyl stretching vibrations. With increasing initial loading pressure, it was found that the reaction can be accelerated efficiently until 0.67 GPa. Above 0.67 GPa, such as 1.26 and 2.34 GPa, the reaction was deaccelerated (Figure 19a). The evolution of ln*k* values for the polymerization reaction as a function of pressure can be best fitted in a quadratic equation. By using this quadratic dependence of ln*k* on *P*, the activation volume at any pressure is obtained (Figure 19b). As can be seen, $\Delta V^{\neq}$ is negative below 0.77 GPa, and the formation of the transition state from the reactants involves a net

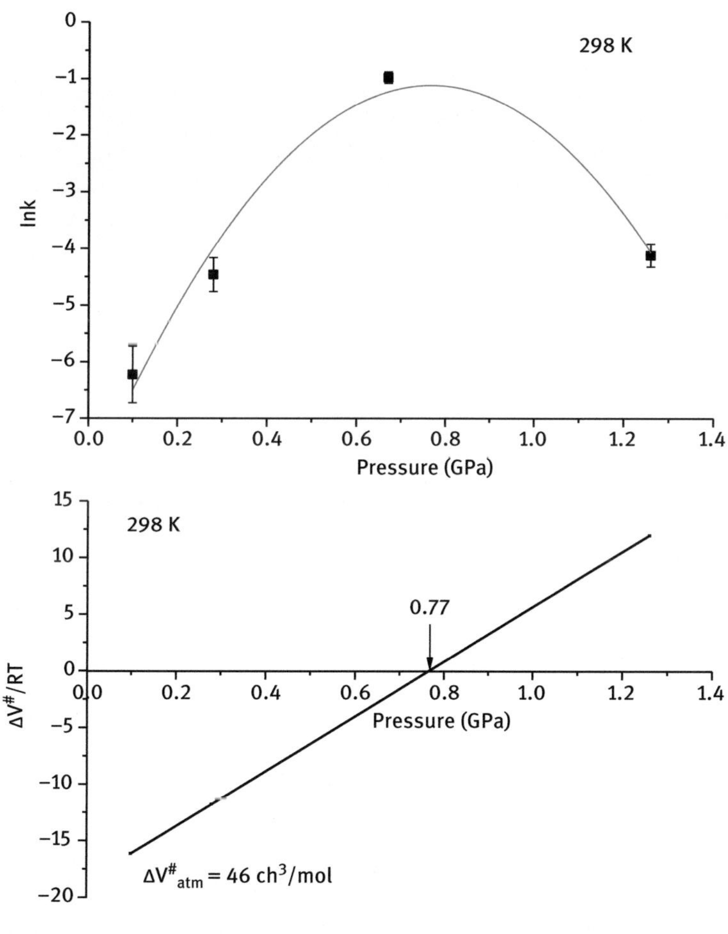

Figure 19: Quadratic fit of the values of ln*k* as a function of pressure at ambient temperature. Reproduced with permission from ref. [22].

contraction in volume, the rate constant increases with pressure. On the other hand, $\Delta V^{\neq}$ is positive above 0.77 GPa, k decreases with increasing pressure.

Photopolymerization of unsaturated compounds is usually explained by excitation of π^* antibonding states. For styrene, the available photochemical data are nevertheless limited to ambient pressure range. A schematic representation of styrene energy levels is proposed in Figure 20. In the case of the UV excitation, the single 350 nm photon (3.5 eV) is not sufficient in energy to excite the first singlet state (S_1) located at 4.43 eV. The UV-visible absorption of liquid styrene in high wavelength region is measured shown in Figure 20. Clearly, a broad absorption onset is 480 nm (2.6 eV), which corresponds to the allowed 0–0 transition to the T_1. Although the energy of S_1 is higher than the energy of one 350 nm photon, two-photon absorption would be able to populate the S_1. Owing to the small energy difference (0.25 eV) between the S_1 and T_2 states, the possible singlet-triplet intersystem crossing (ISC) can lead to the efficient energy transfer to the triplet state.

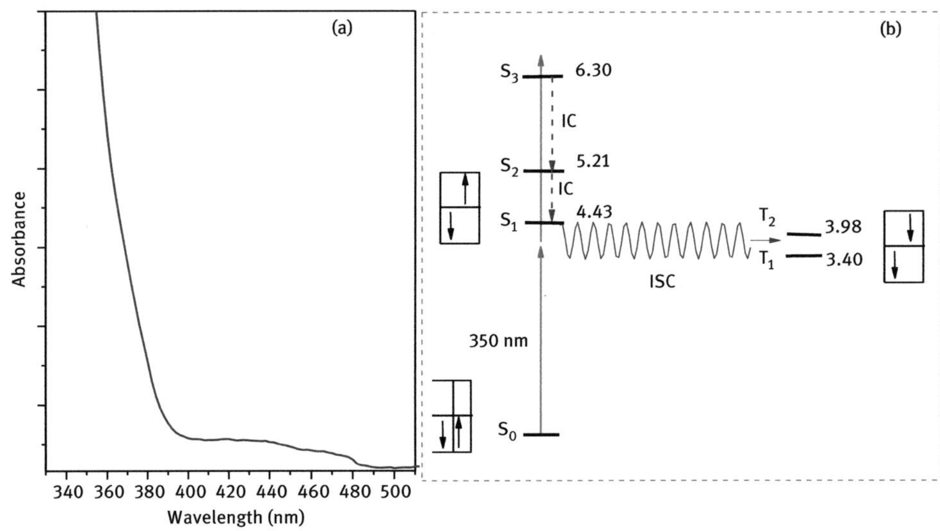

Figure 20: (a) UV-vis spectrum of liquid styrene at ambient condition. (b) The schematic energy levels of styrene. Reproduced with permission from ref. [22].

4.3 2-Butyne

Conjugated polymers are prominent semiconductors that have unique electric conductivity and photoluminescence. Guan et al. reported a high-pressure study on producing conjugate polymers starting from an acetylene derivative, dimethyl acetylene (DMA or 2-butyne) with formula of $H_3CC\equiv CCH_3$ as the monomer characterized by *in situ* FTIR and Raman spectroscopy [23]. The microphotographs (Figure 21) show

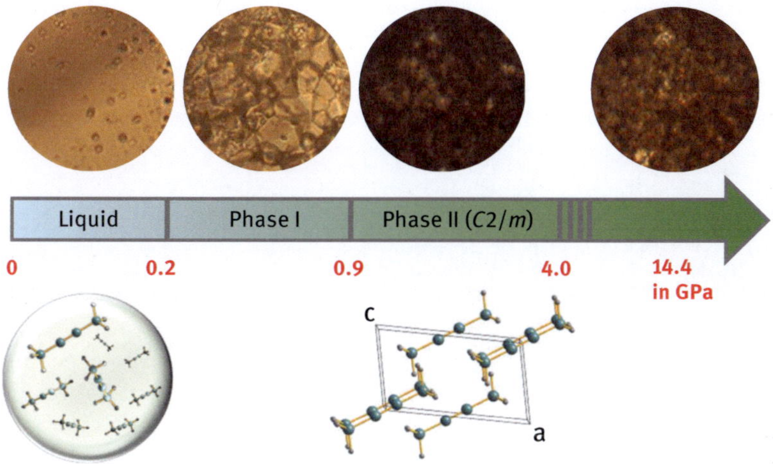

Figure 21: Microphotographs (~200 μm in diameter) of 2-butyne under high pressures: transparent liquid (0–0.2 GPa), transparent solid phase I (0.2–0.9 GPa) and opaque solid phase II (above 0.9 GPa). Graphs below are illustrations of the liquid and the crystal structure of phase II viewed from *b*-axis. In the phase II, the bright color at 14.4 GPa is due to the compression-reduced sample thickness that enhances the transmitted illumination. Reproduced with permission from ref. [23].

obvious physical differences as pressure reaches the phase transition thresholds (0.2 and 0.9 GPa). The Raman and IR results show that phase II crystal structure of 2-butyne is surprisingly stable at high pressure up to 12 GPa, in strong contrast to the stability of acetylene and its other derivatives. Above this pressure, amorphization was initiated and the process was completed at 24.4 GPa. The expected polymeric transformation was not evident upon compression, but only observed upon decompression from a threshold compression pressure (e.g., 14.4 GPa). As shown in Figure 22, IR spectrum shows broader IR peaks at 1,400 and 1,450 cm^{-1} as well as a new IR absorption at 1,650 cm^{-1}, which can be assigned to C = C stretching vibrations indicating the formation of the sp^2 carbons. Finally, the IR spectrum of the recovered solid shows a totally different IR profile than that of initial 2-butyne especially with the evident vibrational mode of C = C bond (labeled with asterisks in Figure 22).

In a subsequent study, Guan et al. reported high-pressure photochemistry between 2-butyne and trace amount of H_2O was investigated at room temperature using multiline UV radiation at $\lambda \approx 350$ nm and monitored by FTIR spectroscopy [24]. As a starting point, multiple loadings of 2-butyne + H_2O with different ratios with different initial pressures are measured with the IR absorption spectra as reference (Figure 23a) prior to the irradiation with UV laser. Pressure dependence of IR modes at 1,440 (ν_{14}), 2,740 ($2\nu_3$), 2,860 ($2\nu_{14}$), 2,920 (ν_1), and 2,960 (ν_{13}) cm^{-1} clearly show two-phase transition boundaries (Figure 23b). The first boundary at around 0.25 GPa suggests the phase transition from liquid to solid phase I. The IR frequency shifts vs pressure and appearance of a new mode at 2,740 cm^{-1} indicate another

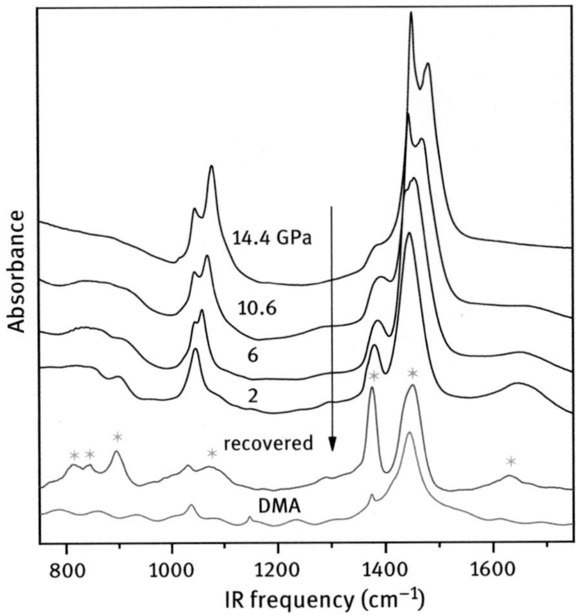

Figure 22: IR spectral evolution collected in the range of 750–1,750 cm^{-1} collected upon decompression from 14.4 GPa. The IR spectrum of the recovered product shows obvious differences (labeled by asterisks) compared to the IR spectrum of original 2-butyne. Reproduced with permission from ref. [23].

phase transition from phase I to phase II at 0.9 GPa. The low-temperature neutron powder diffraction study indicated that phase II has a monoclinic structure (space group C2/m) with two molecules per unit cell (Figure 23b).

Starting from these phases at pressures of 0.27 (liquid-phase I mixture), 0.52 (phase I), 0.90 GPa (phase II), photochemical studies were carried out by irradiating the multiline UV at λ ~350 nm with a power of 250 mW. Photon-induced reactions were observed at 0.27 GPa as evidenced by almost identical IR evolution profiles (Figure 24). Clearly, after 30 min irradiation, the new IR bands appear and are intensified with further radiation. After 100 min, these IR peaks become extremely intense but no more time-dependent changes of the IR profile were observed, indicating the completion of the photochemical reactions under the given conditions. As a result, a brownish solid as the product can be observed under the microscope. The observed IR peaks at 969 cm^{-1} (=C-H bend) and 3,020 cm^{-1} (=C-H stretch) both infers a possible formation of alkenes. By comparing the IR spectra at 30, 45 and 100 min with corresponding characteristic IR peaks of 2-butene, the new peaks at 676, 969, 1,661, 2,862 and 3,023 cm^{-1} exhibits an excellent agreement with the IR bands of solid cis 2-butene ($H_3CCH=CHCH_3$), which absorbs at 682, 972, 1,661, 2,860 and 3,014 cm^{-1}, respectively. In addition to 2-butene, the IR profiles suggest the formation of other

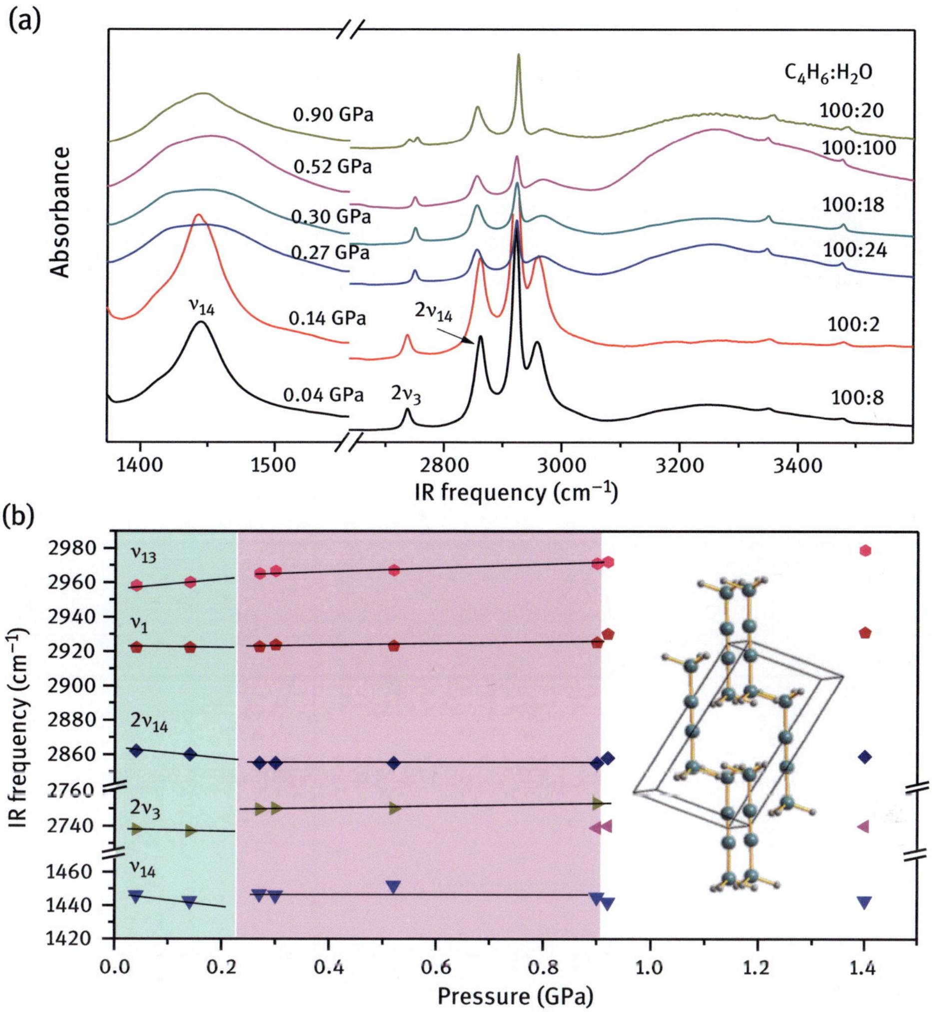

Figure 23: (a) Selected IR spectra of 2-butyne mixed with small amount of water collected at room temperature with different initial loading pressures. (b) Pressure dependence of selected IR active modes of 2-butyne. Three pressure regions with different colors indicate different phases. The inset shows the crystal structure of 2-butyne in phase II. Reproduced with permission from ref. [24].

possible photochemical products, such as solid 2-butanone. The IR signatures for solid 2-butanone become more prominent at 0.52 GPa.

At higher pressure 0.9 GPa, interestingly, the IR spectra suggests that instead of 2-butene, only 2-butanone is produced. The closure of the production channel for 2-butene at this pressure could be primarily due to the fact that H_2O and/or 2-butyne transformed to crystalline phases (Ice VI and phase II, respectively). At room temperature, specifically, the phase transition pressures for both liquid $H_2O \rightarrow$ Ice VI

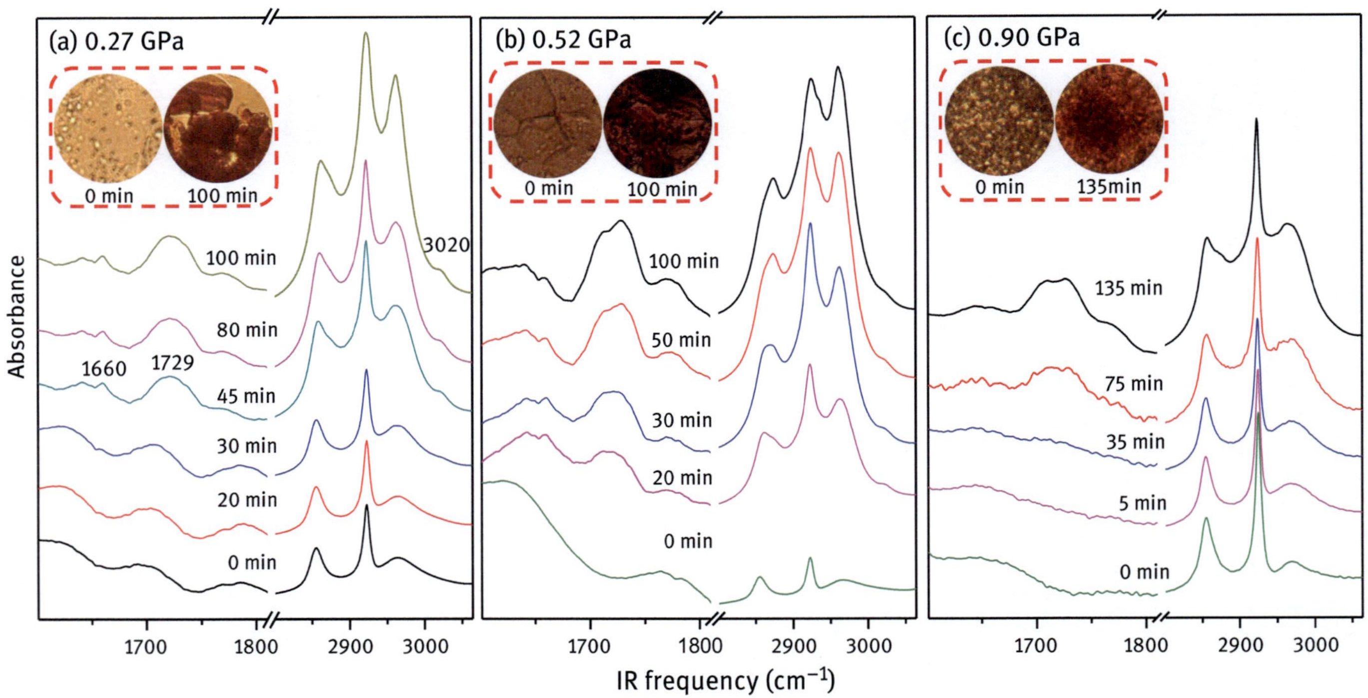

Figure 24: IR spectra of 2-butyne and H_2O at pressure 0.27 (a), 0.52 (b) and 0.90 GPa (c) upon UV irradiation as a function of time in the spectral region of 1,600–1,800 cm^{-1}. The insets show corresponding microphotographs (60 microns in diameter) of 2-butyne/H_2O mixture collected before and after irradiation. Reproduced with permission from ref. [24].

and phase I → phase II of 2-butyne were reported in pressure region of 0.9–1GPa. The crystallization of water would significantly impact the formation of water cages which are essential to preserve the reactive H atoms and thus the formation of 2-butene.

The formation of 2-butene infers a cleavage of the triple carbon-carbon bond of 2-butyne and subsequent reaction with H atoms. The source of H atoms therefore is of fundamental importance for the formation of 2-butene. As shown in Figure 25, the photodissociation of water via the two-photon process allows the production of a pair of radicals within the cage of water molecules. The H and OH radical pair can either escape the water cage and persist separately in solution or diffusively recombine to form the starting molecule. Alternatively, the radical pair could also remain the original water cage and recombine on a much faster timescale than for diffusive recombination. Therefore, water cages play a prominent role to sequestrate reactive OH radicals but release H atoms for subsequent reactions. Ultimately, the light H

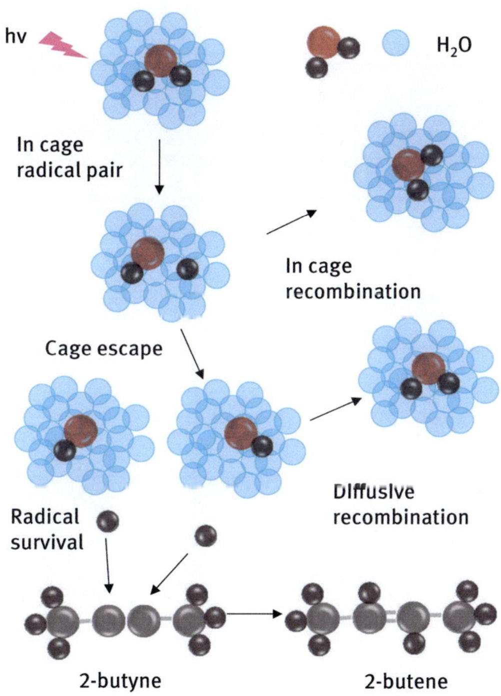

Figure 25: Schematics showing three possible processes following a photodissociation event that produces a pair of H and OH radicals within a water cage. The H and OH radical pair can escape the solvent cage and persist separately in solution or diffusively recombine to form original H_2O molecule. The H and OH radical pair could also remain within the original H_2O cage and recombine on a much faster timescale than that for diffusive recombination. The escaped H atoms are proposed to react with 2-butyne and give rise to 2-butene. Reproduced with permission from ref. [24].

atom with enough kinetic energy survives the recombination process and thus favorably react with the carbon-carbon triple bond of 2-butyne to produce 2-butene.

Finally, the effects of loading pressures on the reaction kinetics were quantitatively examined for in-depth understanding of the reaction mechanisms. The yield of two primary products can be estimated by the integrated intensity of the characteristic IR bands, e.g., 675 cm^{-1} for 2-butene and 1,729 cm^{-1} for 2-butanone (Figure 26). For 2-butene, the relative reaction yield suggests that low pressure (i.e., <0.27 GPa) is favorable for its production whereas relatively high pressure (e.g., 0.52 GPa) significantly reduces the production. Moreover, the production channel for 2-butene is

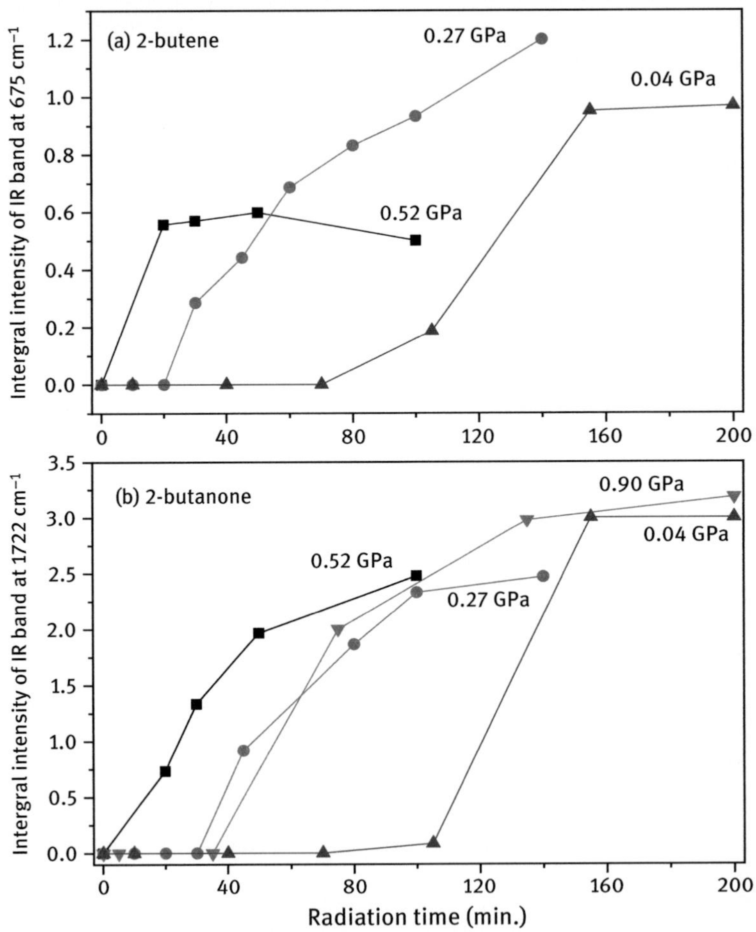

Figure 26: Relative photochemical reaction yields of (a) 2-butene and (b) 2-butanone estimated by integrating the IR absorbance of the respective characteristic IR modes as a function of radiation time for 2-butyne-water mixture with an initial loading pressure of 0.04, 0.27, 0.52, and 0.90 GPa. Reproduced with permission from ref. [24].

completely closed at 0.9 GPa. For 2-butanone, in contrast, the relative reaction yield is less affected by different loading pressures. In particular, this production channel remains open even under high pressure of 0.9 GPa (Figure 26b). The pressure-dependent reaction yield and can be understood in terms of activation volume as well as the production efficiency of radicals.

4.4 Phenylacetylene

Pressure-induced polymerization of unsaturated hydrocarbons has received revitalized interest due to the technological importance of polymer products. High-pressure -tuned polymerization kinetics were examined to elucidate the (photon-assisted) polymerization mechanism of phenylacetylene (PA, $C_6H_5C\equiv CH_3$) under different stimuli including high-pressure, physical-state, and UV radiation effects by using *in situ* FTIR spectroscopy by Guan et al. [25]. It was found that the compression rate is an important factor in controlling the pressure-induced crystallization of liquid samples. In the first attempt, the liquid PA sample (S1) is compressed slowly (~25 MPa/s) from 0.08 GPa to 0.95 GPa, leading to a fast crystallization after 1 min (Figure 27). However, in the second attempt, when the fresh loaded sample (S2) is compressed directly from a pressure below 0.9 GPa to 1.46 GPa, the sample still undergoes the crystallization but remains in a liquid-like state for up to 8 min. Therefore, in the pressure range 1.1–1.46 GPa, the observations suggest that the relaxation (nucleation) time of the crystallization is prolonged as pressure increases. More interestingly, when a freshly loaded sample S3 is directly compressed from a pressure below 0.9 GPa to 1.7 GPa with a much faster compression rate, and the solidification process is found to be completely avoided. This phenomenon suggests that the liquid-like state above 1.1 GPa is metastable and is known as the glass-forming liquid state. Although the pressure-induced formation of the glass-forming state of PA was not reported in previous studies under high pressure, it was observed at low temperatures when PA undergoes a rapid cooling rate. In this case, upon fast compression across freezing pressure of 1.1 GPa, liquid PA falls into the glass-forming liquid (also known as the supercooled liquid). Under the pressure, the metastable glass-forming liquid phase undergoes two competitive physical processes: crystallization and vitrification (glass-transition). This pressure-induced transformation mechanism together with the polymerization process to be discussed below can be understood in the schematic P-V diagram in Figure 28. The threshold pressure (P_g) for the vitrification is expected with much faster compression rates.

In attempt to induce the chemical reactions of PA by applying near-UV (ca. 350 nm) irradiation below 1.70 GPa where no chemical reactions were observed by pure compression, the PA sample is loaded at five different pressures (0.05, 0.55, 0.76, 1.10, and 1.38 GPa) to examine the effects of pressure and physical states on the photochemical reactivity of PA. IR spectral evolution at 0.05, 0.55, and 0.76 GPa show that the photon-assisted polymerization reaction was initiated at all pressures with no

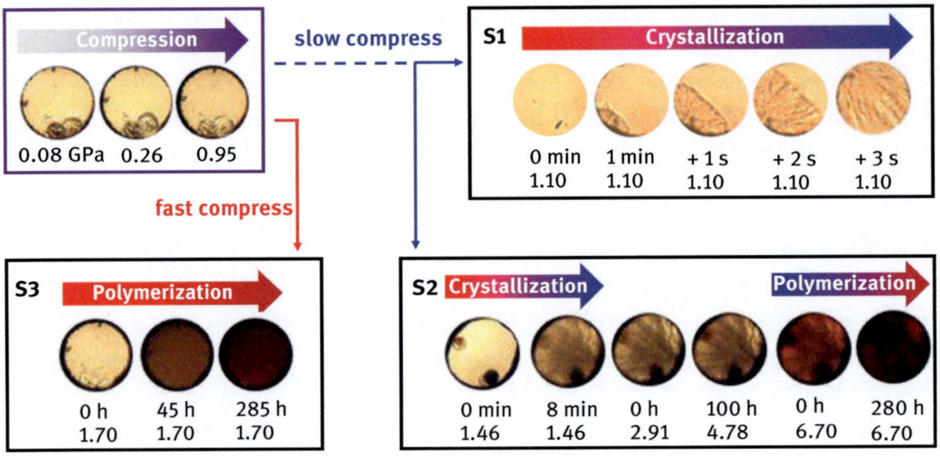

Figure 27: Microphotographs (~200 µm in diameter) of phenylacetylene (PA) samples (S1–S3) upon compression at room temperature. In the slow compression trials, S1, S2 are gradually compressed from 0.08 to 1.10, 1.46 GPa, respectively. In the fast compression trial, S3 is compressed directly from a pressure below 0.95 to 1.70 GPa. The crystalized S2 is further compressed to 2.91, 4.78, and 6.70 GPa and kept at 4.78, 6.70 GPa for 100 h, 280 h, respectively. The S3 was kept at 1.70 GP for more than 280 h. Reproduced with permission from ref. [25].

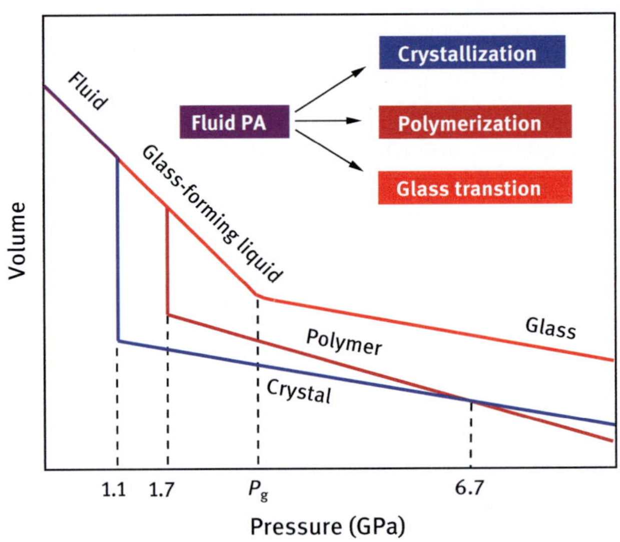

Figure 28: Schematic pressure–volume diagram of liquid PA at room temperature. The crystallization and polymerization pressures are labeled. The glass-transition pressure P_g is expected with a much faster compression rate. Reproduced with permission from ref. [25].

differences in the polymerization product formed but only the varied radiation time required to complete the polymerization. As shown in Figure 29, as pressure increases from 0.05 to 0.75 GPa in the fluid, the reaction rate k is remarkably enhanced from 0.1 h^{-1} to 5,135 h^{-1}. However, the reaction rate and yield are unexpectedly reduced in the case of 1.10 GPa. At this pressure, the reaction is quickly induced by laser radiation before it transforms into the crystal as it is at the freezing pressure threshold. The kinetics at 1.38 GPa (crystalline phase) shows a similar trend. In both cases, the PY_∞ and reaction rate (steepness of the curves within the first half-minute) are comparable, in contrast with the ones obtained in the fluid. As the chain propagation of the polymerization in the crystal is known to be restricted, the results suggest although the reaction in these phases can efficiently induced by laser radiation, the structure of the glass-forming liquid state has a "frozen" nature, which can impose restriction effects on the chain propagation of the polymerization.

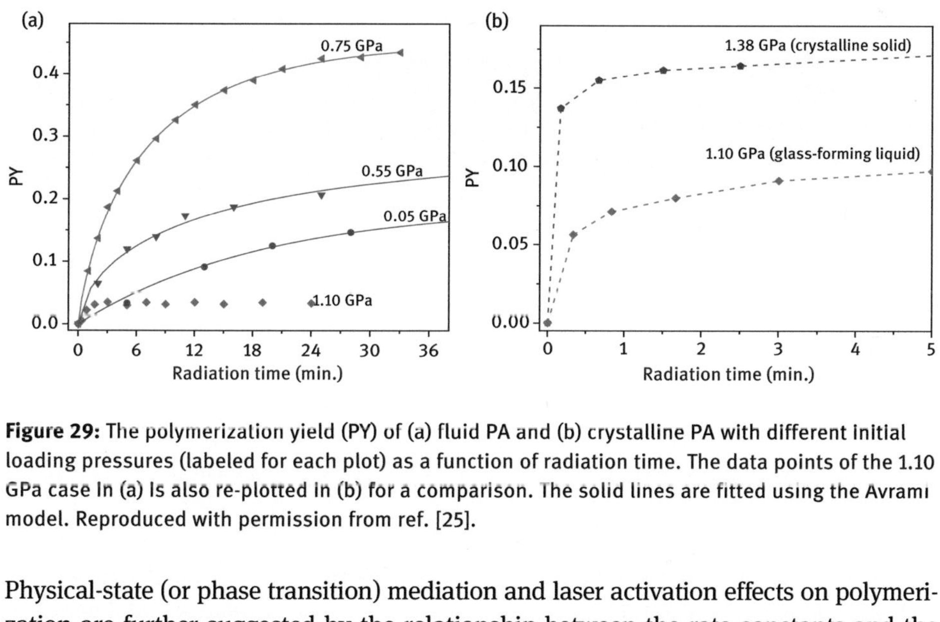

Figure 29: The polymerization yield (PY) of (a) fluid PA and (b) crystalline PA with different initial loading pressures (labeled for each plot) as a function of radiation time. The data points of the 1.10 GPa case in (a) is also re-plotted in (b) for a comparison. The solid lines are fitted using the Avrami model. Reproduced with permission from ref. [25].

Physical-state (or phase transition) mediation and laser activation effects on polymerization are further suggested by the relationship between the rate constants and the isothermal activation volume. To derive the activation volumes $\Delta V^{\neq}$, the logarithm of the rate constant ln$k$ as a function of pressure is plotted in Figure 30. In the different physical states achieved by compression, ln$k$ is assumed to be a linear function of pressure and hence that the isothermal activation volume $\Delta V^{\neq}$ is independent of pressure in the same physical state. At first glance, the polymerization of the fluid under low pressures and UV radiations gives the highest reaction rates. In the fluid phase, no polymerization was observed if the laser radiation is not applied. Thus, without the photoactivation, an infinitely high activation volume $\Delta V^{\neq}$ can be considered in the fluid phase, which is significantly reduced by the photoactivation. Similarly, the $\Delta V^{\neq}$

is decreased from infinity to – 2RT because of the fluid to the glass-forming liquid phase transition. The activation volume of the polymerization of PA in crystalline form was found comparable to the one in the photoactivation experiments.

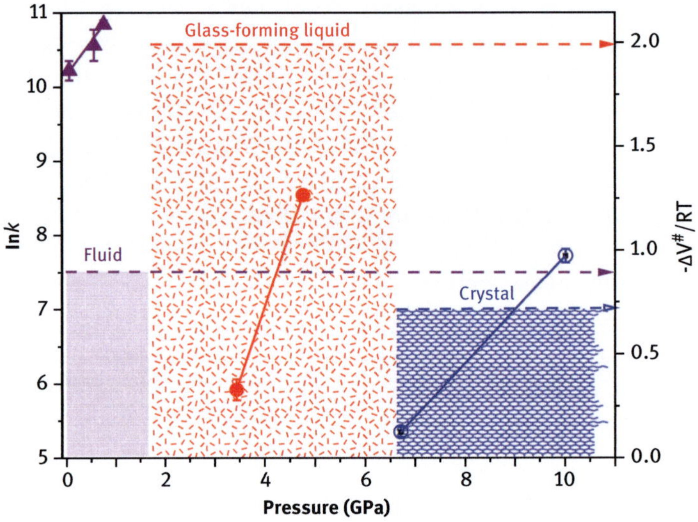

Figure 30: Linear regressions of ln*k* (left axis) as a function of pressure at 298 K for laser-induced polymerization in the fluid (purple), pressure-induced polymerization in the glass-forming liquid (red), and in the crystal (blue). The color bars and arrows show magnitudes of the derived isothermal activation volumes (right axis). Reproduced with permission from ref. [25].

5 Porous materials and guest–host interactions

5.1 ZIF-8

Metalorganic framework (MOF) is an emerging class of porous materials with promising applications in gas storage and catalysis, etc. ZIF-8 is a representative member of the zeolitic imidazolate framework (ZIF) family, a subgroup of MOF. Substantial interest has been focused on the investigation of its structure and properties under different conditions. Pressure tuning has proven an important and effective means to modify the structures and thus the associated properties of porous materials. Therefore, ZIF-8 was investigated under high pressures up to ~39 GPa using *in situ* IR spectroscopy [26]. Upon compression to 1.6 GPa followed by decompression, the pressure effects on ZIF-8 framework were found reversible (Figure 31a). Further compression to higher pressures resulted in irreversible structural transitions to a disordered or amorphous phase (Figure 31b). However, the chemical structure of the framework was found to sustain extreme compression without breaking down. The compression

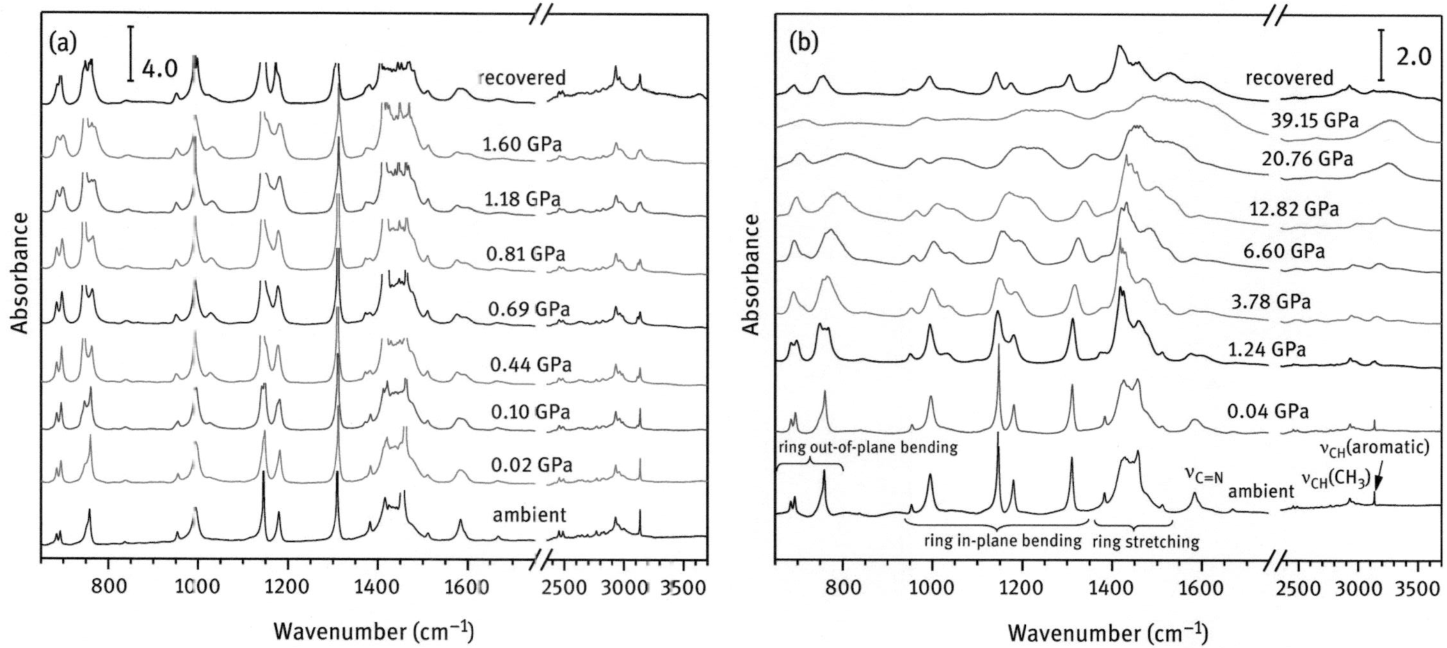

Figure 31: Selected IR spectra of ZIF-8 on compression to a highest pressure of 39.15 GPa and as recovered (a), and to another highest pressure of 1.60 GPa and as recovered (b). Reproduced with permission from ref. [26].

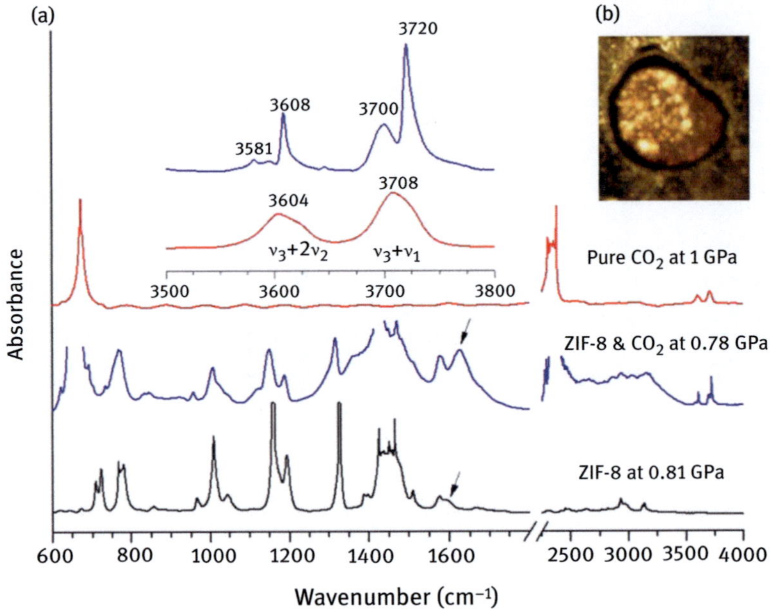

Figure 32: (a) IR spectrum of ZIF-8 loaded with CO_2 (middle) compared with that of pure CO_2 (top) and that of pure ZIF-8 (bottom) at similar pressures. The inset shows the zoomed spectral region for the combination modes of ZIF-8 loaded with CO_2 (top) and pure CO_2 (bottom). (b) Photograph of ZIF-8 loaded with CO_2 taken under an optical microscope. The arrows indicate the positions of the C = C stretching mode of the imidazole ring. Reproduced with permission from ref. [27].

behavior and especially the surprising chemical stability probed by *in situ* IR spectroscopy provide new insight into the storage applications of ZIF-8.

In a subsequent study by Hu et al., ZIF-8 framework was investigated when loaded with CO_2 in a DAC at high pressures to examine the guest–host interactions [27]. Upon loading, two types of CO_2 molecules (inside the framework and outside as bulk medium) can be differentiated by the combination IR bands of CO_2 unambiguously (Figure 32). Pressure was found to play a regulating role in the insertion and extrusion of CO_2 with respect to the framework, even at room temperature. The strong interactions between CO_2 and framework are apparent from the IR features of the framework in the C = C stretching region, providing consistent information about the possible interaction site. As guest molecules, CO_2 can substantially influence the structural stability of the ZIF-8 framework as compared to the empty framework characterized by the synchrotron far-IR data in the lattice region (Figure 33). The enhanced CO_2 storage capacity of ZIF-8 at high pressure provides new insight into the gas capture and storage applications of ZIFs.

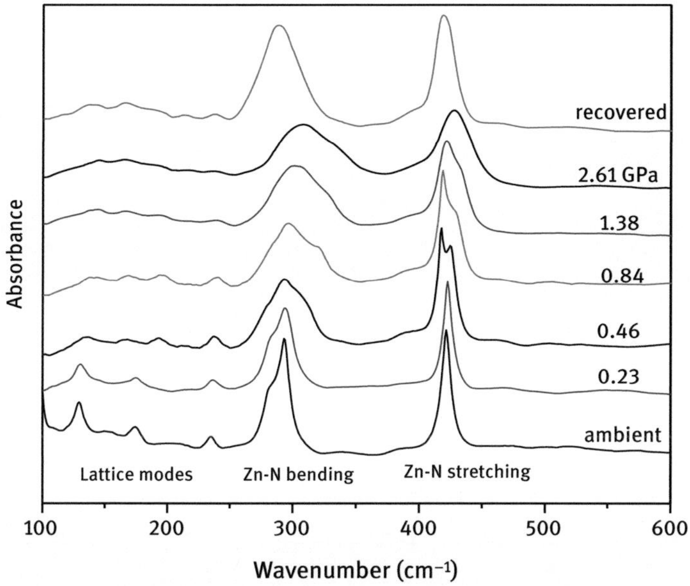

Figure 33: Far-IR spectra of empty ZIF-8 framework upon compression to 2.61 GPa and decompression to ambient pressure. These far-IR spectra suggest that pressure can significantly modify the crystal structures of empty ZIF-8 framework irreversibly. Reproduced with permission from ref. [27].

5.2 MIL-68

Another representative MOF, MIL-68 exhibits unique structural topology with two types of channels with distinct pore sizes, making it a promising candidate for the application of gas storage and separation. The behavior of as-made and activated MIL-68 (In) and their reversibilities under high pressure were investigated by IR spectroscopy [28]. Overall, the structures of both frameworks are highly stable upon compression to 9 GPa, but with some modifications to the local structure, especially the bridging O–H units. Upon the release of external pressure, the structural modifications are found to be completely reversible for as-made MIL-68 (In) while irreversible for the activated framework. The difference in the reversibility of framework is attributed to the solvent DMF molecules contained in the channels. In addition, the stability of the activated framework was studied under hydrostatic compression by using a pressure transmitting medium (PTM). The pressure-induced inclusion of PTM makes the framework more resilient to pressure (18 GPa) and the structural modifications are completely reversible upon decompression (Figure 34a). Furthermore, the performance of MIL-68 (In) for CO_2 adsorption under high pressure was investigated. The results show that the hexagonal pores are readily accessible for CO_2 at relative low pressures (e.g., <0.35 GPa), while the triangular pores become accessible for CO_2 at elevated pressures (e.g.,

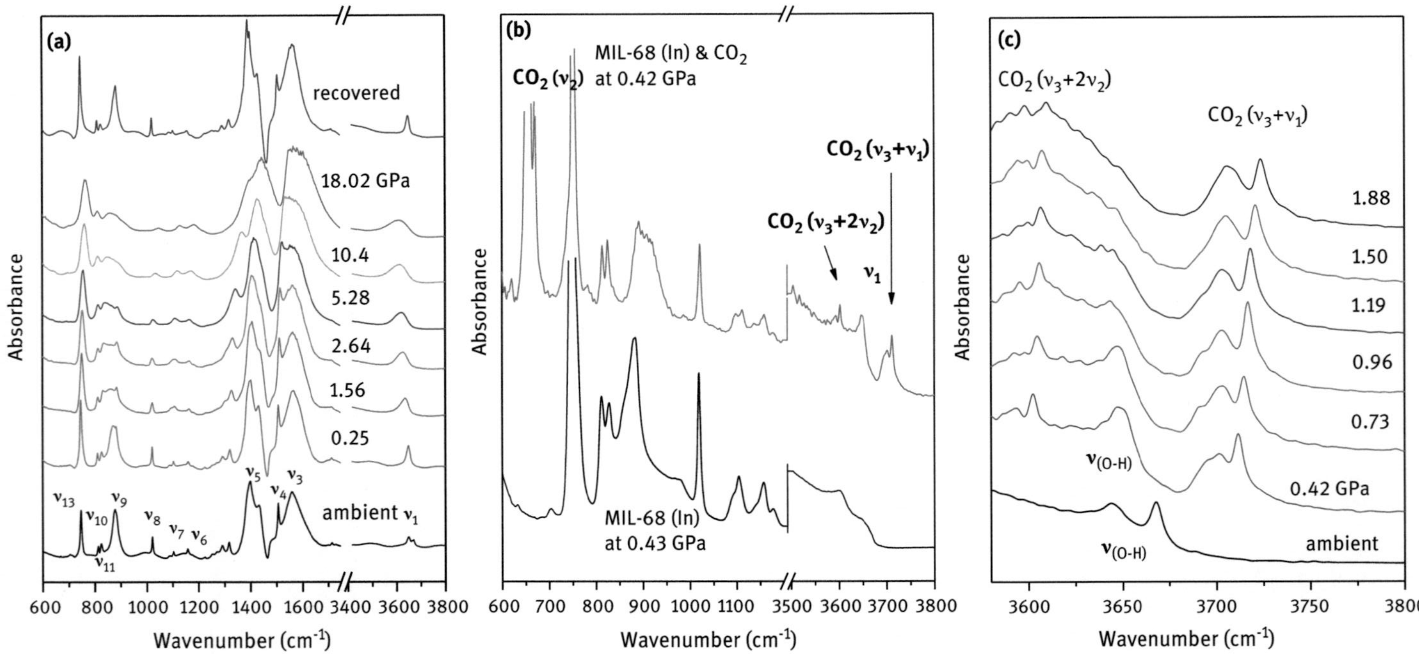

Figure 34: (a) IR spectra of activated MIL-68 (In) with PTM upon compression. (b) IR spectra of activated MIL-68 (In) and MIL-68 (In) loaded with CO_2 at around 0.4 GPa in the frequency region of 600–3,800 cm^{-1}. (c) IR spectra of MIL-68 (In) loaded with CO_2 upon compression. Reproduced with permission from ref. [28].

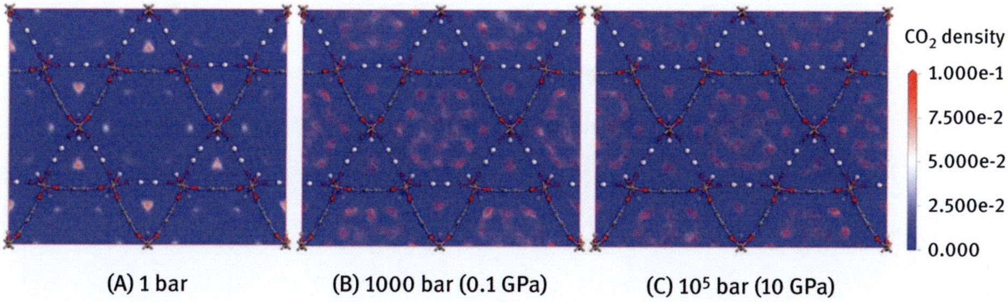

Figure 35: Simulated contour plots of the CO_2 probability density distributions along the hexagonal and triangular channels of MIL-68 (In) framework at (a) 1 bar, (b) 1,000 bar or 0.1 GPa, and (c) 10^5 bar (or 10 GPa). Reproduced with permission from ref. [28].

>1.5 GPa) (Figure 34b). The pressure-regulated CO_2 occupation in different channels of the MIL-68 framework is found to be reversible between compression and decompression (Figure 34c). The unique adsorption behavior of CO_2 in the framework is associated with the OH units that contribute as the primary binding sites through hydrogen bonding with CO_2. These observations are strongly supported by the simulation results (Figure 35). The high stability and enhanced CO_2 adsorption of MIL-68 (In) under high pressure make it a promising candidate for greenhouse gas storage.

5.3 PdSDB and CbSDB

The SDB (SDB = 4,4′-sulfonyldibenzoate)-based MOFs have attracted much attention because of their high CO_2 selectivity over other gas molecules. Although PbSDB and CdSDB are built from the same V-shaped organic ligand (SDB), however, different metal centers result in different coordination environments, thus featuring different topologies, shapes of channels, and pore sizes. Using *in situ* Raman and FTIR spectroscopy as well as synchrotron powder X-ray diffraction (XRD), the structural stabilities, possible phase transitions of activated PdSDB and CdSDB frameworks, and the guest–host interactions in CO_2-loaded PbSDB and CdSDB frameworks at high pressures were investigated by Jiang et al. [29].

As shown in Figure 36a, Raman measurements on PbSDB reviewed prominent pressure-induced structural changes from the crystalline phase to the amorphous phase. Such a change is mostly irreversible upon pressure release. The IR measurements (Figure 36b) suggest the framework topology, and chemical connectivity remain robust despite the reduction of crystallinity. A similar pressure effect was found in activated CdSDB framework, but with substantially higher onset transition pressure for amorphization, indicating that CdSDB is more resilient to structural deformation induced by pressure.

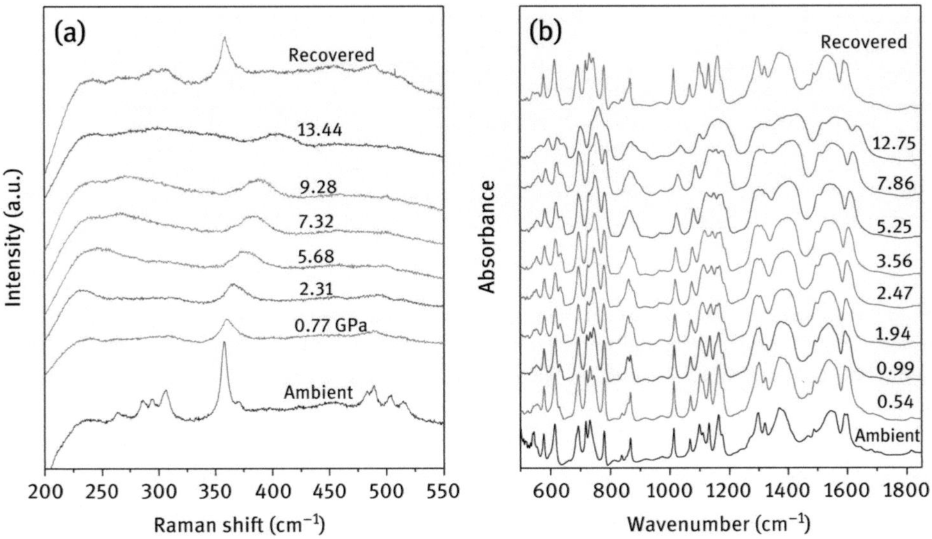

Figure 36: Raman (a) and IR (b) spectra of PbSDB at selected pressures upon compression and recovery. Reproduced with permission from ref. [29].

Then the CO_2 adsorption performance under high pressures was investigated comparatively between CdSDB and PbSDB by examining the CO_2-loaded frameworks using FTIR spectroscopy, as shown in Figures 37 and 38. The splitting of both combination modes (i.e., $\nu_3 + 2\nu_2$ and $\nu_3 + \nu_1$) at 0.63 GPa (Figure 37a) indicates the insertion of CO_2 into the pores of the CdSDB framework resulting in two different types of CO_2 in the system. The lower frequency modes (3,582 and 3,690 cm^{-1}) can be assigned as the CO_2 inside the CdSDB channels while the higher frequency modes (3,603 and 3,712 cm^{-1}) are due to the CO_2 outside the framework (Figure 37c). Further compression to 0.87 GPa resulted in more prominent separation of CO_2 in two different environments due to the solid CO_2 formed from liquid-to-solid phase transition residing outside the framework. The red shift suggests the weakening of the C–O bond in CO_2 as a result of enhanced guest–host interaction. The most interesting observation is the appearance of a new IR mode as a shoulder for the $\nu_3 + \nu_1$ mode of CO_2 inside the framework when compressed to 3.45 GPa (Figure 37a). The additional IR absorption suggests the formation of an additional nonequivalent CO_2 adsorption site within the framework due to compression. Upon decompression, the new CO_2 adsorption remains observable down to 2.5 GPa (Figure 37b), below which the IR profile in this region suggests that CO_2 mainly resides on a single adsorption site in the low-pressure region. These observations suggest that high external pressure is responsible for creating and maintaining the new adsorption site, which is highly dependent on the framework structures that are affected by pressures.

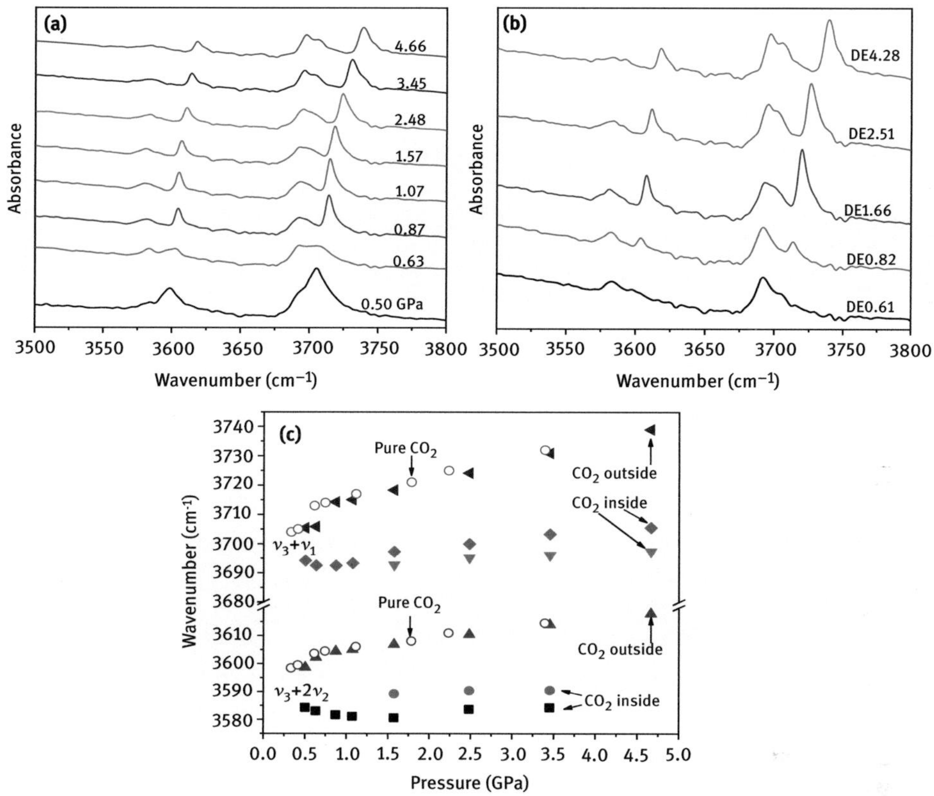

Figure 37: IR spectra of CdSDB loaded with CO_2 upon (a) compression from 0.50 to 4.66 GPa and (b) decompression to 0.61 GPa in the spectral region of 3,500–3,800 cm^{-1}. (c) Pressure dependence of $\nu_3 + 2\nu_2$ and $\nu_3 + \nu_1$ modes of loaded CO_2 in CdSDB. Reproduced with permission from ref. [29].

Compared to the IR spectrum of pure liquid CO_2 collected at 0.33 GPa, interestingly, the adsorbed CO_2 PbSDB framework at similar loading pressure (e.g., 0.29 GPa) exhibited a distinctively different IR profile in both the band shape and particularly the frequencies that are substantially red shifted (Figure 38a). With increasing pressure to 0.59 GPa, the $\nu_3 + \nu_1$ mode exhibits an obvious splitting leading to the observation of a higher frequency component at 3,694 cm^{-1}. The splitting of this mode is most likely associated with the formation of a new, nonequivalent CO_2 adsorption site of the PbSDB framework, similar to the observation of CdSDB framework at 3.45 GPa, although the formation pressure is much lower here. With further compression, strikingly, the intensity of the new higher-frequency component of $\nu_3 + \nu_1$ mode develops rapidly and becomes the dominant mode above 1.68 GPa. At 4.28 GPa, the original low-frequency component of $\nu_3 + \nu_1$ mode is completely depleted with only a single sharp symmetrical IR mode observed at 3,705 cm^{-1}. This low frequency indicates that all the adsorbed CO_2 is confined in PbSDB framework via the strong guest–host

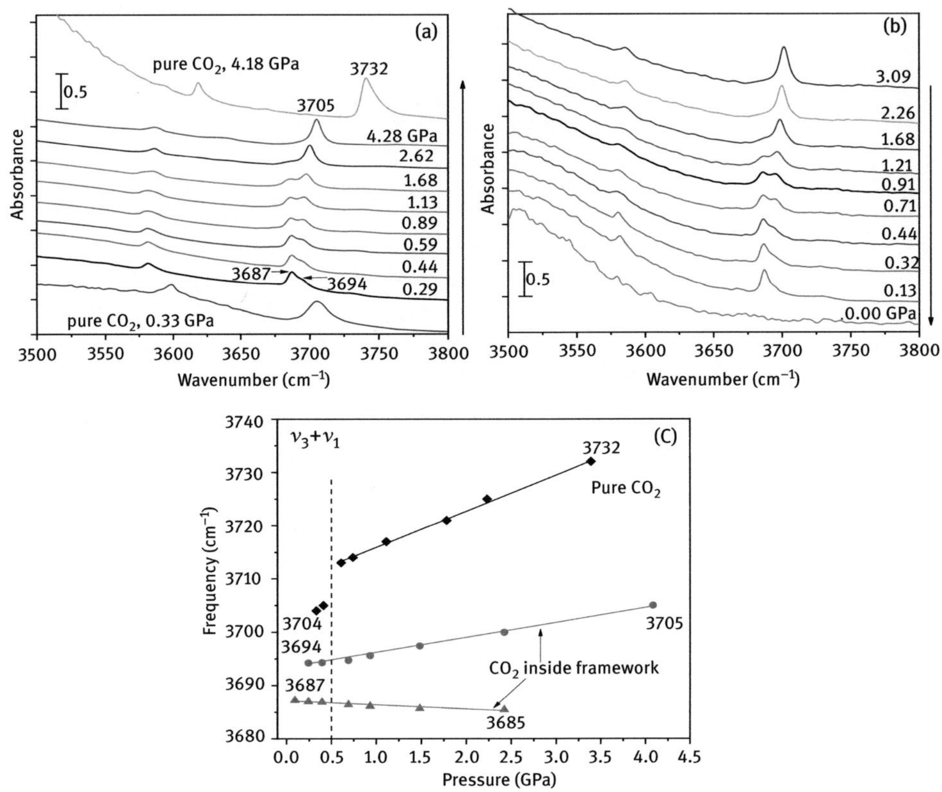

Figure 38: IR spectra PbSDB loaded with CO_2 upon (a) compression from 0.24 to 4.28 GPa in comparison with that of pure CO_2 at representative pressures and (b) decompression to ambient pressure in the spectral region of 3,500–3,800 cm^{-1}. (c) The pressure dependence of the $\nu_3 + \nu_1$ mode of CO_2 loaded with PbSDB (red circles and triangles) and that of pure CO_2 (black squares). Reproduced with permission from ref. [29].

interactions. On decompression (Figure 38b), the spectral revolutions suggest that pressure-regulated formation and new CO_2 adsorption site is completely reversible.

Interestingly, careful examination of the pressure dependence on the two components of the $\nu_3 + \nu_1$ mode reveals that the low-frequency component undergoes a red shift while the high-frequency component shifts to blue (Figure 38c), in strong contrast to the CO_2 loaded CdSDB systems, and thus the mechanism of pressure-regulated adsorption sites could be different in the two systems. Based on structural analysis, CdSDB is expected to exhibit two distinctive adsorption sites spectroscopically. At low loading pressures (e.g., <2.5 GPa) in this experiment, however, only one CO_2 adsorption site was observed as evidenced by the single peak of both $\nu_3 + 2\nu_2$ and $\nu_3 + \nu_1$ combination modes. This can be interpreted as although the two adsorption sites are crystallographically nonequivalent, the corresponding vibrational frequencies are near degenerate due to the small difference between these two sites

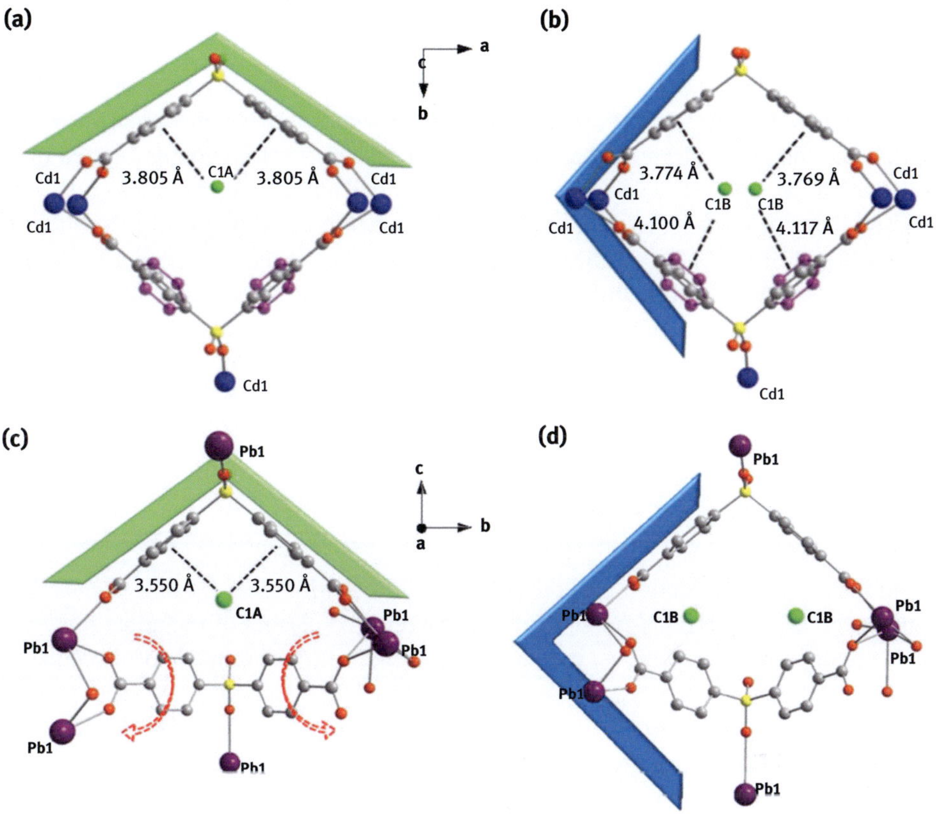

Figure 39: Schematics of CO_2 adsorption sites in CdSDB at ambient pressure and low temperature (a) and (b), and at high pressures (b), and in PbSDB at ambient pressure (c) and high pressures (d). The green balls represent CO_2 molecules inside the pores of CdSDB and PbSDB, with the oxygen atoms omitted. The green and blue brackets are showing two different types of π-pockets in the channels. The dash lines symbolize the interactions between CO_2 and the specific adsorption sites of the frameworks. Reproduced with permission from ref. [29].

(Figure 39a). Upon further compression, however, due to the different extent of the guest–host interaction associated with nonequivalent environment of the SDB framework, the degeneracy of the vibrational frequency is removed, resulting in the observation of a higher frequency component associated with the second adsorption site (Figure 39b).

In contrast, single-crystal X-ray diffraction on CO_2-loaded PbSDB frame shows that only one CO_2 adsorption site is possible given the orientations of the SDB ligands in the cavity. This single adsorption site is similarly constituted by a π-pocket from the phenyl rings of the same SDB linker that is 3.550 Å away from the CO_2 center (Figure 39c). The observation of the single component of the $\nu_3 + \nu_1$ mode at low loading pressures (e.g., 0.29 GPa) is consistent with that only one π-pocket is favorable at low pressures. At high pressures, the clear splitting of the $\nu_3 + \nu_1$ mode strongly suggests the formation of a new CO_2 adsorption site. Therefore, it is necessary that the phenyl rings involving two SDB linkers undergo a pressure-induced change in their orientations. The fact that the original adsorption site (C1A) remains upon compression indicates that the π-pocket involving the two phenyl rings from the same SDB linker is affected very little by the pressure. Thus, it is highly likely that pressure changes the orientation of the phenyl rings in two adjacent SDB linkers substantially (as illustrated in Figure 39c), creating a second π-pocket. As indicated in Figure 39d, the rotation along the axis collinear with the C-S and C-COO bonds can change the phenyl orientation. Such rotation results in the two phenyl rings from the two SDB linkers face each other, effectively creating a new CO_2 adsorption site (C1B, Figure 39d).

6 Summary and outlook

In summary, the elucidation of molecular and crystal structures, as well as chemical and structural transformation mechanisms for selected classes of materials rendered under high-pressure conditions, was achieved using *in situ* vibrational spectroscopy as the essential tool. Specifically, pressure-mediated hydrogen bonding interactions, pressure and photon-tuned chemical reactions, as well as guest–host interactions in nanoporous MOF materials can be efficiently and accurately probed and characterized using *in situ* high-pressure Raman spectroscopy, FTIR spectroscopy, often combined with other local lab-based or synchrotron-enabled techniques. These spectroscopic data provided enormously valuable information for us to understand the pressure-induced phenomena at the microscopic level in-depth. Thus, vibrational micro-spectroscopy can be considered a routine but indispensable technique to obtain first-hand, reproducible spectroscopic data in any high-pressure material research labs.

In addition to extreme pressure, extreme temperatures such as several Kelvin and over a thousand degrees Celsius, and especially their combinations pose experimental challenges yet offer new and unexplored *P*–*T* domains for novel structures and properties of materials to be discovered. For instance, numerous novel stoichiometry compounds have been synthesized at simultaneous high-P and high-T conditions. *In situ* vibrational micro-spectroscopy is expected to play an important role in structural characterization under these tough conditions. Although extremely convenient, one should realize that vibrational spectroscopy itself alone is seldom successful in solving totally unknown structures. Therefore, other experimental techniques such as synchrotron X-ray diffraction and spectroscopy, and especially computational modeling, are also essential to realize the full potential of vibrational spectroscopy on structural characterization for novel materials formed under extreme conditions.

References

[1] Hemley RJ, Ashcroft NW. The revealing role of pressure in the condensed matter sciences. Physics Today 1998;51(8):26.

[2] Hemley RJ. Effects of high pressure on molecules. Annual Review of Physical Chemistry 2000;51:763.

[3] Schettino V, Bini R. Molecules under extreme conditions: Chemical reactions at high pressure. Physical Chemistry Chemical Physics 2003;5(10):1951.

[4] McMillan PF. Chemistry at high pressure. Chemical Society Reviews 2006;35(10):855.

[5] Grochala W, Hoffmann R, Feng J, Ashcroft NW. The chemical imagination at work in very tight places. Angewandte Chemie-International Edition 2007;46(20):3620.

[6] Schettino V, Bini R. Constraining molecules at the closest approach: Chemistry at high pressure. Chemical Society Reviews 2007;36(6):869.

[7] Song Y, Manaa MR. New trends in chemistry and materials science in extremely tight space. Journal of Physical Chemistry C 2012;116(3):2059.

[8] Bini R, Schettino V. Materials Under Extreme Conditions. Imperial College Press, London, UK 2014.

[9] Mao HK, Chen B, Chen JH, Li K, Lin JF, Yang WE, Zheng HY. Recent advances in high-pressure science and technology. Matter Radiat Extremes 2016;1(1):59.

[10] Song Y. In: Stauffer MT., editors. Applications of Molecular Spectroscopy to Current Research in the Chemical and Biological Sciences. Intech, 2016, doi:http:/dx.doi.org/10.5772/64617 http:/dx.doi.org/10.5772/64617.

[11] Mao HK, Chen XJ, Ding Y, Li B, Wang L. Solids, liquids, and gases under high pressure. Reviews of Modern Physics 2018;90(1):55.

[12] Song Y. New perspectives on potential hydrogen storage materials using high pressure. Physical Chemistry Chemical Physics 2013;15(35):14524.

[13] Mao HK, Bell PM, Shaner JW, Steinberg DJ. Specific volume measurements of copper, molybdenum, palladium, and silver and calibration of the ruby R1 fluorescence pressure gauge from 0.06 to 1 Mbar. Journal of Applied Physics 1978;49(6):3276.

[14] Guan J. Thesis, University of Western Ontario, 2018.

[15] Dong Z. Thesis, University of Western Ontario, 2012.

[16] Dong Z, Song Y. Transformations of cold-compressed multiwalled boron nitride nanotubes probed by infrared spectroscopy. Journal of Physical Chemistry C 2010;114(4):1782.

[17] Murli C, Lu N, Dong Z, Song Y. Hydrogen bonds and conformations in ethylene glycol under pressure. Journal of Physical Chemistry B 2012;116(41):12574.

[18] Zhou L, Shinde N, Hu A, Cook C, Murugesu M, Song Y. Structural tuning of energetic material bis(1H-tetrazol-5-yl)amine monohydrate under pressures probed by vibrational spectroscopy and x-ray diffraction. Journal of Physical Chemistry C 2014;118(46):26504.

[19] Wang P, Guan JW, Galeschuk DTK, Yao YS, He CF, Jiang S, Zhang SJ, Liu Y, Jin ML, Jin CQ, et al. Pressure-induced polymorphic, optical, and electronic transitions of formamidinium lead iodide perovskite. Journal of Physical Chemistry Letters 2017;8(10): 2119.

[20] Guan RF, Wang P, Song Y, Staroverov VN. Pressure-induced polymorphic transformations of ethylenediamine bisborane. Journal of Physical Chemistry C 2021;125(34):18614.

[21] Guan J, Song Y. Pressure selected reactivity and kinetics deduced from photoinduced dissociation of ethylene glycol. Journal of Physical Chemistry B 2015;119(8):3535.

[22] Guan JW, Song Y. Highly efficient and selective pressure-assisted photon-induced polymerization of styrene. Journal of Chemical Physics 2016;144(21):8.

[23] Guan JW, Daljeet R, Kieran A, Song Y. Pressure-induced amorphization and reactivity of solid dimethyl acetylene probed by *in situ* FTIR and Raman spectroscopy. Journal of Physics: Condensed Matter 2018;30(22):8.

[24] Guan JW, Daljeet R, Song Y. Pressure-selected reactivity between 2-butyne and water induced by two-photon excitation. Canadian Journal of Chemistry 2017;95(11):1212.

[25] Guan JW, Wang P, Song Y. Mechanistic insights of multiplexed effects of pressure, physical states, and laser on the polymerization kinetics of phenylacetylene probed by *in situ* FTIR spectroscopy. Journal of Physical Chemistry B 2021;125(16):4169.

[26] Hu Y, Kazemian H, Rohani S, Huang Y, Song Y. *In situ* high pressure study of ZIF-8 by FTIR spectroscopy. Chemical Communication 2011;47(47):12694.

[27] Hu Y, Liu Z, Xu J, Huang Y, Song Y. Evidence of pressure enhanced CO2 storage in ZIF-8 probed by FTIR spectroscopy. Journal of the American Chemical Society 2013;135(25):9287.

[28] Hu Y, Lin B, He P, Li Y, Huang Y, Song Y. The structural stability of and enhanced CO_2 storage in MOF MIL-68 (In) under high pressures probed by FTIR spectroscopy. Chemistry – A European Journal 2015;21:18739.

[29] Jiang S, Hu Y, Chen SS, Huang YN, Song Y. Elucidation of the structural origins and contrasting guest-host interactions in CO2-loaded CdSDB and PbSDB metal-organic frameworks at high pressures. Chemistry-a European Journal 2018;24(72):19280.

Ian S. Butler

Diamond-anvil cell molecular spectroscopy of solid materials

1 Introduction

The effect of "extreme conditions" on the structures of materials is of considerable interest to materials chemists, physicists, geologists, and engineers. This term implies subjecting materials to very high temperatures and/or high pressures. In this chapter, we will focus chiefly, however, on the effects of high pressures on solid inorganic materials.

One of the simplest ways of achieving high pressures in the laboratory is through the use of diamond-anvil cells (DACs). In a DAC, materials are squeezed together between the parallel faces of two opposed gemstone-quality diamonds. Two excellent histories on the DAC are available [1, 2]. The first DACs were developed by Charles E. Weir, Ellis R, Lippincott, Alvin Van Walkenburg, and Elmer N. Bunting at the National Bureau of Standards (NBS) in the USA in 1958 [3]. Their research built on the pioneering work of the 1946 Nobel Laureate in Physics, Percy W. Bridgeman, who had succeeded in squeezing materials between two opposed anvils made of tungsten carbide [4]. Professor Bridgeman was particularly famous for being the first person to show that the protein in an egg could be denatured by the application of pressure at room temperature [5]. The NBS researchers focused their work on the effects of high pressures on X-ray diffraction and infrared spectroscopic measurements. Shortly afterwards, other university researchers became interested and, eventually, boutique companies such as High Pressure Diamond Optics in Tucson, Arizona, and, more recently, Almax-easyLab in the UK, were established for the commercial production of DACs [6]. The Tucson company was founded by Alvin Van Valkenburg, another former NBS scientist. These commercial DACs can achieve pressures of up to about 20 GPa (~200 kbar). The culet sizes of the two gemstone-quality diamonds are typically 100–250 μm and the smaller the diamonds, the higher the pressures that can be achieved.

Acknowledgments: I would like to thank all the undergraduate and graduate students, postdoctorals, and visiting scientists, who have been involved in this high-pressure work over the past 35 years, for their efforts. In addition, I would like to acknowledge the generous research funding from the following agencies: NSERC (Canada), FQRNT (Quebec), IUPAC, and NATO.

Ian S. Butler, Department of Chemistry, Otto Maass Building, Room 002, McGill University, 801 Sherbrooke Street West, Montreal, Quebec, Canada H3A 0B8

https://doi.org/10.1515/9783110668612-002

An initial problem with the DAC was how does one determine the actual pressure between the two diamond faces? It was quickly realized that there was a need for a thin metal gasket between the two diamonds in the DAC to ensure a uniform pressure gradient between them. These gaskets are usually made of stainless steel or Inconel with approximately a 300-μm hole drilled in the center. Originally, these holes were drilled using the expertise of Swiss watchmakers. Nowadays, the holes are laser drilled. The next problem was how could the pressure being applied be determined? The breakthrough came in 1972 when some of the original NBS scientists and others discovered that the intense R_1 fluorescence of ruby red-shifted systematically with increasing pressure [7]. The NBS researchers continued their work and finally succeeded in establishing that a small ruby chip provided a convenient calibration measure for the pressure in a DAC. This calibration proved to be linear up to almost 200 kbar [8]. Modifications of the original DAC design quickly led to high-pressure studies being performed using both powder X-ray [9] and single-crystal X-ray diffraction, as well as infrared and Raman spectroscopy [10]. Even viscosity measurements can be undertaken at high pressures [11].

2 Experimental methods

Most of our own infrared and Raman high-pressure studies over the past 35 years were undertaken using DACs from High Pressure Diamond Optics (Tucson, Arizona, USA) (Figure 1). Infrared measurements were made on a Bruker IFS-48 FTIR or a Nicolet FTIR instrument. The Raman work was performed on either a Bruker IFS-88 FTIR spectrometer coupled to a Bruker FRA-105 Raman module and on a Renishaw In Via spectrometer. We almost always used the R_1 fluorescence line of a ruby chip as the pressure calibrant and determined the pressure using the following equation:

$$P = 1.775Xv - \left(0.795ve^{-Xv/78}\right)$$

where Xv is the difference in cm^{-1} between the peak position at pressure P and at atmospheric pressure.

Occasionally, we used an alternative technique to determine pressures up to about 62 kbar. This technique involved monitoring the behavior of the T_{2g} phonon mode of the type-IIA diamonds in a DAC using FT-Raman spectroscopy and 1,064.1 nm laser excitation (Figures 2 and 3) [12]. By making measurements at the sample–diamond interface, we found that there is a linear relationship between the phonon peak position and the applied pressure, so that $d\upsilon/dP = 0.16\ cm^{-1}\ kbar^{-1}$.

Figure 1: High-pressure diamond-anvil cell from High Pressure Diamond Optics, Tucson, Arizona, USA. Photograph courtesy of Tristan Borchers, Department of Chemistry, McGill University, Montreal, Quebec, Canada.

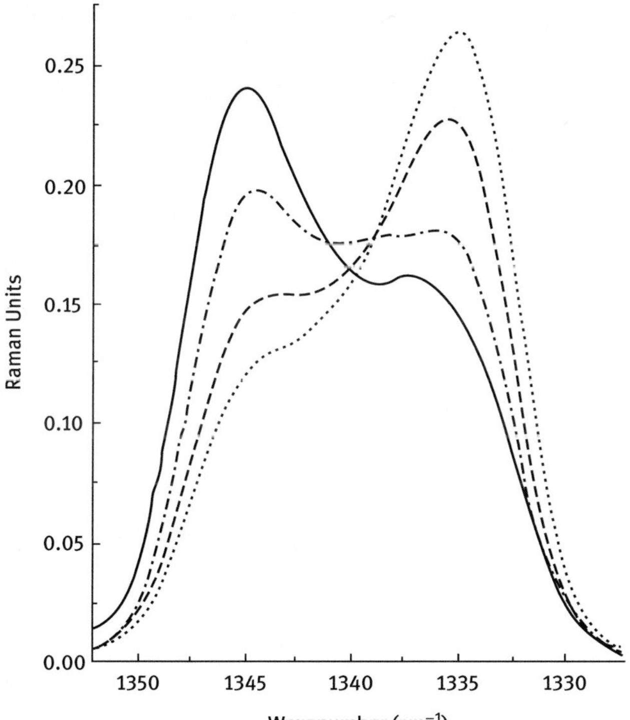

Figure 2: FT Raman spectra of the T_{2g} phonon mode of the DAC at 6.5 kbar pressure focused at various depths in the DAC at optimum depth resolution and fitted bands at these depths. (—) sample–diamond interface; (- – -) center (reproduced with permission from Butler et al. [12]. Copyright @ 1997, Sage Publications).

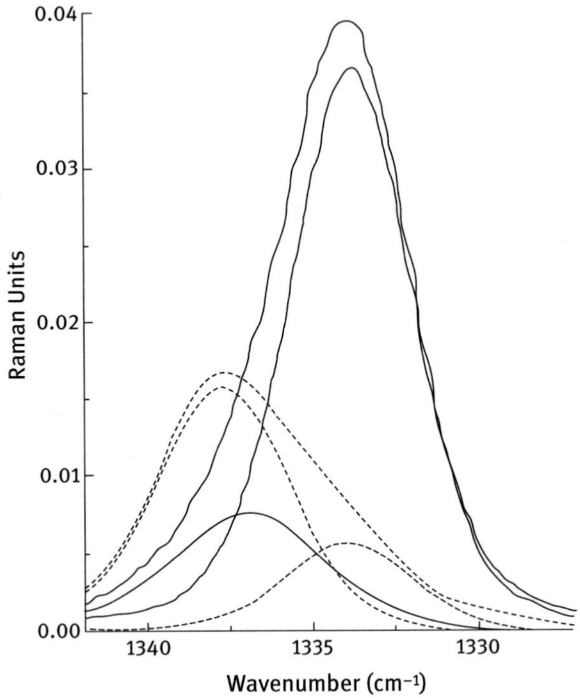

Figure 3: Variation in fitted band positions with pressure at optimum depth resolution at focused at the sample–diamond interface. The line is the least-squares fit of the highest energy band at each pressure (reproduced with permission from Butler et al. [12]. Copyright @ 1997, Sage Publications).

3 High-pressure studies on solid materials

We originally started to extend our initial vibrational spectroscopy work into the realm of solid materials in 1987 by repeating a high-pressure Raman spectroscopic study on mercuric cyanide, $Hg(CN)_2$ [13]. Our work was apparently the first time that anyone had located a DAC cell on a micro-Raman spectrometer (Instruments SA U-1000 Ramanor). This particular instrument was equipped with a Nachet optical microscope and we used green laser excitation (514.5 nm, Spectra Physics Model 164–5 W) to study phase transitions in solid materials. The DAC used in this case was purchased from Diacell Products (Leicester, UK), which was founded by Dr. David M. Adams. It employed two type-IIA diamonds that were specially selected from a diamond merchant in Amsterdam because of their low fluorescence. *n*-Octane was usually used as a hydrostatic pressure-transmitting fluid between the diamonds to ensure a uniform pressure across the faces of the gemstones. We chose $Hg(CN)_2$ as the test material for this setup because Wong [14] had reported a high-pressure Raman study of this material about eight years earlier for pressures up to about 45 kbar in a DAC using a conventional optical bench arrangement. He presented evidence of structural phase transitions at 20 and 40 kbar. In our work, we examined the changes occurring in the Raman spectra in the CN stretching region – about 30 spectra were recorded at different pressures up to 45 kbar at 20 min intervals over a 12 h period. As shown in Figure 4, only one ν(CN) peak was observed initially but, at about 16 kbar, this peak began to be replaced by two new ones. This peak eventually completely disappeared at 20 kbar. The two new peaks, which were separated by about 10 cm^{-1}, continued to be observed until the higher one vanished at 40 kbar (Figure 5). Our micro-Raman-DAC data confirmed the two pressure-induced phase transitions occurring in $Hg(CN)_2$ reported by Wong [14] and showed that excellent high-pressure results could be readily obtained with our instrumental arrangement.

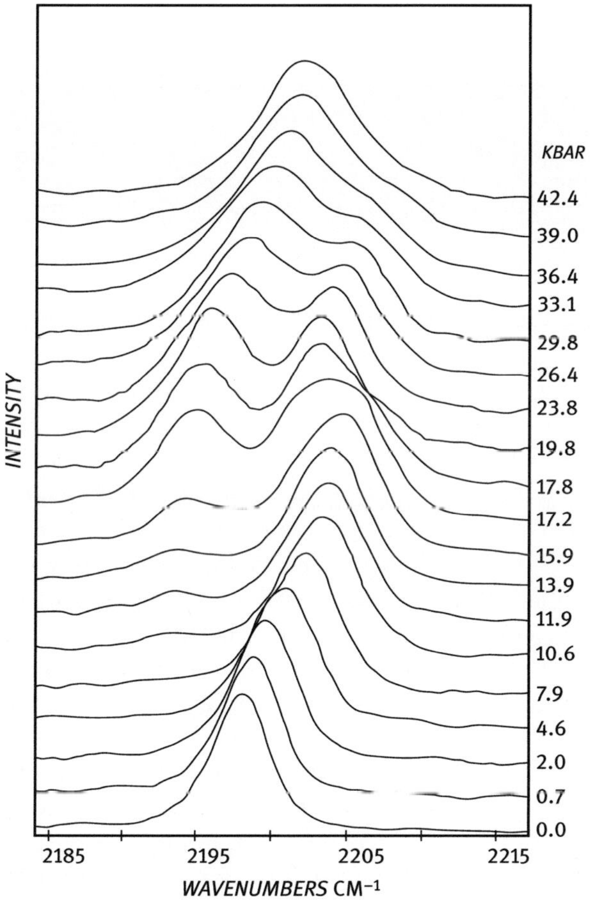

Figure 4: Micro-Raman spectra of solid $Hg(CN)_2$ in the CN stretching region at different pressures (reproduced from Benham et al. [13]. Copyright @ 1987, Sage Publications).

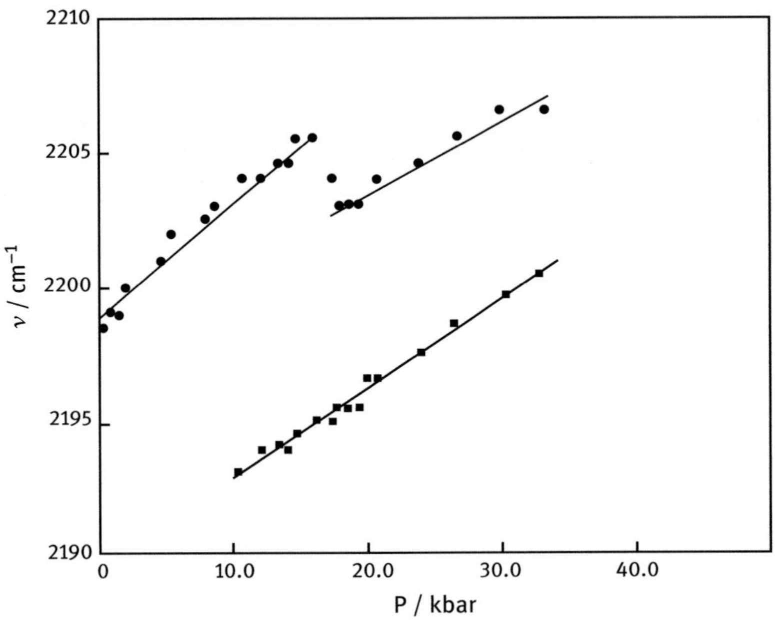

Figure 5: Pressure dependences of the υ(CN) modes of $Hg(CN)_2$ (reproduced from Benham et al. [13]. Copyright @ 1987, Sage Publications).

Our next high-pressure study, using the same micro-Raman spectrometer mentioned above, was on the cesium salt of the icosahedral $[B_{12}H_{12}]^{2-}$ anion [13]. Earlier X-ray analyses had shown that this material crystallizes in a face-centered cubic structure ($Fm3$; T_h^3) with four molecules per unit cell [15, 16]. A group theoretical analysis revealed that there are two Raman-active B–H stretching modes expected, a_g and h_g, which were observed in our work at atmospheric pressure at 2506 and 2471 cm^{-1}, respectively. Even though a factor group treatment suggests that the h_g B–H stretching mode should split under T_h site symmetry into an $e_g + t_{2g}$ doublet, this splitting was not observed. Both the a_g and the h_g B–H stretching modes did, however, shift gradually to higher wavenumbers as the pressure was increased in the DAC (Figure 6). A phase transition was detected at 8.7 kbar. The pressure coefficients, dv/dP for the a_g and h_g modes in the low- and high-pressure regimes were 1.6 and 1.9 cm^{-1} $kbar^{-1}$ and 2.1 and 2.3 cm^{-1} $kbar^{-1}$, respectively.

Once we had established that our DAC-micro-Raman spectroscopic approach was completely satisfactory, we moved on to other materials. We investigated several cage structures, e.g., o-$B_{10}C_2H_{12}$, 1-, and 2-chloroadamantane. These globular molecules, which are sometimes known as "plastic crystals," were known to undergo order–disorder phase transitions in the solid state at low temperatures [15, 16]. We rationalized that it might be possible to induce such transitions under high pressures. This indeed proved to be the case and the vibrational data revealed

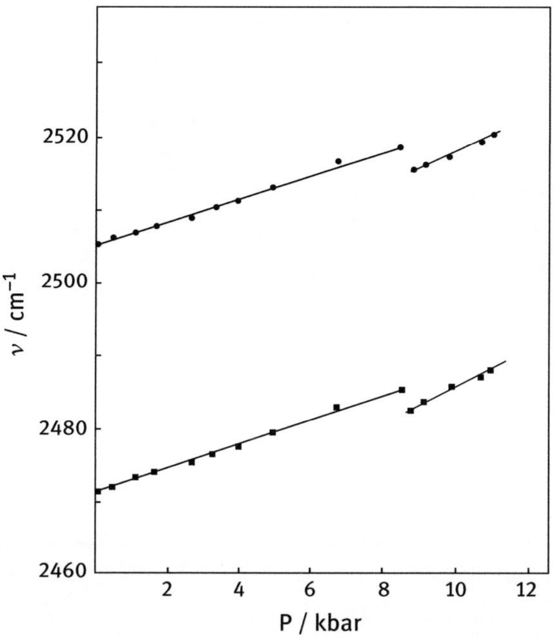

Figure 6: Pressure dependences of the a_g and h_g B–H stretching modes of $Cs_2[B_{12}H_{12}]$ (reproduced from Benham et al. [13]. Copyright @ 1987, Sage Publications).

that the high-pressure phases produced mirrored those spectroscopically obtained at low temperatures [17]. The 2-chloroadamantane material was also investigated by differential scanning calorimetry. Two transitions were detected at 231 and 178 K on cooling, and at 242 and 227 K on heating. For the change of phase I to phase II, the enthalpy and entropy were 8.3 kJ mol^{-1} and 25 K^{-1} mol^{-1}, while for the change of phase II to phase I, the enthalpy and entropy were 0.47 kJ mol^{-1} and 2.3 K^{-1} mol^{-1}.

Following this work, we next moved on to study the effect of high pressure on the order–disorder phase transition in 2-adamantanone at pressures up to ~21 kbar [18]. There was indeed a pressure-induced phase transition and a large hysteresis was observed in both the compression and decompression cycles with the order–disorder transition being observed at ~10.0 and ~6.9 kbar, respectively. The Raman spectra of the high-pressure ordered phase was examined carefully and the peak shifts and the small splittings obtained were similar to those observed for the ordered phase formed at low temperature (150 K) suggesting the production of the same crystalline phase. The mode Grüneisen parameters were determined for the ordered and disordered phases as $0.031 < x_i < 0.057$ and $0.011 < x_i < 0.046$, respectively, that is, somewhat lower than those known for the parent adamantane molecule.

It had been established earlier that the seven-membered cage hydrocarbons, norbornane (C_7H_{12}), norbornylene (C_7H_{10}), and norbornadiene (C_7H_8), undergo phase transitions from high-temperature, disordered structures to low-temperature ordered ones [19, 20]. A preliminary variable-temperature study by us on the related cage molecule quadricyclane (C_7H_8), suggested that an order–disorder phase transition might also occur for this material at low temperature [21]. Consequently, we decided to investigate more thoroughly the effect of both temperature and pressure on quadricyclane [22], especially since this particular molecule apparently has considerable potential in solar-to-thermal energy conversion processes [23]. On the basis of the observed vibrational spectra, a factor group analysis was undertaken to predict the possible space group symmetries of the ordered low-temperature phase. The pressure-induced order–disorder phase transition occurred at 9.3 ± 0.5 kbar on compression and ~7 kbar on decompression (Figures 7 and 8). Upon decompression, the ordered phase transformed back to the disordered one. The volume change at the transition [24] was estimated to be 4.0 cm^3 mol^{-1}. This value is larger than that for both norbornylene (3.1 cm^3 mol^{-1}) [25] and adamantane (2.2 cm^3 mol^{-1}) [26]. Based on the observed variable-temperature infrared and Raman spectra and the band splittings detected, we concluded that the most probable structures of the ordered phase were monoclinic (C_s) with $Z = 2$, orthorhombic (D_2 or C_{2v}) also with $Z = 2$, or tetragonal (D_{2d}) with $Z = 4$.

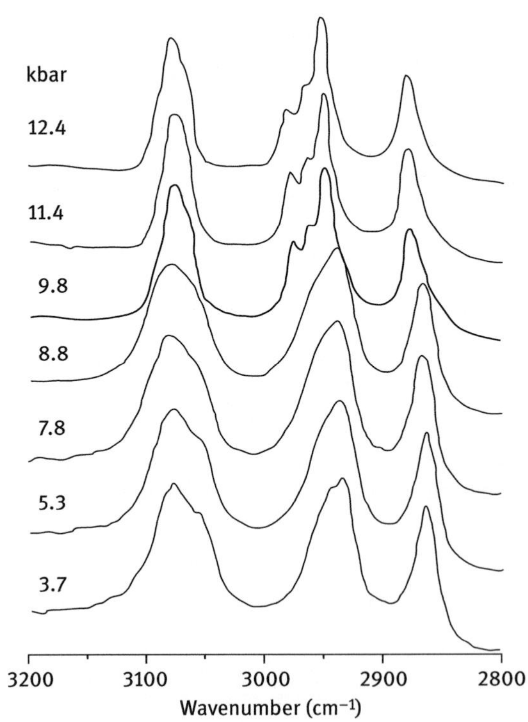

Figure 7: Infrared spectra in the C–H stretching region of quadricyclane as a function of pressure on compression (reproduced with permission from Kawai et al. [22] Copyright @ 1992, American Chemical Society).

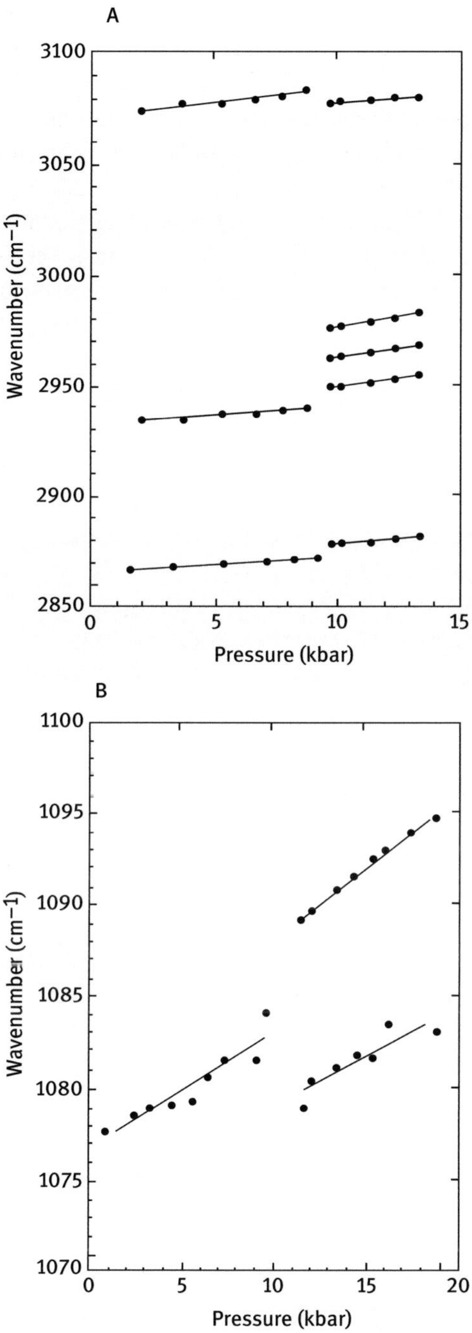

Figure 8: Pressure dependences of various quadricyclane vibrations in infrared (A) and Raman (B) spectra (reproduced with permission from Kawai et al. [22] Copyright @ 1992, American Chemical Society).

Our next steps in studying the effect of high pressures on solid materials involved investigations of simple inorganic salts and various organometallic complexes. We first looked at potassium hexabromostannate(IV), K_2SnBr_6 [27]. A pressure-induced phase transition was detected by Raman spectroscopy at around 19 kbar. All the observed bands exhibited blue shifts with increasing pressure (Figure 9). Grüneisen parameters were calculated from the pressure dependences and these were greater for the high-pressure phase than for the low-pressure monoclinic one indicating that the former is denser and less compressible than the low-pressure one. A year after we published our work, Adams and Norman [28] reported a synchrotron energy-dispersive X-ray powder diffraction study of K_2SnBr_6, which showed a transformation at 14 kbar and above from the parent monoclinic cell to a cubic structure of symmetry Fm3m, which is stable to 120 kbar. Another phase appears to exist above 120 kbar.

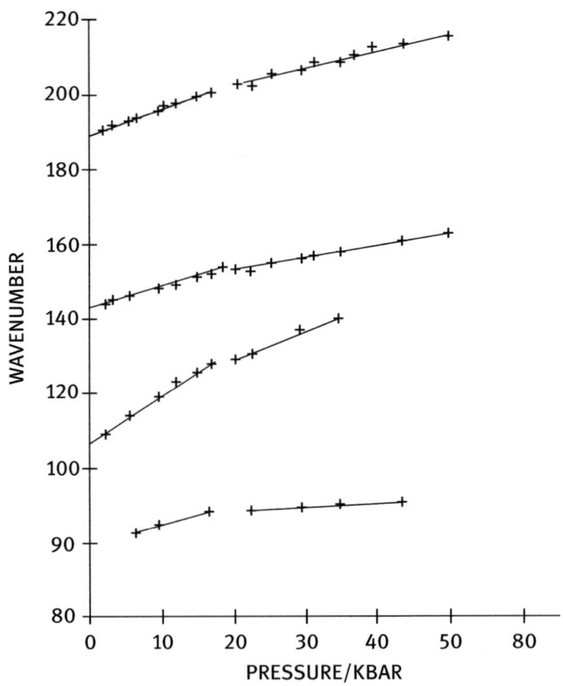

Figure 9: Pressure dependence of selected peaks in the Raman spectra of K_2SnBr_6 (reproduced from Butler and Huang [27]. Copyright @ 1990, Wiley Analytical Science).

We then moved on to look at the micro-Raman spectra of potassium chromate(VI), K_2CrO_4, and tungstate(VI), K_2WO_4, at pressures approaching 45 kbar [29]. The chromate complex did not exhibit any evidence for a phase change but the tungstate complex did at about 13 kbar. The observed pressure dependences of the vibrational bands were comparable to those reported for typical *spinels* but slightly larger than those for *scheelites*. K_2CrO_4 was also subjected to a high-pressure, angle-dispersive X-ray study in a DAC [30]. Although it is a highly compressive pseudo-*scheelite*, there is no pressure-induced phase transition observed up to 52 kbar. A gradual decrease in the relative intensities of some of the diffraction peaks was observed above 16 kbar, which from calculations are consistent with disordering of the CrO_4^{2-} ions. This step appears to be the first one in amorphization of the material.

Moving on to organometallics, we examined the vibrational spectroscopic effects of high pressures on the phase behavior of the prototype metal carbonyls, $CH_3Mn(CO)_5$ and $CH_3Re(CO)_5$ [31]. Both the infrared and Raman spectra were investigated for pressures up to about 18 and 40 kbar, respectively. The two complexes exhibited phase transitions (most probably second order) – $CH_3Mn(CO)_5$ at ~ 9 and $CH_3Re(CO)_5$ at ~22 kbar, respectively. Of particular interest was the fact that the pressure-wavenumber plots for the CO stretching modes revealed that the π-bonding interactions between the central Mn and Re atoms and the CO groups (especially the axial ones *trans* to the CH_3 groups) were appreciably strengthened with increasing high external pressures resulting in *negative* pressure dependences, that is, −0.82 for the Mn and −0.15 cm^{-1}/kbar for the Re complexes, respectively (Figure 10). This observation was completely in accord with the earlier observation by Adams and Ekejiuba [32] in their high-pressure Raman study of $Mn(CO)_5Br$. The application of pressure appears to enhance the π-backbonding between a transition metal and a CO group. We found a similar effect on Re-CO π-backbonding for $Re(CO)_5Cl$ [33].

A subsequent micro-Raman study of crystalline $(\eta^5\text{-}C_5H_5)Re(CO)_3$ at pressures up to 40 kbar also indicated some negative discontinuities in the CO stretching mode pressure–wavenumber plots ($d\nu/dP$ = −0.16 and −0.45 cm^{-1}/kbar), as well as a phase change, at about 9 kbar (Figures 11 and 12) [34]. Again, this situation indicates that the π-backbonding interactions are strengthened with increasing pressure, most probably in this case partly as the result of increased charge transfer from the π-cyclopentadienyl ring to the Re(I) ion.

Following our work on several other types of transition metal carbonyls, we thought it would be interesting to expand our research into the structurally related transition metal thiocarbonyls. We knew from our earlier work that the CS and CSe ligands are better π-acceptor ligands than CO [35]. Consequently, this situation should also be reflected in high-pressure vibrational measurements. Initially, we looked at $Cr(CO)_5(CS)$ and $(\eta^6\text{-}C_6H_6)Cr(CO)_2(CS)$ [36]. The infrared spectra (4,300–400 cm^{-1}) of both complexes were recorded at room temperature for pressures up to about 30 kbar, while the micro-Raman spectrum of $Cr(CO)_5(CS)$ were examined in the 450–300 cm^{-1} region at pressures up to 25 kbar. The $d\nu/dP$ values of the IR-active binary (CS) overtone and combination modes were determined and used to calculate the $d\nu/dP$ values for the parent (CS) modes, which could not be obtained directly because these absorptions were buried beneath the intense absorption of the diamonds in the DAC that is centered at 1332 cm^{-1}. The $d\nu/dP$

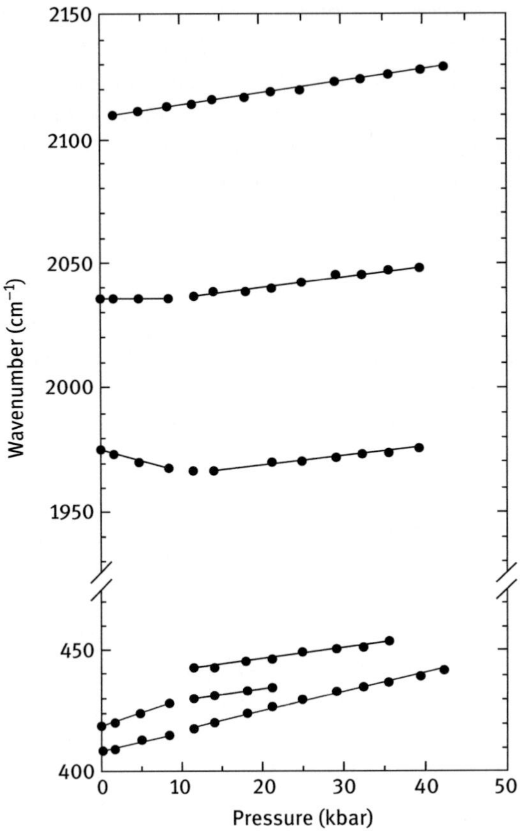

Figure 10: Pressure dependences of some of the observed Raman bands of $CH_3Mn(CO)_5$ (reproduced from Huang et al. [31]. Copyright @ 1991, American Chemical Society).

values obtained for the (CO) and (CS) modes do indeed suggest that in both complexes the π-backbonding is enhanced by increasing pressure, as we expected.

Our another organometallic study included an examination of the crystal structure and high-pressure micro-Raman spectra of $(\eta^5\text{-}C_5Me_5)Re(CO)_3$ up to 50 kbar [37]. The complex crystallizes in the triclinic space group P1 (C_i^1, $Z = 2$) rather than the $P2_1/n$ (C_{2h}^5, $Z = 4$) space group for the related $(\eta^5\text{-}C_5H_5)Re(CO)_3$ complex. The high-pressure measurements revealed a possible pressure-induced phase change at about 9 kbar. We have also investigated the utility of near-IR FT-Raman microspectroscopy for high-pressure work, e.g., on $Cr(CO)_4(bipy)$ (bipy = 2,2′-bipyridyl) [38]. This study was achieved by mounting for the first time a DAC on the optical stage of FT-Raman microprobe equipped with a long-range 4× objective an using a near-IR Nd:YAG (1,064.1 nm) laser to excite the sample. We resorted to this approach because the $Cr(CO)_4(bipy)$ complex decomposed when using a DAC with visible laser

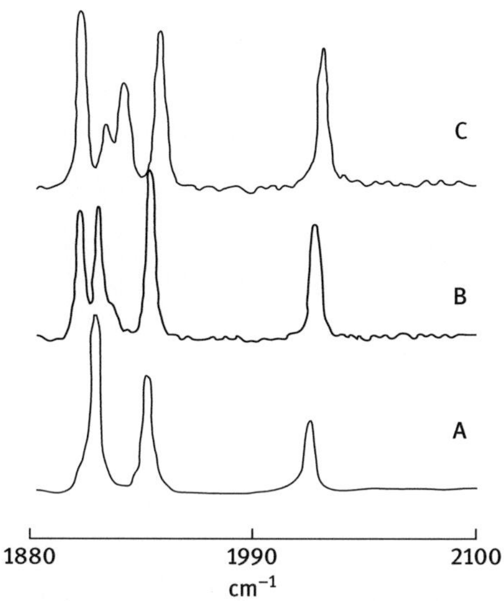

Figure 11: Micro-Raman spectra of (η^5-C_5H_5)Re(CO)$_3$ in the ν(CO) region at (A) ambient pressure, (B) 12.8 kbar, and (C) 23.3 kbar (reproduced from Huang et al. [32]. Copyright @ 1991, American Chemical Society).

excitation. Previous bipy ring mode assignments could be verified from an examination of the pressure-induced changes in the vibrational spectra. The electron-donating ability of the bipy ligand to the central chromium atom was clearly modulated by the application of pressure [39].

Since it was becoming clear to us that high-pressure vibrational spectroscopy did indeed provide an alternative way of probing the π-backbonding ability of various organometallic π-acid ligands, we also started to look into the effect of high external pressures on ligands such as isocyanides and ethylene. For example, the two crystalline chromium(0) isocyanide complexes, Cr(CO$_5$(CNBu-*t*) and *cis*-Cr(CO)$_4$ (CNBu-*t*)$_2$, were examined using the DAC method [40]. The high-pressure measurements were performed at ambient temperature for pressures up to 45 kbar. Both complexes underwent possible second-order phase transitions at about 11 and 6 kbar, respectively. The Cr–CO and Cr–CN backbonding interactions were increased upon the application of high pressures and it could be concluded that the CNBu-*t* ligand is a weaker π-acceptor than the CO ligand. The (CO) modes involving the vibrations of the CO groups *trans* to the CNBu-*t* ligands had higher pressure dependences. In the case of the ethylene ligand, we investigated the pressure dependences of the IR and Raman spectra of the two Zeise's complexes, K[Pt(η^2-C_2H_4)Cl$_3$] and [Pt(η^2-C_2H_4)Cl$_2$]$_2$, for pressures up to ~33 kbar [41]. No phase changes were detected for either complex. The observed $d\nu/dP$ values ranged from −0.13 to 0.79 cm^{-1} kbar^{-1}. The negative values helped in identifying more clearly the positions of the C=C stretching modes of the Pt-C_2H_4 groups. Another alkene complex that we investigated for its

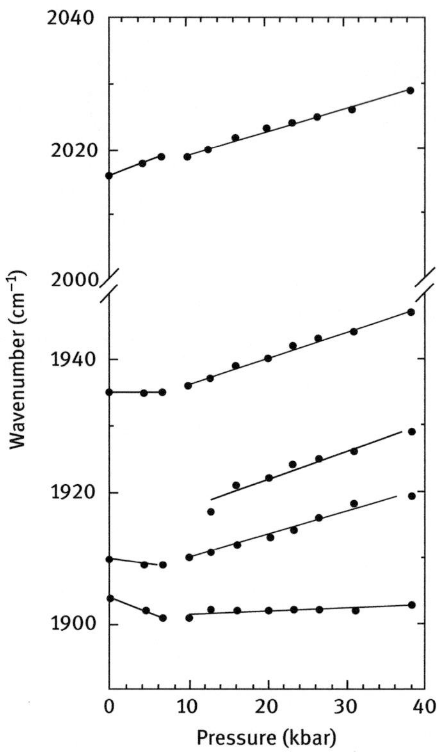

Figure 12: Pressure dependences of the observed bands of (η^5-C_5H_5)Re$(CO)_3$ in the ν(CO) region (reproduced from Huang et al. [32]. Copyright @ 1991, American Chemical Society).

pressure dependence is dichloro(1,5-cyclooctadiene)platinum(II), Pt(COD)Cl_2 (COD = η^4-C_8H_{12}) [42]. This square-planar complex did not exhibit any pressure-induced phase changes up to ~30 kbar and the observed $d\upsilon/dP$ values ranged from –0.18 to 0.79 cm^{-1} $kbar^{-1}$. The value observed for the IR band at 1,426 cm^{-1} was –0.18 cm^{-1} $kbar^{-1}$ suggesting that this band is associated with the C=C stretching mode of the COD group. Our success with alkene complexes also led us to examine the pressure effects on the vibrational spectra of some metal carbyne complexes [43].

Among the other metal complexes that we have examined for their response to high external pressures are *cis*-$(Ph_3P)_2Pt(SH)_2$ [44], the molybdenum-sulfur cluster $[Mo_3(S_2)_6S]^{2-}$ [45], *trans*-Pt$[PPh(CF_3CF_2)_2]_2Cl_2$ [46], CuCN [47], the three Magnus' Green salts, $Pt(NH_3)_4][PtCl_4]$, $[Pt(ND_3]_4][PtCl_4]$ and $[Pt(NH_3)_4][PtBr_4]$ [48], *trans*-Ni $(PhCH_2)_3P]_2(NCS)_2$ [49], the iron cyanide complexes, Prussian blue, $Fe_4[Fe(CN)_6]_3$, $K_3[Fe(CN)_6]$ and $Na_2[Fe(CN)_5NO]\cdot 2H_2O$ [50], (η.^{6}-$C_6H_5CO_2CH_3$)Cr$(CO)_2$(CX) (X = S, Se) [51], CH_3ReO_3 [52], M_2RuH_6, and M_2RuD_6 (M = Ca, Sr, Eu) [53], A_2MH_5, and A_2MD_5 (A = Ca, Sr, Eu; M = Ir, Rh) [54], $KMHP_2O_7\cdot 2H_2O$ (M = Co, Mn, Zn, Mg) [55], Ph_4M (M = Si, Ge, Sn, Pb) [56], $CaHPO_4\cdot 2H_2O$ and $CaHPO_4$ [57], and $[Si(H)CH_2Ph]_6$.[58], α-Si_3N_4

[59], $Cs_2[B_{10}Cl_{10}]$ [60] and buckminsterfullerene, C_{60} [61]. We will only discuss, however, a few of these studies here.

High-pressure IR and FT-Raman spectra of the homogeneous catalyst candidate *cis*-dimercaptobis(trphenylphosphine)platinum(II), *cis*-$[(PPh_3)_2Pt(SH)_2]$, were measured up to pressures of 55 kbar [44]. A phase transition was detected at around 15 kbar. The most pressure-sensitive modes were the v(Pt-S) and the C–H stretching of the phenyl rings with dv/dP values of about 1.0 cm^{-1} $kbar^{-1}$. The pressure-induced phase transition was attributed to reorientation of the phenyl rings. In the case of the three Magnus' Green salts, $Pt(NH_3)_4][PtCl_4]$, $[Pt(ND_3]_4][PtCl_4]$, and $[Pt(NH_3)_4][PtBr_4]$, pressure-tuning IR spectra were recorded up to ~40 kbar [48]. There was evidence of restricted rotation of the NH_3 groups about the Pt–N bonds, most possibly owing to weak hydrogen bonding to the halogens [N–H ⋯ X (X = Cl, Br)]. In another example, sodium nitroprusside, $Na_2[Fe(CN)_5NO]{\cdot}2H_2O$, is well known in medical circles as a source of nitric oxide. Under pressures up to ~45 kbar, there was no evidence of a phase change [50]. The a_1 ν(Fe-C_{ax}) mode of the C_{4v} symmetry $[Fe(CN)_5NO]^{2-}$ ion was the most pressure sensitive with a dv/dP value of 0.74 cm^{-1} $kbar^{-1}$. Moreover, a pressure-sensitive Fermi resonance was detected for the a_1 ν(Fe-N) and δ(FeNO) modes located at 654 and 662 cm^{-1}, respectively. The Raman spectra of the calcium, strontium and europium salts of hexahydridoruthenate(II). $[RuH_6]^{4-}$, were also examined for pressures up to 50 kbar [53]. No phase transitions were observed. The bulk moduli (129–188 kbar) and mode Grüneisen parameters were determined for these materials. In a related study, the vibrational spectra of the ternary metal hydrides and deuterides, A_2MH_5, and A_2MD_5 (A = Ca, Sr, Eu; M = Ir, Rh) were assigned on the basis of Raman, IR, photoacoustic IR, and inelastic neutron scattering spectroscopy together with density functional theory calculations [53]. It was determined that the positions of the IR-active stretching vibrations of these hydrides and deuterides depended markedly on the ionization energies of the cation and the central metal in the anions. In addition, a pressure- and temperature-induced phase transition was present for $CaIrH_5$.

Another study involved the investigation of the effect of high pressures on $KMnO_4$ trapped in KBr and $KClO_4$ matrices up to 50 kbar [62]. The permanganate ion, MnO_4^-, exhibits resonance Raman effects with different laser lines with up to eight overtones being observed for the a_1 ν(Mn–O) mode of the T_d symmetry MnO_4^- ion. This work enabled us to determine the pressure dependences of the anharmonicity constants and the harmonic frequencies of the a_1 ν(Mn–O) vibration in the two matrices. In both cases, the anharmonicity constants decreased with increasing pressure, while the harmonic frequencies increased steadily. The phase transition of KBr from its face-centered structure to the body-centered one took place at about 24 kbar. The perchlorate matrix did not exhibit any evidence of a phase transition under pressure.

We have also undertaken some hydrothermal DAC (HDAC) work using a special high-pressure spectroscopic cell designed by William Bassett et al. at Cornell University

[63]. For instance, we undertook a HDAC Raman microscopy study of the manganese carbonyls, $Mn_2(CO)_{10}$, $(\eta^5\text{-}C_5H_5)Mn(CO)_3$ and $Mn(CO)_5Br$ [64]. We were able to investigate the effect of water at elevated pressures and temperatures on these complexes both visually and spectroscopically. The measurements were performed at 22.1 MPa and around 374 °C. All three organometallic complexes dissolved in subcritical water at a pressure of 22.1 MPa at different temperatures: $Mn_2(CO)_{10}$ (150 °C, $(\eta^5\text{-}C_5H_5)Mn(CO)_3$ (125°C), and $Mn(CO)_5Br$ (225 °C). It also appeared that $Mn(CO)_5Br$ dimerized to the bridged $Mn_2(CO)_8Br_2$ complex at 425 °C.

Much of our later high-pressure vibrational spectroscopic DAC work has been on inorganic artists' pigments. This research has shown that the pigments investigated are remarkably stable under high-pressure and high-temperature conditions [65]. For example, we have studied, both *α*- and *β-copper phthalocyanine* polymorphs [66], *wulfenite* ($PbMoO_4$) [67], *ultramarine blue* [68] and *Prussian blue* ($K_4[Fe(CN)_6]_3$) [50]. The metastable *β-phthalocyanine* polymorph did not undergo a phase transition at pressures approaching 50 kbar. However, the more thermodynamically stable *α*-polymorph did exhibit a reversible phase transition at 20 kbar. Two spectral regions of the Raman spectra were particularly pressure sensitive, 800–900 and 1,100–1,200 cm^{-1}, and these regions are indicative of the two polymorphic forms. The mineral *wulfenite*, lead molybdate ($PbMoO_4$), is present in the artists' pigment, *orange molybdate*. Both variable-temperature and high-pressure Raman measurements were performed on *wulfenite* [67]. No phase changes were detected by either spectroscopic method, but some amorphization did take place at ~600 °C. The stability of *wulfenite* under such extreme conditions bodes well for it continued use in *orange molybdate* as an artists' pigment. The resonance Raman spectra of the overtone progressions of the ν_1 fundamental and combination bands of S_3^- and the ν_1 fundamental of S_2^- in *ultramarine* were studied as a function of pressure. No phase transitions were detected up to ~50 kbar, and the ν_1 fundamental of S_3^- proved to be almost completely harmonic.

4 Conclusions

It is clear from our studies that considerable structural information can indeed be gleaned from high-pressure vibrational spectroscopic measurements on solid inorganic materials using DACs and HDAC equipment. Further chapters in this monograph will focus on other important high-pressure techniques, such as synchrotron X-ray and neutron diffraction.

References

[1] Bassett WA. Diamond anvil cell, 50th birthday. High Pressure Research 2009;29(2):163–186.

[2] Piermarini GJ. High-pressure X-ray crystallography with the diamond cell at NIST/NBS. Journal of Research of the National Institute of Standards and Technology 2001;106:889–920.

[3] Weir CE, Lippincott ER, Van Valkenburg A, Bunting EN. Infrared studies in the 1- to 15-micron region to 30,000 atmospheres. Journal of Research of the National But Standards and Technology 1959;63(A):55–62.

[4] For a recent example on the use of tungsten carbide anvils in high-pressure studies, see Z. Ying, C. Xiping, S. Guangai, Lu Yuping, G. Jian and H. Duanwei, Optimization of Tungsten Carbide Opposite Anvils Used in *In Situ* High-Pressure Loading Apparatus, Article ID 60720, 2014, 5. http://dx.doi.org/10.1155/2014/607520.

[5] Bridgman PW. The coagulation of albumen by pressure. Journal of Biological Chemistry 1914;19:511–512.

[6] For more information on these two companies, see High Pressure Diamond Optics, 2954 N. Campbell Ave., Tucson, AZ 85719, USA (Tel: 2-520-544-9338) and www.almax-easylab.com/.

[7] Forman RA, Piermarini GJ, Barnett JD, Block S. Pressure measurement made by the utilization of ruby sharp-line luminescence. Science 1972;176:284–284.

[8] Piermarini GJ, Block S, Barnett JD, Forman RA. Calibration of the pressure dependence of the R_1 ruby fluorescence to 195 kbar. Journal of Applied Physics 1975;46:2774–2780.

[9] Piermarini GJ, Weir CE. A diamond cell for X-ray diffraction studies at high pressures. Journal of Research of the National But Standards 1962;66(A):325–331.

[10] Weinstein BA, Piermarani GJ. High-pressure single crystal studies of ice VI. Science 1965;148:947–948.

[11] Piermarani GJ, Forman RA, Block S. Viscosity measurements in the diamond anvil pressure cell. Review of Scientific Instruments 1978;49:1061–1066.

[12] Butler IS, Edwards CM, Markwell RD. Pressure-tuning FT-Raman spectroscopic study of the T_{2g} phonon mode of a diamond-anvil cell. Spectrochimica Acta – Part A 1997;53:2253–2259.

[13] Benham V, Lord G, Butler IS, Gilson DFR. High-pressure diamond-anvil-cell micro-Raman spectra of mercuric cyanide, $Hg(CN)_2$, and cesium dodecahydroborate, $Cs_2[B_{12}H_{12}]$. Applied Spectroscopy 1987;41(5):915–917.

[14] Wong PTT. The effect of pressure on the molecular structure of mercury(II) cyanide and the anomalous pressure dependence of the phonon frequencies. Journal of Physical Chemistry 1980;72(12):6721–6725.

[15] Uspenskaya SI, Solntsev KA, Kusnetsov NT. The X-ray diffraction study of alkali metal dodoceahydroclosododecaborates. Zh Strukt Khim 1973;14(1):166–167.

[16] Uspenskaya SI, Solntsev KA, Kusnetsov NT. Preliminary X-ray diffraction of rubidium and cesium dodecahydroclosododecaborates and their double salts of $M_2B_{12}H_{12}$.MCl-type. Zh Strukt Khim 1975;16:482–484.

[17] Butler IS, Gilson DFR, Paroli R, Kawai NT, Lord G. Application of high-pressure micro-Raman spectroscopy in the detection of phase transitions in orientationally-disordered Solids. In: Proc. 23rd Microbeam Anal. Meeting, San Francisco, CA, USA, 1988, 165–166.

[18] Harvey PD, Butler IS, Gilson DFR. Raman study of the effect of high pressure on the order-disorder phase transition in 2-Adamantanone. Journal of Physical Chemistry 1986;90 (19):44546–44549.

[19] Parsonage NG, Staveley LAK. Disorder in Crystals. Oxford, UK, Clarendon Press, 1978.

[20] Westrum EF. In: Carter RS, Rush JJ., editors, Molecular Dynamics and the Structure of Solids. Washington, DC, USA, National Bureau of Standards, 1969.

[21] Kawai NT, Gilson DFR, Butler IS. The order-disorder phase transition in the cage hydrocarbon quadricyclane. Molecular Crystals and Liquid Crystals Science Technology 1992;211:59–61.

[22] Kawai NT, Gilson DFR, Butler IS. Variable-temperature and -pressure studies of the vibrational spectra and phase transition in quadricyclane. Journal of Physical Chemistry 1992;96 (21):8556–8561.

[23] Hammond GS, Turro NJ, Fischer A. Photosensitized cycloaddition reactions. Journal of the American Chemical Society 1961;83:4674–4675.

[24] Smith EB. Phase coexistence and hysteresis. Journal of Physics and Chemistry of Solids 1959;9:182–183.

[25] Kawai NT, Butler IS, Gilson DFR. Vibrational spectroscopic study of the temperature- and pressure-induced phase transition of norbornylene. Journal of Physical Chemistry 1991;95 (2):634–639.

[26] Hara K, Katou Y, Osuhi J. Pressure-induced phase transition and comprehensive behavior of substituted adamantanes. Bulletin of the Chemical Society of Japan 1981;54(3):687–691.

[27] Butler IS, Huang Y. High-pressure micro-Raman spectra of potassium hexabromostannate(IV), K_2SnBr_6. Journal of Raman Spectroscopy 1990;21(4):273–275.

[28] Adams DM, Norman AJ, A High-pressure synchrotron X-ray powder diffraction study of K_2SnBr_6. Journal of Physics Condensed Matter 1991;3:3211.

[29] Huang Y, Butler IS. High-pressure micro-Raman spectra of potassium chromate(VI) and potassium tungstate(VI). Applied Spectroscopy 1990;44(8):1326–1328.

[30] Edwards CM, Haines J, Butler IS, Leger J-M. High-pressure X-ray diffraction study of potassium chromate: pressure-induced orientational disorder and its implications for amorphization in A_2BX_4 compounds. Journal of Physics and Chemistry of Solids 1999;60 (4):529–538.

[31] Huang Y, Butler IS, Gilson DFR, Lafleur D. Vibrational study of the effects of high external pressures on the structure and phase behavior of pentacarbonyl(methyl)-manganese(I) and pentacarbonyl(methyl)rhenium(I). Inorganic Chemistry 1991;30(1):117–120.

[32] Huang Y, Butler IS, Gilson DFR, High-pressure micro-raman A. Study of crystalline tricarbonyl (^{5}-cyclopentadienyl)rhenium(I), (^{5}-C_5H_5)$Re(CO)_3$. Inorganic Chemistry 1991;30(5):1098–1102.

[33] Adams DM, Ekejiuba IOC. A second-order phase transition in $Mn(CO)_5Br$. Journal of Chemical Physics 1982;77:4793–4795.

[34] Huang Y, Butler IS, Gilson DFR, Pressure-tuning Micro-raman A. Study of crystalline pentacarbonyl(chloro)rhenium(I), $Re(CO)_5Cl$. Spectrochimica Acta 1991;47A(7):909–913.

[35] Butler IS. Coordination chemistry of the thiocarbonyl ligand. Pure and Applied Chemistry 1988references therein;60(8):1241–1244.

[36] Huang Y, Butler IS, Gilson DFR. High-pressure infrared and micro-Raman spectra of crystalline pentacarbonyl(thiocarbonyl)chromium(0) and (^{6}-benzene)-dicarbonyl(thiocarbonyl)chromium (0). Inorganic Chemistry 1992;31(2):303–305.

[37] Huang Y, Butler IS, Gilson, DFR. Crystal structure and high-pressure micro-Raman spectra of tricarbonyl(^{5}-pentamethylcyclopentdienyl)rhenium(I), (^{5}-C_5Me_5)$Re(CO)_3$. Inorganic Chemistry 1992;31(23):4762–4765.

[38] Barnett SM, Markwell RD, Brienne SHR, Butler IS, Gilson DFR, Kawai NT, Vlcek A Jr. Near-infrared Fourier transform Raman microspectroscopy at high external pressures. Journal of Raman Spectroscopy 1993;24(7):471–473.

[39] Barnett SM, Vlcek A Jr., Butler IS. Pressure-tuning infrared and Raman spectroscopic study of (2,2′-bipyridine)etetracarbonylchromium(0). Canadian Journal of Chemistry 1998;54A (12):1869–1878.

[40] Li H, Butler IS. Effect of high external pressures on the infrared and micro-Raman spectra of crystalline, octahedral mixed carbonyl-*tert*-Butyl isocyanide complexes of chromium(0). Inorganic Chemistry 1995;34(5):1193–1197.
[41] Baldwin JA, Gilson DFR, Butler IS. Effect of high external pressures on the vibrational spectra of Zeise's complexes, $K[Pt(^2-C_2H_4)Cl_3]$ and $[Pt^2-C_2H_4)Cl_2]_2$. Journal of Organometallic Chemistry 2005;690(13):3165–3168.
[42] Baldwin JA, Butler IS, Gilson DFR. Effect of high external pressures on the vibrational spectra of crystalline dichloro(1,5-cyclooctadiene)platinum(II), $Pt(^4-C_8H_{14})Cl_2$. Inorganica Chimica Acta 2006;359(9):3079–3083.
[43] Xu Z, Butler IS, Mayr A. The pressure-tuning Raman and IR spectral studies on the multnuclear metal carbyne complexes. Spectrochimica Acta – Part A: Molecular and Biomolecular Spectroscopy 2005;61A(5):995–1000.
[44] Song Y, Butler IS, Shaver A. High-pressure vibrational study of the catalyst candidate *cis*-Dimercaptobis(triphenylphosphine)platinum(II), *cis*-$[(Ph_3P)_2Pt(SH)_2]$. Spectrochimica Acta, Part A: Molecular and Biomolecular Spectroscopy 2002;58A(12):2581–2587.
[45] Butler IS, Höfstatter M, Müller A. Fourier transform Raman spectroscopic study of pressure-induced ligand bond activation in a molybdenum sulfur cluster. Journal of Raman Spectroscopy 2002;33(3):213–217.
[46] Xu Z, Kapoor PN, Butler IS. Pressure-tuning infrared and Raman spectra of trans-dichlorobis [(diperfluoroethyl)phenylphosphine)platinum(II), *trans*-$Pt[PPh(CF_3CF_2)_2]_2Cl_2$. Inorganica Chimica Acta 2006;359(12):4121–4124.
[47] Romao C, Barsan MM, Butler IS, Gilson DFR. A high-pressure micro-raman spectroscopic study of copper cyanide, CuCN. Journal of Materials Science 2010;45(9):2518–2520.
[48] Edwards CM, Butler IS. Pressure-tuning infrared spectra of the three magnus' salts, $[Pt(NH_3)_4]$ $[PtCl_4]$, $[Pt(ND_3)_4][PtCl_4]$ and $[Pt(NH_3)_4][PtBr_4]$. Inorganica Chimica Acta 2011;377(1):155–158.
[49] Xu Z, Christendat D, Butler IS. Pressure-tuning infrared spectroscopic study of *trans*-Di-N-thiocyanato-bis(tribenzylphosphine)nickel(II), *trans*-$Ni[PhCH_2)_3P]_2(NCS)_2$. Journal of Molecular Structure 2010;980(1-3):114–116.
[50] Barsan MM, Butler IS, Fitzpatrick J, Gilson DFR. High-pressure studies of the micro-Rmaan specctra of iron cyanide complexes: Prussian blue $(Fe_4)[Fe(CN)_6]_3)$, potassium ferricyanide (K_3 $[Fe(CN)_6]$) and sodium nitroprusside, $(Na_2)[Fe(CN)_5(NO)].2H_2O$. Journal of Raman Spectroscopy 2011;42(9):1820–1824.
[51] Edwards CM, Moore RJ, Barsan MM, Butler IS. Pressure-tuning Raman spectra of the arene chromium(0) chalcocarbonyl complexes $(^6-C_6H_5CO_2CH_3)-Cr(CO)_2(CX)$ (X = O, S). Spectroscopy Letters 2011;45(1):40–44.
[52] Barsan MM, Gilson DFR, Reber C, Butler IS. High-pressure micro-Raman spectroscopic study of methyltrioxorhenium(VII). Journal of Raman Spectroscopy 2006;37(12):1321–1326.
[53] Barsan MM, Moyer RO, Butler IS, Gilson DFR. Pressure dependence of the raman spectra of salts of hexahydridodiruthenate(II), M_2RuH_6 (M = Ca, Sr, Eu) and their deuteride analogues. Journal of Alloys and Compounds 2006;424(102):73–77.
[54] Barsan MM, Butler IS, Gilson DFR, Moyer RO, Zhou W, Wu H, Udovic TJ, Raman FTIR. Photoacoustic-infrared, and inelastic neutron scattering spectra of ternary metal hydride salts A_2MH_5 (A = Ca, Sr, Eu; M = Ir, Rh) and their deuterides. Journal of Physical Chemistry A 2012;116(10):2490–2496.
[55] Baril M, Assaaoudi H, Butler IS. Pressure-tuning Raman microspectroscopic study of cobalt (II), Manganese(II), Zinc(II) and Magnesium(II) pyrophosphate dihydrates. Journal of Molecular Structure 2005;751(1-3):168–171.

[56] Warner SD, Butler IS, Wharf I. Pressure-tuning infrared and Raman spectroscopy of group 14 Tetraphenyl compounds. Spectrochimica Acta – Part A: Molecular and Biomolecular Spectroscopy 2000;56A(3):453–466.
[57] Xu J, Butler IS, Gilson DFR. FT-Raman and high-pressure infrared spectroscopic studies of dicalcium phosphate dihydrate ($CaHPO_4.2H_2O$) and anhydrous dicalcium phosphate ($CaHPO_4$). Spectrochimica Acta – Part A: Molecular and Biomolecular Spectroscopy 1999;55A:2801–2809.
[58] Li H, Butler IS, Harrod JF. Effect of high pressure on the internal vibrations of all-*trans*-Hexaphenylcyclohexasilane. Journal of Physical Chemistry 1994;98(9):2266–2270.
[59] Huang Y, Butler IS. Infrared study of the effect of high external pressures on crystalline -silicon nitride. Applied Spectroscopy 1992;46(8):1303–1304.
[60] Huang Y, Butler IS. High-pressure vibrational spectoscopic study of cesium decachloroborate, $Cs_2[B_{10}Cl_{10}]$. Inorganica Chimica Acta 1992;192(1):5–7.
[61] Barsan MM, Butler IS, Gilson DFR. Vibrational spectroscopy at high pressure of the permanganate anion trapped in potassium bromide and potassium perchlorate matrices. Journal of Physical Chemistry B 2006;110(18):9291–9297.
[62] Huang Y, Gilson DFR, Butler IS. High-pressure infrared study of Buckminsterfullerene, C_{60}. Journal of Physical Chemistry 1991;95(15):5723–5725.
[63] Bassett WA, Stern AH, Bucknum M, Ming CI. A new diamond anvil cell for hydrothermal studies to 2.5 GPa and from 190 to 1200 C. Review of Scientific Instruments 1993;64:2340.
[64] Calahoo C, Barsan MM, Thomas ML, Kozinski JA, Butler IS. Hydrothermal raman microscopy studies of metal carbonyls. Vibrational Spectroscopy 2011;57(1):2–7.
[65] Bayne JM, Butler IS. Effect of temperature and pressure on selected artists' pigments. New Journal of Chemistry 2013;37(12):3833–3839.
[66] Beaulieu-Houle G, Gilson DFR, Butler IS. Pressure-tuning micro-raman spectra of artists' pigments: – and -copper phalocyanine polymorphs. Spectrochimica Acta – Part A: Molecular and Biomolecular Spectroscopy 2014;117:61–64.
[67] Bayne JM, Butler IS. Variable-temperature and high-pressure micro-raman of inorganic artists' pigments: crystalline wulfenite, lead(II) molybdate(VI), $PbMoO_4$. Spectroscopy Letters 2014;47(8):616–620.
[68] Barsan MM, Butler IS, Gilson DFR. High-pressure resonance raman spectroscopic study of ultramarine blue pigment. Spectrochimica Acta – Part A: Molecular and Biomolecular Spectroscopy 2012;98:457–459.

Janice L. Musfeldt

Introduction to molecule-based multiferroics

Single-phase materials that simultaneously expose ferroelectricity and ferromagnetism have been challenging to realize. This is due to the mutually exclusive mechanisms for these two properties in cubic ABX_3 perovskites: standard ferroelectricity requires empty d orbitals whereas ferromagnetism derives from partially filled d orbitals [1]. Both proper and improper ferroelectrics offer paths forward [2], and when combined with weak ferromagnetism arising from a non-collinear magnetic state, both inversion and time-reversal symmetries can be broken [2–18]. The mechanisms that allow magnetoelectric coupling to emerge are best established in oxides. That said, molecule-based multiferroics offer a number of attractive opportunities – retaining many of the salient physical problems observed in oxides while offering simpler chemistry, flexible structures, and overall lower energy scales [19–35]. Their modular nature, with easily interchangeable metal centers, ligands, and counter-ions, also provides a platform for the development of structure–property relationships [36–51]. Molecular materials offer a particular edge to researchers interested in the fundamental aspects of magnetism [24, 34, 39, 44, 52–80]. The smaller exchange interactions, for instance, allow magnetization to be saturated in experimentally available fields [27, 28, 46, 78, 81–89]. Ferroelectricity is also observed in molecule-based materials – a result of flexible lattices, hydrogen-bond networks, and the ability to stabilize charge-separated states [15, 57, 66, 75, 90–101]. Other consequences of a soft lattice include high compressibility, the tendency to own photo-induced states, and processability [33, 35, 66, 71, 94, 99, 101–112]. In molecule-based materials, competing interactions and delicately balanced mixing processes give rise to elaborate $P-T-B-\hbar\nu$ phase diagrams. Small external perturbations can change important energy and length scales, driving the system into new states with very different properties [113]. Pressure is different from magnetic field and temperature in that it directly alters local structure (bond lengths and angles) and hybridization [114]. New energy scales and areas of phase space can be revealed under compression in a manner that is much cleaner than chemical doping. In materials with spin-containing centers, distortions can trigger new magnetic properties [115]. In some cases, the competing states and different functionalities can be harnessed in technologies with widespread impact. Multiferroics hold promise for ultra-low power memory and logic devices – if coupling is strong enough [116].

Acknowledgments: Research at Tennessee is supported by the National Science Foundation (DMR-1707846). A. J. Clune's assistance with the figures is very much appreciated. Useful conversations with A. J. Clune, N. S. Dalal, K. D. Hughey, Z. Liu, K. R. O'Neal, H. Xiang, and J. Yang are gratefully acknowledged.

Janice L. Musfeldt, Departments of Chemistry and Physics, University of Tennessee, Knoxville, TN 37996, USA

https://doi.org/10.1515/9783110668612-003

1 $[(CH_3)_2NH_2]Mn(HCOO)_3$: a highly tunable multiferroic

$[(CH_3)_2NH_2]Mn(HCOO)_3$ is a molecule-based multiferroic with coupling across broad energy and time scales and an outstanding platform with which to concentrate our discussion. As we shall see, this system has overall low-energy scales and, as a result, is extraordinarily responsive to external stimuli. Figure 1 displays the crystal structure of $[(CH_3)_2NH_2]Mn(HCOO)_3$ [33, 91, 117, 118]. This metal-organic framework material consists of octahedral MnO_6 building blocks connected by formate ligands. Dimethylammonium cations fill the anion pockets. The amine is disordered at high temperatures. Below $T_{o/d} = 185$ K, the counterion orders to create an extensive hydrogen bonding network between the amine and framework [91, 93, 95, 117, 119–123]. Ferroelectricity is triggered by the accompanying $R\bar{3}c \rightarrow Cc$ structural distortion [91, 93, 119, 120, 122–124]. The electric polarization P is 1.5 $\mu C/cm^2$ in the low-temperature phase. It is the ordered state that is shown in the structural rendering below. The domain walls have been visualized [74].

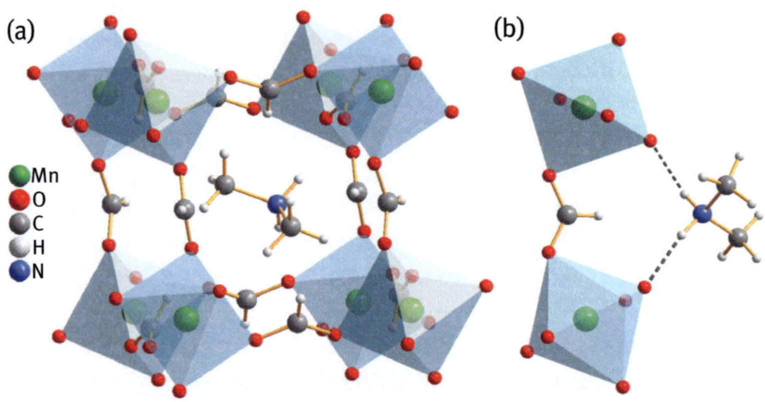

Figure 1: (a) Crystal structure of $[(CH_3)_2NH_2]$ $Mn(HCOO)_3$ below $T_{o/d}$ [91]. (b) Close-up view of the amine inside the anion pocket and the intermolecular hydrogen bonding that develops in the low-temperature phase (reprinted with permission from Clune et al. [135]. Copyright 2020 American Chemical Society).

Figure 2 summarizes the series of temperature-, magnetic field-, and pressure-induced transitions in $[(CH_3)_2NH_2]Mn(HCOO)_3$. In addition to ferroelectricity below the 185 K order–disorder transition, there is a transition to a weakly canted antiferromagnetic state at $T_N = 8.5$ K [56, 91, 125]. The canting angle is 0.08°. This non-collinear state arises from the non-centrosymetric nature of the formate bridge [56, 91, 117, 125–127]. Cross-coupling between the electric and magnetic orders has been observed [119–121, 123, 128], and low-temperature phase separation leads to quantum tunneling of the magnetization [129]. The exchange interaction J is −0.64 K. Magnetic field drives $[(CH_3)_2NH_2]Mn(HCOO)_3$ across a spin-flop transition at 0.31 T and a magnetic quantum phase transition [130] to the fully saturated state at 15.1 T [131]. Interestingly, electric polarization is enhanced under magnetic field due to magneto-electric coupling [74, 121, 132]. Magneto-infrared spectroscopy reveals that distortions of the formate bending mode link temperature- and field-driven ferroic states [133]. Room-temperature Raman spectroscopy under pressure and prior X-ray scattering [134] also reveals how symmetry is broken across the 4.2 GPa transition. This is thought to be a $R\bar{3}c \rightarrow Cc$ transition [134] – although there are other candidate space groups for the high-pressure phase.

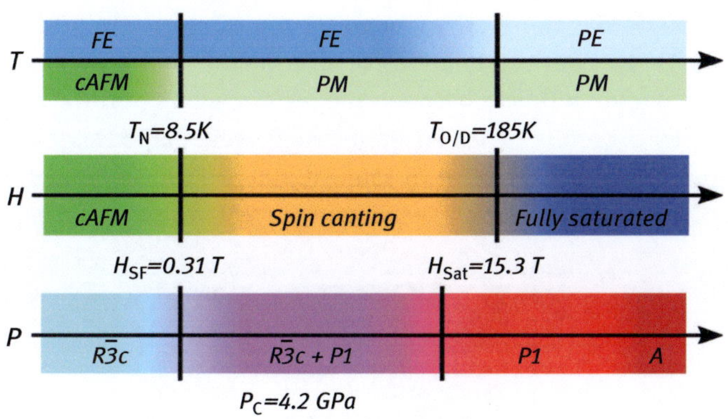

Figure 2: Schematic view of the temperature-, magnetic field-, and pressure-induced transitions in $[(CH_3)_2NH_2]Mn(HCOO)_3$. The properties are summarized as well. T_N, Néel temperature; $T_{o/d}$, order–disorder transition temperature; H_{SF}, spin-flop field; H_{Sat}, critical field; P_c, critical pressure; A, amorphous (reprinted with permission from Clune et al. [135]. Copyright 2020 American Chemical Society).

In the remainder of this chapter, we review how variable pressure Raman scattering reveals the pressure–temperature–magnetic field ($P-T-H$) phase diagram of the molecule-based multiferroic $[(CH_3)_2NH_2]Mn(HCOO)_3$ [135]. We discuss how these findings resolve controversy concerning the symmetry of the high-pressure phase and, at the same time, unveil the presence of an additional polar phase that is accessible under compression (and very likely strain) at room temperature. At this time, it is not known whether the polar phase is ferroelectric (switchable) or pyroelectric (not switchable), but we anticipate that this question will motivate future work.

2 Methods

2.1 Raman scattering of molecule-based multiferroics

For the work on multiferroic [(CH3)$_2$NH$_2$]Mn(HCOO)$_3$ discussed here, we carried out synchrotron-based infrared absorption and Raman scattering measurements at the COMPRES 22-IR-1 beamline at the National Synchrotron Light Source II at Brookhaven National Laboratory. Based upon a comparison of the infrared and Raman spectra, we identified a handful of features to track under compression. There are several criteria that must be met to assure a successful outcome. It is best to focus on peaks that have a very clear chemical and physical interpretation. In this case, we elected to track the behavior of the HCOO$^-$ bending and C–N–C stretching modes of [(CH$_3$)$_2$NH$_2$]Mn(HCOO)$_3$. These Raman-active vibrational modes reveal a great deal about the superexchange pathway between the metal centers and the hydrogen bonding experienced by the amine. We also want strong, well-shaped spectral features that persist over the full pressure range of interest and are sensitive to the process under investigation. A room-temperature test usually reveals where to focus.

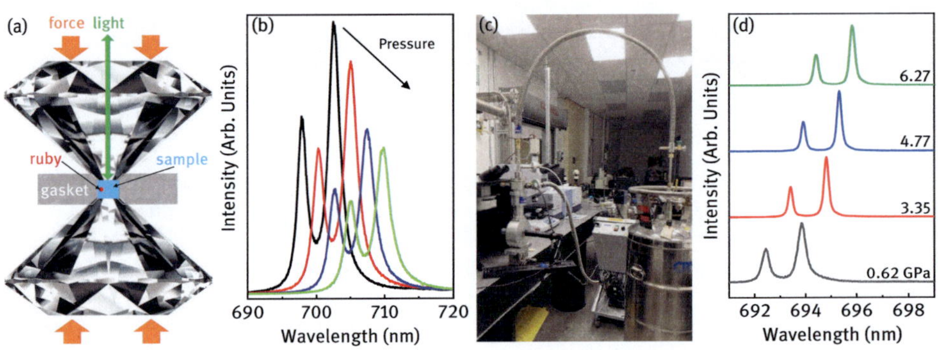

Figure 3: (a) Schematic of a diamond-anvil cell. (b) Ruby fluorescence under pressure at 300 K. (c) Photograph of the cryostat setup at the COMPRES beamline at the National Synchrotron Light Source II, Brookhaven National Laboratory. (d) Examining the ruby fluorescence line shapes is a great way to confirm that your sample is hydrostatic at elevated pressure. These fluorescence spectra were taken at 150 K. The fact that the line shapes have not broadened or become deformed is good evidence of isotropic compression.

The Raman scattering measurements presented here were performed between 70 and 3,600 cm^{-1} using a 532 nm diode pumped solid state laser, 1 mW power, and a 1200 g/mm grating. Small single crystals, on the order of 10 μm, were loaded into a diamond-anvil cell (with ultra-low fluorescence diamonds and 400 or 500 μm culet anvils depending on the run) for high-pressure spectroscopy along with an annealed ruby ball to determine the pressure [136, 137]. A schematic of our setup and an example of how we use the ruby fluorescence to reveal pressure is shown in

Figure 3(a,b). It is customary to add a pressure-transmitting medium to the cell in order to achieve hydrostatic compression of the sample. Due to the softness of our target material, a neat load was used – although we typically carry out a number of tests with vacuum grease, KBr, KCl, and liquid Ar to determine which pressure medium gives the best results. Of course, the "best" quality spectrum can be highly subjective, but it generally involves the appearance of peak shapes that are not inhomogeneously broadened and a sharp transition region. That said, it is well known that molecule-based materials are highly compressible and can function as their own pressure medium, so a neat load is very convenient if there are no optical density concerns. This is clearly the case for Raman scattering but not for infrared absorption. For these scans, we employed integration times up to 15 s for both room- and low-temperature work, averaging as appropriate.

The diamond-anvil cell was loaded into an open-flow helium cryostat for variable temperature experiments. A photo of the setup is shown in Figure 3(c). Pressure was increased *in-situ* to 8 or 10 GPa in small increments at each temperature of interest (between 300 and 75 K) and fully released after each cycle. Typically, we warm up to room temperature to be sure that pressure is released. Because we need to close the diamond-anvil cell securely, the initial starting pressure tends to increase with decreasing temperature. This can prevent investigation of the lowest pressure regime. In this effort, all distortions were reversible upon release of the pressure. Achieving isotropic compression can be an even greater challenge at low temperatures. We employ the shape of the ruby fluorescence spectrum as a reliable indicator of hydrostaticity (Figure 3(d)).

To complement the high-pressure work, we performed variable temperature Raman scattering measurements at ambient pressure conditions using a Horiba Jobin-Yvon T64000 Raman spectrometer (10–5,000 cm^{-1}) equipped with a 50× objective, 1800 g/mm grating, 532 nm laser, and an open-flow helium cryostat. Laser power was kept below 1 mW to avoid sample damage. Standard peak fitting procedures were employed as appropriate.

2.2 Complementary experimental methods

It is useful to combine high-pressure vibrational spectroscopies with complementary measurement techniques. Our team often couples Raman scattering and synchrotron-based infrared absorption spectroscopies due to the different selection rules for low symmetry materials. For studies of multiferroics, we like to add dielectric, susceptibility, magnetization, and heat capacity where possible. To illustrate the benefits of this approach, we include the susceptibility of $[(CH_3)_2NH_2]Mn(HCOO)_3$ in Figure 4. These measurements reveal how T_N evolves under compression, and we as shall see below, the data can be included on the $P-T-H$ phase diagram to illustrate spin behavior.

A similar approach was taken to reveal the $P–T$ phase diagram of $Co[N(CN)_2]_2$. In this case, the local lattice distortion triggers a magnetic crossover [138].

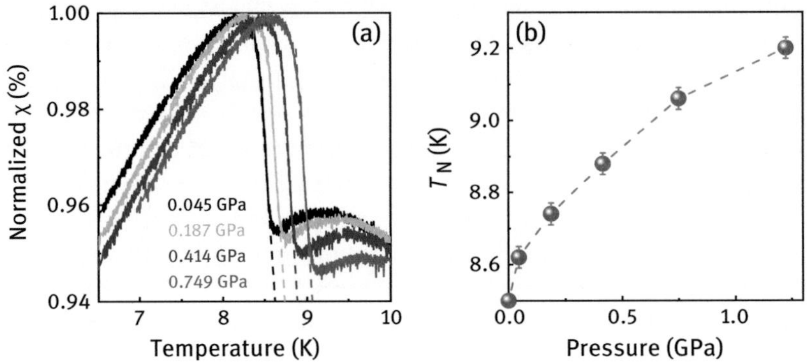

Figure 4: (a) Magnetic susceptibility of $[(CH_3)_2NH_2]$ $Mn(HCOO)_3$ under compression. (b) T_N versus pressure. Here, we concentrate on the temperature range of the magnetic ordering transition (reprinted with permission from Clune et al. [135]. Copyright 2020 American Chemical Society).

2.3 Complementary lattice dynamics calculations and a correlation group analysis

High-pressure infrared and Raman spectroscopies are nicely complemented by a detailed symmetry analysis, first-principles theory, and simulations [139–141]. There is a natural connection between vibrational spectroscopy and lattice dynamics because the calculations provide resonance frequencies, mode symmetries, displacement patterns, and force constants. A combined analysis of vibrational mode trends and displacement patterns can reveal local lattice distortions, phase transition mechanisms, and charge ordering patterns [43, 142–149]. As a result, lattice dynamics calculations are very useful for assigning modes, extracting symmetries, uncovering subgroup relationships, and revealing phase transition mechanisms [133]. Without complementary calculations, mode assignments are mostly guesswork – relying on both experience and similarities with chemically related compounds.

An important strategy for resolving subgroup relationships in molecule-based multiferroics involves energy calculations as a function of pressure. By examining the relative stability of a compound relaxed under various structural constraints as a function of some external stimuli, one can predict which phase – out of several viable candidates – is more stable under the conditions of interest [135, 150]. This allows one to (i) estimate where the crossover between different ground states takes place, (ii) predict relative phase stability and sometimes the phase transition mechanism, and (iii) determine whether there is a clear ground state or a metastable phase nearby. An analysis of relative stability can also reveal the properties of the low-symmetry phase [141]. Similar approaches have been used to simulate the relative importance of certain exchange interactions under compression [151, 152]. In $(NH_4)_2FeCl_5.H_2O$, for instance, tuning the J ratios promotes the development of collinear vs. non-collinear magnetic states.

3 Results and discussion

3.1 Assigning modes and identifying phase boundaries

Figure 5 summarizes the Raman scattering response of $[(CH_3)_2NH_2]Mn(HCOO)_3$ as a function of pressure at room temperature [135]. We focus on three different mode clusters in order to analyze the site-specific responses of the material. The key excitations of interest include: (i) the low-frequency Mn-containing lattice modes (ii) the formate bending mode, and (iii) the CNC stretching mode. These features are local probes of the Mn environment, the superexchange ligand, and amine behavior, and they are assigned based upon complementary lattice dynamics calculations [133]. As we shall see below, the formate bending mode is a particularly sensitive indicator of symmetry breaking in this multiferroic.

In order to reveal local lattice distortions under compression, we plot frequency versus pressure for each set of modes (Figure 5(b, d, f)). There is a broad crossover region with mixed character centered at a critical pressure of 4.2 GPa, in excellent agreement with prior work [134]. The breadth of this crossover region is indicative of first-order character. There are substantial frequency shifts in the Mn-containing lattice modes and extensive splitting of the $HCOO^-$ bending and CNC stretching modes. The symmetry breaking of the formate bending mode is particularly revealing. It appears as a weak doublet at ambient conditions and develops into a triplet with two additional features near 775 cm^{-1}. Originally, the development of this cluster was attributed to a highly distorted formate cage and a change in space group from $R\bar{3}c \rightarrow$ mixed phase $\rightarrow Cc$ [118, 134]. More recently, an alternate sequence has been proposed in which pressure drives the system from $R\bar{3}c \rightarrow$ mixed phase $\rightarrow P1$ at room temperature [135]. We will show evidence for the latter below. Signatures of amorphous character appear in the Mn-containing phonons (Figure 5(a)) near 6.74 GPa and above.

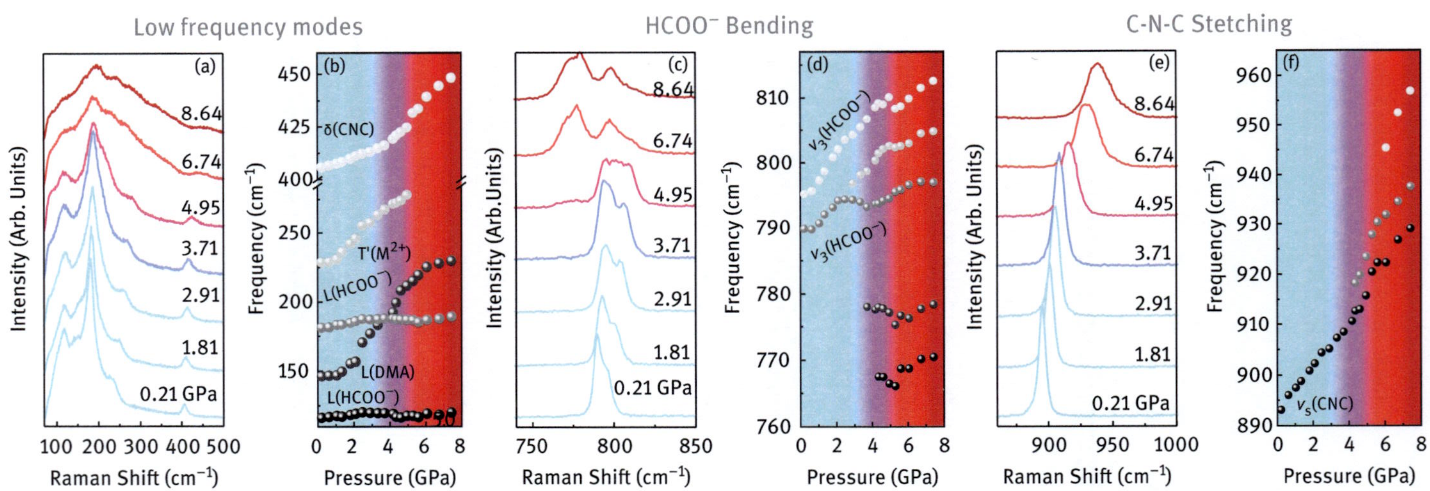

Figure 5: (a, c, e) Raman scattering spectra of $[(CH_3)_2NH_2]Mn(HCOO)_3$ and (b, d, f) the associated frequency versus pressure plots showing the crossover region near 4.2 GPa [135]. All data are at room temperature. Blue, $R\bar{3}c$ space group; purple, mixed $R\bar{3}c$ + P1 phase; red, P1 space group; deep red, amorphous region (reprinted with permission from Clune et al. [135]. Copyright 2020 American Chemical Society).

3.2 What happens across the order–disorder transition?

We employed variable temperature Raman scattering spectroscopy to clearly identify the signature of *Cc* space group symmetry in the low-temperature phase of $[(CH_3)_2NH_2]Mn(HCOO)_3$ at ambient pressure conditions. This type of complementary data – without the diamond-anvil cell – is useful for disentangling temperature and pressure effects. Again, we focus on the formate bending mode of this molecule-based multiferroic. The $HCOO^-$ ligand is extremely sensitive to hydrogen bonding which changes dramatically across the order–disorder transition at 185 K and is responsible for the occurrence of ferroelectricity [91, 119, 120]. Moreover, the formate ligand creates the superexchange pathway between metal centers, making it pertinent to the development of the canted antiferromagnetic state below $T_N = 8.5$ K as well.

Figure 6 displays a close-up view of the Raman scattering response of $[(CH_3)_2NH_2]Mn(HCOO)_3$ in the vicinity of the formate bending mode. The spectral behavior across the temperature-driven $R\bar{3}c \rightarrow Cc$ transition is particularly interesting. Above 185 K, the formate bending mode displays a well-isolated doublet. This is the standard signature of the $R\bar{3}c$ phase. Below $T_{o/d}$, the amine is ordered, the long-range hydrogen bonding network is formed, and the system becomes ferroelectric with space group *Cc* [33, 91, 118–120, 124].

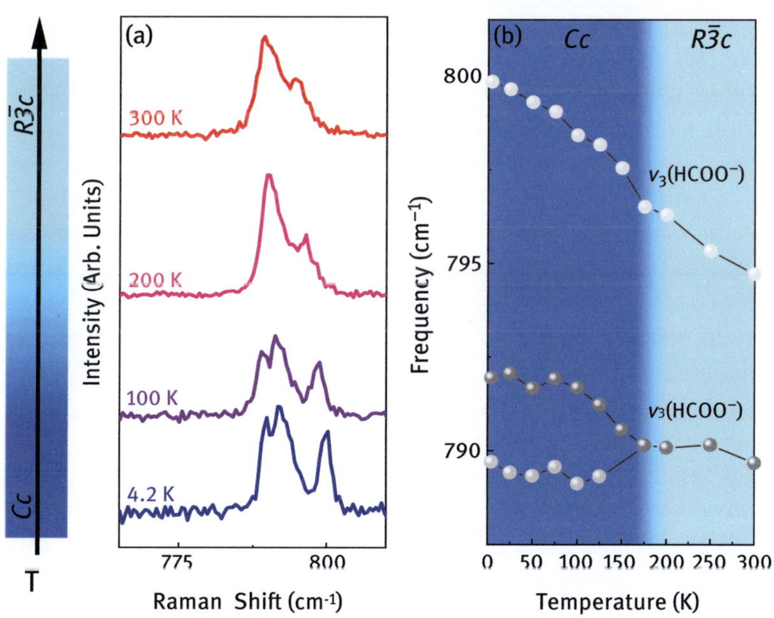

Figure 6: (a) Close-up view of the Raman-active formate bending mode in $[(CH_3)_2NH_2]Mn(HCOO)_3$ at selected temperatures. (b) Peak position versus temperature, showing how the formate bending mode splits across the order–disorder transition (reprinted with permission from Clune et al. [135]. Copyright 2020 American Chemical Society).

Here, the formate bending mode contains three well-isolated peaks. There are no other vibrational modes nearby. The frequency versus temperature plot emphasizes how the upper branch of this structure hardens across $T_{o/d}$ whereas the lower frequency branch splits into two features. The formate bending mode in the low-temperature *Cc* phase is therefore characterized by a clear triplet structure. As we discuss below, this triplet structure does not appear in isolation under compression suggesting that the high-pressure space group is not *Cc* as previously supposed [134].

3.3 Pressure dependence of the Raman-active $HCOO^-$ bending and C–N–C stretching modes at temperatures above and below the order–disorder transition

Figure 7 summarizes the behavior of the formate bending and C–N–C stretching modes of $[(CH_3)_2NH_2]Mn(HCOO)_3$ under compression at several different temperatures. In the high-temperature phase, we find $P_C = 4.2$ GPa as discussed previously. This critical pressure marks the center of a broad sluggish region which is clearly observed in the vibrational response [134]. The 200 K spectra are very similar to those at room temperature with a comparable type of mixed-phase regime. We begin by analyzing the symmetry of the high-temperature phase as a function of pressure. Prior Raman scattering and X-ray work suggests an $R\bar{3}c$ → mixed phase → *Cc* transition under compression [134]. In this case, we expect to observe doublet → triplet splitting of the formate bending mode just like that in Figure 6. Instead, we find that, in addition to the characteristic triplet centered at approximately 795 cm^{-1}, we have broad doublet near 775 cm^{-1}. These features develop in tandem suggesting that they are related. The broad, lower frequency doublet cannot be accounted for within the *Cc* space group, implying that the true symmetry of the high-temperature, high-pressure phase of $[(CH_3)_2NH_2]Mn(HCOO)_3$ is lower than *Cc*. The intermediate-pressure mixed phase must therefore consist of $R\bar{3}c$ + another space group lower than *Cc*. The two-component nature of this intermediate phase along with evidence for amorphous character at the highest pressures is in agreement with prior work [134].

We carried out similar measurements in the low-temperature ferroelectric phase. Figure 7(b,c) summarizes the development of the formate bending and C–N–C stretching modes at 125 and 75 K. Both are well below the order–disorder transition. In the low-pressure regime, the formate bending mode displays the characteristic triplet structure in both sets of spectra. This pattern is indicative of *Cc* symmetry, consistent with prior X-ray and Raman scattering work [122, 124]. P_C is still centered near 4.2 GPa, although the critical pressure regime broadens considerably as temperature decreases. At 75 K, the broadening of the two-phase regime is extreme. As before, we can employ the splitting pattern of the formate bending mode (and also that of the amine-related mode [135]) as microscopic indicators of symmetry in the high-pressure phase of $[(CH_3)_2NH_2]Mn(HCOO)_3$. In pressure sweeps at 125 and 75 K, we find that the

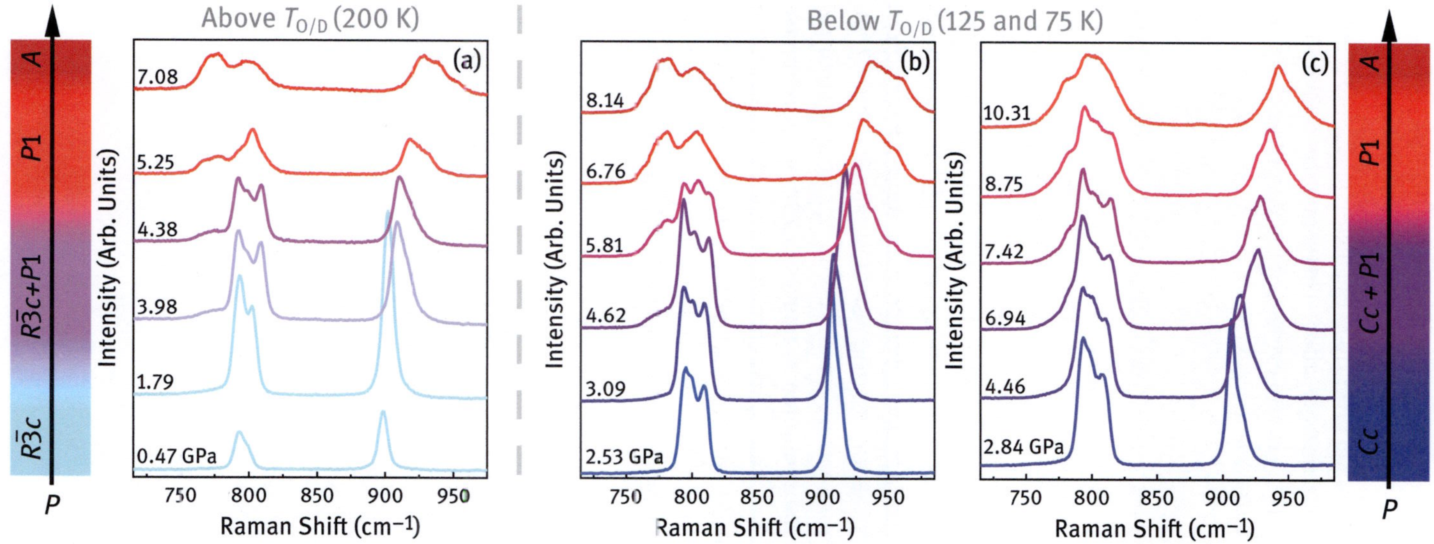

Figure 7: Pressure dependence of the Raman-active $HCOO^-$ bending and C–N–C stretching modes of $[(CH_3)_2NH_2]Mn(HCOO)_3$ at temperatures above and below $T_{o/d}$ [135]. Reprinted with permission from Clune et al. [135]. Copyright 2020 American Chemical Society.

characteristically sharp triplet centered at 795 cm^{-1} develops a broad doublet near 775 cm^{-1} that continues to strengthen and evolve under compression. This pattern is identical to what is observed above the critical pressure in the high-temperature phase. We therefore conclude that the symmetries are alike. The sequence is therefore $R\bar{3}c$ → mixed phase → lower than *Cc*. There is also evidence for amorphous character at the highest pressures – especially in the low-frequency Mn-containing modes.

3.4 Vibrational spectroscopy as a microscopic probe of symmetry

A closer look at the group–subgroup relationship for the $R\bar{3}c$ space group is useful at this point in the process because it allows us to determine the symmetry of the high-pressure phase (Figure 8). We already know that the low-temperature ambient pressure phase has *Cc* symmetry [91, 95, 119, 120, 124, 123]. Using the branch of the symmetry subgroup tree that contains $P\bar{3}c1$ (rather than $P3_221$) is the only route by which the system can end up in the *Cc* space group. Examining the spectra in Figure 7, we see that the low-temperature Raman response of $[(CH_3)_2NH_2]Mn(HCOO)_3$ actually breaks symmetry under compression. There is only one subgroup of *Cc*, which is *P*1. We therefore see that the sequence of phases is $R\bar{3}c$ → mixed phase → *P*1. Importantly, *Cc* is a polar space group, which means that *P1* is also polar [141]. The low-temperature *Cc* phase is already known to be ferroelectric with a polarization of 1.5 μC/cm^2. It is not yet known whether the more distorted *P1* phase is ferroelectric (switchable) or pyroelectric (not switchable). Interestingly, the formate bending mode displays a similar splitting pattern under pressure at

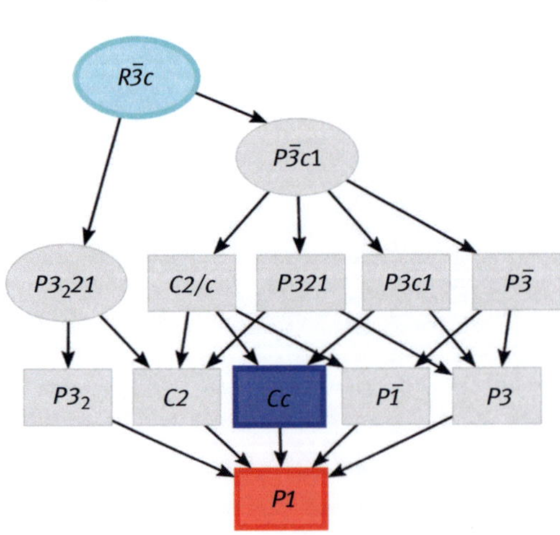

Figure 8: Schematic diagram showing how the $R\bar{3}c$ space group breaks symmetry. Both *Cc* and P1 are polar space groups [141] (reprinted with permission from Clune et al. [135]. Copyright 2020 American Chemical Society).

room temperature as well [Figure 5]. We therefore conclude that the symmetry of the high-pressure, high-temperature phase is also *P1*. Compression beyond this point broadens many of the low-frequency features involving Mn-related displacements and changes some of the line shapes – a signature of amorphous character.

3.5 Order–disorder transition under compression

The 185 K order–disorder transition in $[(CH_3)_2NH_2]Mn(HCOO)_3$ [91, 95, 119, 120, 122–124, 133] also depends upon pressure, and unveiling whether it rises or falls determines whether the polar or non-polar state is stabilized. In many materials, $T_{o/d}$ satisfies $dT_{o/d}/dP > 0$ due to pressure-induced enhancement of dipole interactions [153–155]. Therefore, should shift positively (or negatively in some cases) along the pressure axis. The best strategy for revealing this effect is to carry out a series of variable temperature Raman scattering runs at constant pressure – starting near 0.5 GPa and working up. Analysis of the mode splitting patterns should then reveal the conditions that stabilize the polar state and how the order–disorder transition evolves.

In the absence of a set of constant pressure measurements, one can assemble quasi-isobaric trends from a series of constant temperature experiments [135]. This technique has a number of challenges including variations in the spot measured or the sample used in different runs, the need to renormalize intensities from different runs to make the spectra comparable, the requirement for selecting an acceptable range of pressures rather than a particular pressure, and the potentially different calibrations from different measurements. We solved the majority of these problems by (i) normalizing each data set to the ambient pressure Raman spectrum, (ii) selecting an acceptable pressure range such as 1.9 ± 0.15 GPa, and examining frequency differences between features of the formate bending mode rather than absolute frequency trends under compression [135]. Obviously, the error bars on the position of the paraelectric ↔ ferroelectric transition are not ideal, but the main trend – which is for the order–disorder transition to rise (and for the polar phase to be stabilized) under compression – is not obscured.

Figure 9 summarizes the splitting of the formate bending mode at several different characteristic pressures in $[(CH_3)_2NH_2]Mn(HCOO)_3$. The splitting is small (5–7 cm^{-1}) in the non-polar $R\bar{3}c$ phase, whereas it is large (8 cm^{-1} or more) in the polar *Cc* phase. An analysis of precisely where the crossover occurs reveals $T_{o/d}$ as a function of pressure. At ambient conditions, $\Delta\omega$ rises gradually across the 185 K order–disorder transition (Figure 9(a)). The frequency shift is on the order of 5 or 6 cm^{-1} in the $R\bar{3}c$ phase and significantly larger in the *Cc* phase – although there is no sharp jump. We therefore see that $\Delta\omega$ can distinguish between the $R\bar{3}c$ and *Cc* phases. At ≈ 1.0 GPa, the order–disorder transition temperature has not moved very much [135]. Important changes begin to occur at 1.9 GPa. Based upon the sharp jump in the $\Delta\omega$ versus temperature trend, we estimate that the order–disorder transition rises to approximately

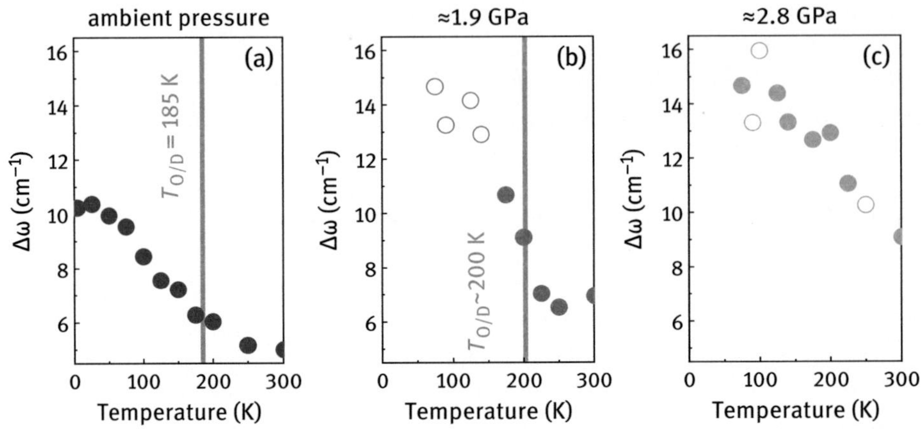

Figure 9: Splitting of the formate bending mode versus temperature at fixed pressures: (a) ambient conditions, (b) 1.9 GPa, and (c) 2.8 GPa (reprinted with permission from Clune et al. [135]. Copyright 2020 American Chemical Society).

200 K (Figure 9(b)). It continues to rise at 2.15 GPa [135]. At ≈ 2.8 GPa, the system is distorted at all temperatures – probably due to proximity to the edge of the mixed-phase boundary (Figure 9(c)). Thus we see that the order–disorder transition increases with pressure until it merges with the mixed-phase boundary line [135]. This is why the order–disorder transition does not persist above the mixed-phase region. The rising nature of the order–disorder transition also explains why the high-pressure phase of $[(CH_3)_2NH_2]Mn(HCOO)_3$ has $P1$ symmetry at all temperatures.

3.6 Developing the phase diagram of multiferroic $[(CH_3)_2NH_2]Mn(HCOO)_3$

In order to reveal the competition between different states of matter, we brought these results together to develop the pressure–temperature ($P-T$) phase diagram of $[(CH_3)_2NH_2]Mn(HCOO)_3$. As discussed above, we see a progression from $R\bar{3}c \rightarrow$ mixed phase $\rightarrow P1 \rightarrow$ amorphous character at room temperature. The strong hysteresis is indicative of first-order behavior. The mixed-phase region broadens significantly below the order–disorder transition, and the progression of the different pressure-induced phases changes slightly. Here, it goes from $Cc \rightarrow$ mixed phase $\rightarrow P1$ and then probably to amorphous under greater compression. Although the error bars are significantly larger, we also find that the order–disorder transition temperature rises with increasing pressure, merging with the boundary that defines the onset of the mixed phase. This is likely the result of enhanced dipolar interactions [135]. High-pressure susceptibility measurements (Figure 4) reveal the behavior of T_N under compression.

By bringing this type of experiment together with Raman scattering work, we can reveal the relationship between the local lattice distortions and the magnetically ordered state. Taken together, the $P-T$ phase diagram illustrates the remarkable competition between charge, structure, and magnetism in $[(CH_3)_2 NH_2]Mn(HCOO)_3$.

Extending the phase diagram to three dimensions provides an even deeper and more complex picture of the interplay between charge, structure, and magnetism. To develop the magnetic field–temperature ($H-T$) part of the diagram, we combine susceptibility, magnetization, and infrared results from several different efforts [91, 126, 131, 133]. Thus, we see how the magnetic ordering transition rises under both pressure and applied magnetic field as well as how the high-pressure branch of the spin-flop transition develops into the magnetic quantum phase transition at 15.3 T. It is worth mentioning that the variety of techniques and overall range of external stimuli employed in this work is remarkable. Too often magnetic field work stops at 5 or 7 T and compression studies go no further than 1.5 or 3 GPa. The $P-T-H$ phase diagram of $[(CH_3)_2 NH_2]Mn(HCOO)_3$ certainly shows the value of extending further from equilibrium in search of new functionalities. The polar space group at 300 K is particularly attractive in this regard.

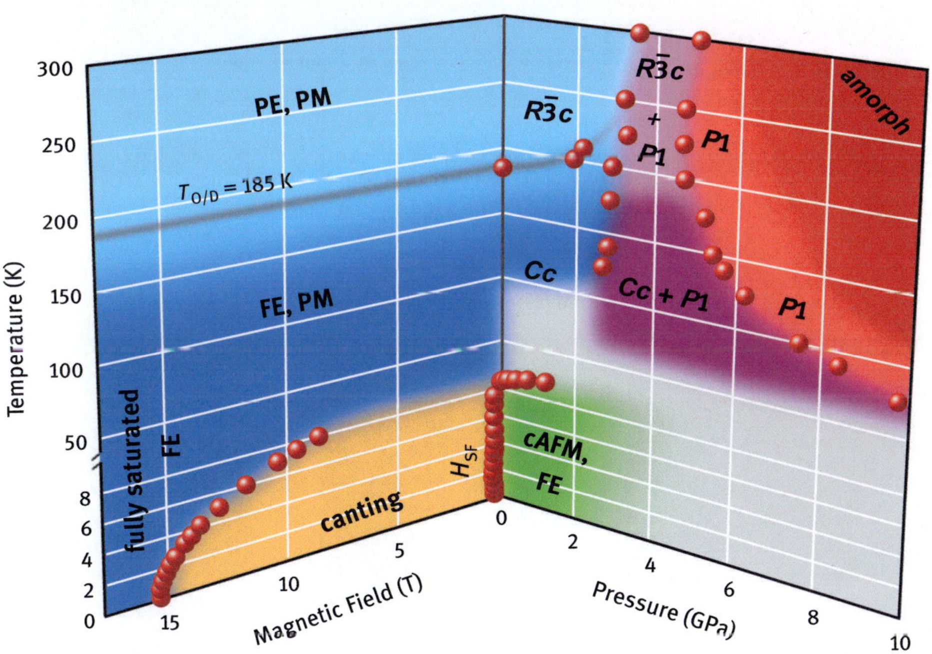

Figure 10: Pressure–temperature–magnetic field ($P-T-H$) phase diagram of multiferroic $[(CH_3)_2NH_2]Mn(HCOO)_3$ [135]. PE, paraelectric; FE, ferroelectric; cAFM, canted antiferromagnetic; PM, paramagnetic; $T_{o/d}$, order–disorder transition; amorph, amorphous; H_{SF}, spin-flop transition (0.31 T) [91, 126, 131]. The gray areas of the phase diagram are inaccessible in our measurements (reprinted with permission from Clune et al. [135]. Copyright 2020 American Chemical Society).

3.7 Structure–property relations

We can also use high-pressure vibrational spectroscopies to unravel structure–property relations. In order to focus our efforts, we concentrate on metal site tunability in the $[(CH_3)_2NH_2]M\,(HCOO)_3$ (M = Mn, Ni, Co) family of materials, although modifications to the amine counterion and ligand anion can also be used to modulate properties. One nice aspect of metal site substitution is the isostructural nature of the process. The room-temperature $R\bar{3}c$ crystal structure is maintained, as are the order–disorder aspects that give rise to ferroelectricity (although the exact values of T_C differ). On the other hand, metal site substitution changes local structure, the spin state, as well as the filling of the d-manifold. The magnetic energy scales are significantly larger in the

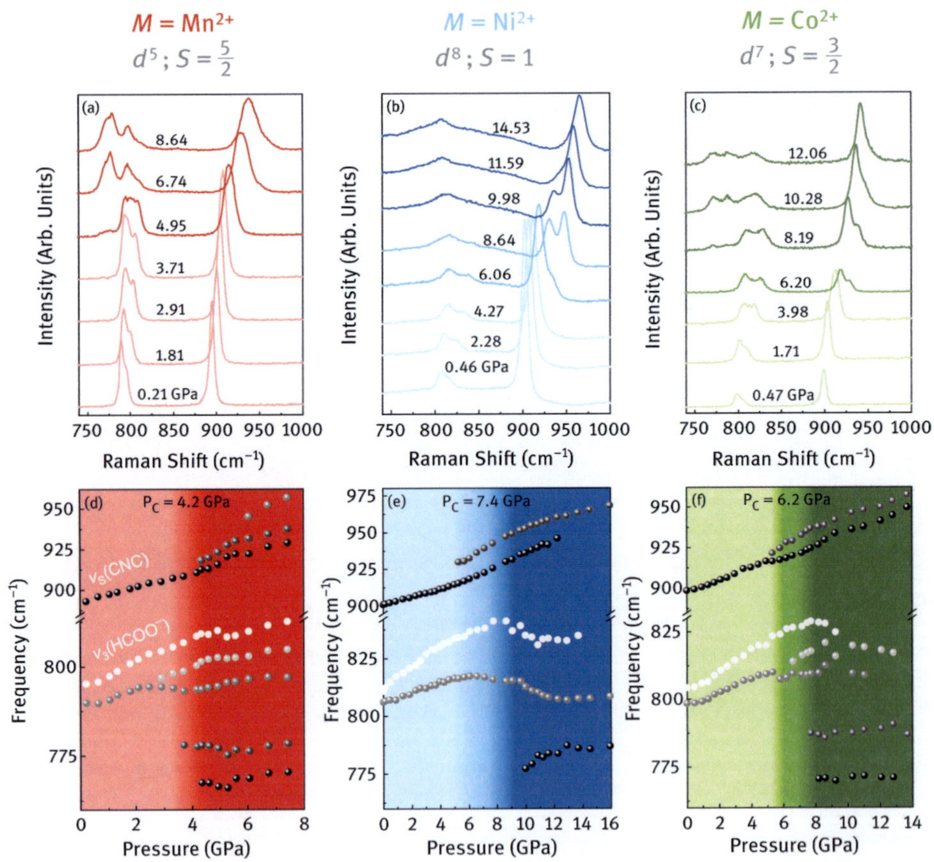

Figure 11: (a–c) Close-up view of the Raman-active formate bending mode in multiferroic $[(CH_3)_2NH_2]Mn(HCOO)_3$ under compression [135] compared with similar data in the Ni and Co analogs. All data are at room temperature. (d–f) Peak position versus pressure showing how symmetry breaking develops in these materials. The critical pressures are indicated.

Ni and Co analogs compared with those in $[(CH_3)_2NH_2]Mn(HCOO)_3$. For instance, spin-flop fields rise to 8 and 11 T in the Ni and Co compounds, respectively, and saturation fields are estimated to be on the order of 80 and 120 T [156]. These energy scales are one of the most important reasons that the Mn material receives so much attention. All of the relevant transitions can be studied using readily accessible temperatures, magnetic fields, and pressures (Figure 10).

Figure 11 summarizes the behavior of the $HCOO^-$ bending and C–N–C stretching modes of the $[(CH_3)_2NH_2]$ M $(HCOO)_3$ (M = Mn, Ni, Co) family of materials. As discussed in prior sections, the Mn compound has a 4.2 GPa critical pressure at room temperature, above which the system enters the $P1$ symmetry phase and then shows amorphous character under additional compression [134, 135]. A similar situation exists in the Ni and Co materials, although the critical pressures are higher. In $[(CH_3)_2NH_2]Ni(HCOO)_3$, the critical field is centered at 7.4 GPa, although the first signs of the mixed phase appear near 5 GPa and diminish above 9 GPa. The 6.2 GPa critical pressure of $[(CH_3)_2NH_2]Co(HCOO)_3$ is also significantly larger than that of the Mn analog. As before, the sluggish behavior around this critical pressure is a signature of two-phase mixing [134]. The mixed-phase regime is quite wide in all of these materials, and in the Mn compound, it broadens significantly below the order–disorder transition (Figure 10) [135]. It is likely that the Ni and Co compounds will show similar behavior.

The possibility of strain-controlled ferroelectricity at room temperature in $[(CH_3)_2NH_2]Mn(HCOO)_3$ is particularly exciting. There is precedent for this effect in $TbMnO_3$ which demonstrates a 10-fold increase in the electric polarization under compression [157]. Here, it is useful to point out that Cc is a well-known polar space group, and a subgroup of a polar space group is also polar [141]. If $[(CH_3)_2NH_2]Mn(HCOO)_3$ is ferroelectric, it is reasonable to expect that a lower symmetry group like $P1$ will have a larger polarization due to the greater distortion – even if we do not know its exact value. Of course, we do not yet know whether $[(CH_3)_2NH_2]Mn(HCOO)_3$ is ferroelectric (switchable) or pyroelectric (not switchable) in the high-pressure phase. This question will surely motivate high-pressure polarization studies of the $[(CH_3)_2NH_2]M$ $(HCOO)_3$ (M = Mn, Ni, Co) family of materials. Based upon the Raman signature of the formate bending and C–N–C stretching modes (Figure 11), the pattern of additional symmetry breaking in the Co analog is most promising in this regard. That said, the higher critical pressure will still make this system less desirable than the Mn analog – unless the value of the high-pressure/high-strain polarization is significantly enhanced.

4 Summary and outlook

This review chapter summarizes our efforts to explore molecule-based multiferroics under compression. These materials are superb platforms with which to unravel competing processes and test strategies for enhancing magnetoelectric coupling due to their overall low energy scales, tunable chemical architectures, and flexible lattices. To concentrate on our efforts, we focused on the high-pressure vibrational properties of multiferroic $[(CH_3)_2NH_2]Mn(HCOO)_3$ which provide microscopic insights into the symmetry of each phase and, when combined with magnetic properties work, support the development of an exquisite $P-T-H$ phase diagram. The latter summarizes – in graphical form – the interplay between charge, structure, and magnetism and, by so doing, may reveal new opportunities for deterministic property control with pressure and/or strain.

References

[1] Hill NA. Why are there so few magnetic ferroelectrics? The Journal of Physical Chemistry: B 2000;104:6694–6709. DOI: https://doi.org/10.1021/jp000114x

[2] Khomskii D. Classifying multiferroics: Mechanisms and effects. Physics 2009;2:20–28. DOI: 10.1103/Physics.2.20

[3] Willett RD, Galeriu C, Landee CP, Turnbull MM, Twamley B. Multiferroic and magnetoelectric materials. Inorganic Chemistry 2004;43:3804–3811. DOI: 10.1021/ic030317a

[4] Fiebig M. Revival of the magnetoelectric effect. Journal of Physics D: Applied Physics 2005;38:R123–R152. DOI: https://doi.org/10.1088/0022-3727/38/8/R01

[5] Eerenstein W, Mathur ND, Scott JF. Multiferroic and magnetoelectric materials. Nature 2006;442:759–765. DOI: https://doi.org/10.1038/nature05023

[6] Cheong SW, Mostovoy M. Multiferroics: A magnetic twist for ferroelectricity. Nature Materials 2007;6:13–20. DOI: https://doi.org/10.1038/nmat1804

[7] Tokura Y, Seki S. Multiferroics with spiral spin orders. Advanced Materials 2010;22:1554–1565. DOI: https://doi.org/10.1002/adma.200901961

[8] Rogez G, Viart N, Drillon M. Multiferroic materials: The attractive approach of metal-organic frameworks (MOFs). Angewandte Chemie International Edition in English 2010;49:1921–1923. https://doi.org/10.1002/anie.200906660

[9] Benedek NA, Fennie CJ. Hybrid improper ferroelectricity: A mechanism for controllable polarization-magnetization coupling. Physical Review Letters 2011;106:107204. DOI: 10.1103/PhysRevLett.106.107204

[10] Benedek NA, Fennie CJ. Polar octahedral rotations: A path to new multifunctional electronic, magnetic, and orbital materials. Journal of Solid State Chemistry 2012;195:11–20. DOI: 10.1016/j.jssc.2012.04.012

[11] Pardo E, Tain C, Liu J, Chamoreau L-M, Dkhil B, Boubekeur K, Lloret F, Nakatani J, Tokoro J, Ohkoshi S-I, Berdaguer M. Multiferroics by rational design: Implementing ferroelectricity in molecule-based magnets. Angewandte Chemie International Edition 2012; 51:8356–8360. DOI: https://doi.org/10.1002/anie.201202848

[12] Tokura Y, Seki S, Nagaosa N. Multiferroics of spin origin. Reports on Progress in Physics 2014;77:076501. DOI: https://doi.org/10.1088/0034-4885/77/7/076501

[13] Rodriguez-Velamazán JA, Fabelo O, Millan A, Campo J 1, Johnson RD, Chapon L. Magnetically-induced ferroelectricity in the $(ND_4)_2$ $[FeCl_5(D_2O)]$ molecular compound. Scientific Reports 2015;5:14475. DOI: 10.1038/srep14475

[14] Fiebig M, Lottermoser T, Meier D, Trassinm M. The evolution of multiferroics. Nature Reviews Materials 2016;1:16046. DOI: 10.1038/natrevmats.2016.46

[15] Xu K, Lu X-Z, Xiang H-J. Designing new ferroelectrics with a general strategy. Npj Quantum Materials 2017;2:1. DOI: 10.1038/s41535-016-0001-8

[16] Cheong S-W, Talbayev D, Kiryukhin V, Saxena A. Broken symmetries, non-reciprocity, and multiferroicity. Npj Quantum Materials 2018;3:19. DOI: 10.1038/s41535-018-0092-5

[17] Peng J, Zhang Y, Lin L-F, Lin L, Liu M, Liu J-M, Dong S. New iron-based multiferroics with improper ferroelectricity. Journal of Physics D 2018;51:243002. https://doi.org/10.1088/1361-6463/aac345

[18] Cheong S-W. SOS: Symmetry-operational similarity. Npj Quantum Materials 2019;4:53. https://doi.org/10.1038/s41535-019-0193-9

[19] Blundell SJ, Pratt FL. Organic and molecular magnets. Journal of Physics: Condensed Matter 2004;16:R771–R828. DOI: https://doi.org/10.1088/0953-8984/16/24/R03

[20] Ward MD. Directing assembly of molecular crystals. MRS Bulletin / Materials Research Society 2005;30:705–712. DOI: https://doi.org/10.1557/mrs2005.206

[21] Miller JS. Magnetically ordered molecule-based assemblies. Dalton Transactions 2006;23:2742–2749. DOI: 10.1039/B515974A

[22] Gaspar AB, Munoz MC, Real JA. Dinuclear iron(II) spin crossover compounds: Singular molecular materials for electronics. Journal of Materials Chemistry 2006;26:2522–2533. DOI: 10.1039/B603488H

[23] Hong T, Kim YH, Hotta C, Takano Y, Tremelling G, Turnbull MM, Landee CP, Kang H-J, Christensen NB, Lefmann K, Schmidt KP, Uhrig GS, Broholm C. Field-induced Tomonaga-Luttinger liquid phase of a two-leg spin-1/2 ladder with strong leg interactions. Physical Review Letters 2010;105:137207. DOI: https://doi.org/10.1103/PhysRevLett.105.137207

[24] Islam MF, Nossa JF, Canali CM, Pederson MR. First-principles study of spin-electric coupling in a Cu_3 single molecule magnet. Physical Review B 2010;82:155446. DOI: https://doi.org/10.1103/PhysRevB.82.155446

[25] Train C, Gruselle M, Verdaguer M. The fruitful introduction of chirality and control of absolute configurations in molecular magnets. Chemical Society Reviews 2011;40:3297. DOI: 10.1039/C1CS15012J

[26] Zapf VS, Sengupta P, Batista CD, Nasreen F, Wolff-Fabris F, Paduan-Filho A. Magnetoelectric effects in an organometallic quantum magnet. Physical Review B 2011;83:140405(R). DOI: https: //doi.org/10.1103/PhysRevB.83.140405. Erratum Physical Review B 2018, 97:219902. DOI: https: //doi.org/10.1103/PhysRevB.97.219902.

[27] Ghannadzadeh S, Möller JS, Goddard PA, Lancaster T, Xiao F, Blundell SJ, Maisuradze A, Khasanov R, Manson JL, Tozer SW, Graf D, Schlueter JA. Evolution of magnetic interactions in a pressure-induced Jahn-Teller driven magnetic dimensionality switch. Physical Review B 2013;87:241102. DOI: 10.1103/PhysRevB.87.241102

[28] Lancaster T, Goddard PA, Blundell SJ, Foronda FR, Ghannadzadeh S, Miller JS, Baker PJ, Pratt FL, Baines C, Huang L, Wosnitza J, McDonald RD, Modic KA, Singleton J, Topping CV, Beale TAW, Xiao F, Schlueter JA, Barton AM, Cabrera RD, Carreiro KE, Tran HE, Manson JL. Controlling magnetic order and quantum disorder in molecule-based magnets. Physical Review Letters 2014;112:207201. DOI: https://doi.org/10.1103/PhysRevLett.112.207201

[29] Zvyagin SA, Kamenskyi D, Ozerov M, Wosnitza J, Ikeda M, Fujita T, Hagiwara M, Smirnov AI, Soldatov TA, Ya. Shapiro A, Krzystek J, Hu R, Ryu H, Petrovic C, Zhitomirsky ME. Direct determination of exchange parameters in Cs_2CuBr_4 and Cs_2CuCl_4: High-field electron-spin-resonance studies. Physical Review Letters 2014;112:077206. DOI: https://doi.org/10.1103/PhysRevLett.112.077206

[30] Stone MB, Chen Y, Reich DH, Broholm C, Xu G, Copley JRD, Cook JC. Magnons and continua in a magnetized and dimerized spin-1/2 chain. Physical Review B 2014;90:094419. DOI: https://doi.org/10.1103/PhysRevB.90.094419

[31] Landee CP, Turnbull MM. Review: A gentle introduction to magnetism: Units, fields, theory, and experiment. Journal of Coordination Chemistry 2014;67:375–439. DOI: https://doi.org/10.1080/00958972.2014.889294

[32] Zapf VS, Jaime M, Batista CD. Bose-Einstein condensation in quantum magnets. Reviews of Modern Physics 2014;86:563. DOI: https://doi.org/10.1103/RevModPhys.86.563

[33] Collings IE, Goodwin AL. Metal-organic frameworks under pressure. Journal of Applied Physics 2019;126:181101. https://doi.org/10.1063/1.5126911

[34] Coronado E. Molecular magnetism: From chemical design to spin control in molecules, materials, and devices. Nature Reviews Materials 2020;5:87–104. DOI: https://doi.org/10.1038/s41578-019-0146-8

[35] Long J, Ivanov MS, Khomchenko VA, Mamontova E, Thibaud J-M, Rouquette J, Beaudhuin M, Granier D, Ferreira RAS, Carlos LD, Donnadieu B, Henriques MSC, Paixao JA, Guari Y, Larionova J. Room temperature magnetoelectric coupling in a molecular ferroelectric ytterbium(III) complex. Science 2020;367:671–676. DOI: 10.1126/science.aaz2795

[36] Tasiopoulos AJ, Vinslava A, Wernsdorfer W, Abboud KA, Christou G. Giant single-molecule magnets: A Mn_{84} torus and its supramolecular nanotubes. Angewandte Chemie International Edition 2004;16:2117–2121. DOI: 10.1002/anie.200353352

[37] Hill S, Anderson N, Wilson A, Takahashi S, Petukhov K, Chakov NE, Murugesu M, North JM, Del Barco E, Kent AD, Dalal NS, Christou G. A comparison between high-symmetry Mn_{12} single-molecule magnets in different ligand/solvent environments. Polyhedron 2005;24:2284–2292. DOI: https://doi.org/10.1016/j.poly.2005.03.079

[38] Baddeley C, Yan Z, King G, Woodward PM, Badjic JD. Structure-function studies of modular aromatics that form molecular organogels. Journal of Organic Chemistry 2007;72:7270–7278. DOI: https://doi.org/10.1021/jo071159y

[39] Kurmoo M. Magnetic metal-organic frameworks. Chemical Society Reviews 2009;38:1353–1379. DOI: 10.1039/b804757j

[40] Auban-Senzier P, Pasquier CR, Jerome D, Suh S, Brown SE, Mizire C, Batail P. Phase diagram of quarter-filled band organic salts [EDT-TTF-$CONMe_2$] $_2$ X, X=AsF_6 and Br. Physical Review Letters 2009;102:257001. DOI: https://doi.org/10.1103/PhysRevLett.102.257001

[41] Bechlars B, D'Alessandro DM, Jenkins DM, Iavarone AT, Glover SD, Kubiak CP, Long JR. High-spin ground states via electron delocalization in mixed-valence imidazolate-bridged divanadium complexes. Nature Chemistry 2010;2:362–368. DOI: https://doi.org/10.1038/nchem.585

[42] Lahti PM. Structure-property relationships for metal-free organic magnetic materials Adv. Physical Organic Chemistry 2011;45:93–169. DOI: 10.1016/B978-0-12-386047-7.00003-5

[43] Olson C, Heth CL, Lapidus SH, Stephens PW, Halder GJ, Pokhodnya K. Spectroscopic study of (two-dimensional) molecule-based magnets: [MII(TCNE)$(NCMe)_2$][SbF_6] (M = Fe, Mn, Ni). Journal of Chemical Physics 2011;135:024503. DOI: https://doi.org/10.1063/1.3599047

[44] Manson J, Lapidus S, Stephens P, Peterson P, Carreiro K, Southerland H, Lancaster T, Blundell S, Steele A. Structural electronic and magnetic properties of quasi 1D quantum magnets [Ni(HF_2)(pyz) $_2$]X (pyz = pyrazine;) exhibiting Ni-FHF-Ni and Ni-pyz-Ni spin interactions. Inorganic Chemistry 2011;50:5990–6009. DOI: https://doi.org/10.1021/ic102532h

[45] Zhang W, Xiong R-G. Ferroelectric metal-organic frameworks. Chemical Reviews 2012;112:1163–1195. DOI: https://doi.org/10.1021/cr200174w

[46] Goddard PA, Manson JL, Singleton J, Franke I, Lancaster T, Steele AJ, Blundell SJ, Baines C, Pratt FL, McDonald RD, Ayala-Valenzuela OE, Corbey JF, Southerland HI, Sengupta P, Schlueter JA. Dimensionality selection in a molecule-based magnet. Physical Review Letters 2012;108:077208. DOI: https://doi.org/10.1103/PhysRevLett.108.077208

[47] Hao J, Davidson RA, Kavand M, van Schooten KJ, Boehme C, Miller JS. Hexacyanobutadienide-based frustrated and weak ferrimagnets: $M(HCBD)_2 \cdot^z CH_2Cl_2$ (M = V, Fe). Inorganic Chemistry 2016;55:9393–9399. https://doi.org/10.1021/acs.inorgchem.6b01565
[48] Escalera-Moreno L, Baldova JJ, Gaita-Arito A, Coronado E. Exploring the high-temperature frontier in molecular nanomagnets: From lanthanides to actinides. Inorganic Chemistry 2019;58:11883–11892. https://doi.org/10.1021/acs.inorgchem.9b01610
[49] Cooper TM, Haley JE, Krein DM, Burke AR, Stwart DJ, Fore JL, Slagle JE. Triplet-state structure-property relationships in a series of platinum acetylides: Effect of chromophore length and end cap electronic properties. Physical Chemistry Chemical Physics 2019;21:26420–26429.
[50] Zhang Z, Rath NP, Cain JM, Meisel MW, Holmes SM. Structure-property studies of a new one-dimensional Fe(III)/Mn(II) chain. Polyhedron 2020;179:114376. DOI: 10.1016/j.poly.2020.114376
[51] Tang S, Yusov A, Li Y, Tan M, Hao Y, Li Z, Chen Y-S, Hu CT, Kahr B, M. D. Ward ROY. confined in hydrogen-bonded frameworks: Coercing conformation of a chromophore. Materials Chemistry Frontiers 2020. Advance Article. DOI: 10.1039/D0QM00200C
[52] de Jongh LJ, Miedema AR. Experiments on simple magnetic model systems. Advances in Physics 1974;23:1–260. DOI: https://doi.org/10.1080/00018739700101558
[53] Manriquez JM, Yee GT, McLean RS, Epstein AJ, Miller JS. A room temperature molecular/organic-based magnet. Science 1991;252:1415–1417. DOI: 10.1126/science.252.5011.1415
[54] de Jongh LJ, Miedema AR. Experiments on simple magnetic model systems. Advances in Physics 2001;50:947–1170. DOI: https://doi.org/10.1080/00018730110101412
[55] Shum WW, Her J-H, Stephens PW, Lee T, Miller JS. Observation of the pressure dependent reversible enhancement of T_c and the loss of the anomalous constricted hysteresis for $[Ru_2(O_2CMe)_4]_3Cr(CN)_6]$. Advanced Materials 2007;19:2910–2913. DOI: https://doi.org/10.1002/adma.200602751
[56] Baker PJ, Lancaster T, Franke I, Hayes W, Blundell SJ, Pratt FL, Jain P, Wang Z-M, Kurmoo M. Muon spin relaxation investigation of magnetic ordering in the hybrid organic-inorganic perovskites $[(CH_3)_2NH_2]M(HCOOO)_3$ (*M*= Ni, Co, Mn, Cu). Physical Review B 2010;82:012407. DOI: https://doi.org/10.1103/PhysRevB.82.012407
[57] Horiuchi S, Tokunaga Y, Giovannetti G, Picozzi S, Itoh H, Shimano R, Kumai R, Tokura Y. Above-room-temperature ferroelectricity in a single-component molecular crystal. Nature 2010;463:789–792. DOI: https://doi.org/10.1038/ncomms2322
[58] Fishman RS, Shum WW, Miller JS. Pressure-induced phase transition in a molecule-based magnet with interpenetrating sublattices. Physical Review B 2010;81:172407. DOI: https://doi.org/10.1103/PhysRevB.81.172407
[59] Miller JS. Magnetically ordered molecule-based materials. Chemical Society Reviews 2011;40:3266–3296. DOI: 10.1039/C0CS00166J
[60] Lancaster T, Pratt FL, Blundell SJ, Steele AJ, Baker PJ, Wright JD, Watanabe I, Fishman RS, Miller JS. Local magnetism in the molecule-based metamagnet $[Ru_2(O_2CMe)_4]_3[Cr(CN)_6]$ probed with implanted muons. Physical Review B 2011;84:092405. DOI: https://doi.org/10.1103/PhysRevB.84.092405
[61] Winter SM, Datta S, Hill S, Oakley RT. Magnetic anisotropy in a heavy atom radical ferromagnet. Journal of the American Chemical Society 2011;133:8126–8129. DOI: https://doi.org/10.1021/ja202156u
[62] Demir S, Zadrozny JM, Nippe M, Long JR. Exchange coupling and magnetic blocking in bipyrimidyl aradical-bridged dilanthanide complexes. Journal Of the American Chemical Society 2012;134:18546–18549. DOI: https://doi.org/10.1021/ja308945d

[63] Winter SM, Oakley RT, Kovalev AE, Hill S. Spin-orbit effects in heavy atom organic radical ferromagnets. Physical Review B 2012;85:094430. DOI: https://doi.org/10.1103/PhysRevB.85.094430
[64] Wriedt M, Yakovenko AA, Halder GJ, Prosvirin AV, Dunbar JR, Zhou H-C. Reversible switching from antiferro- to ferromagnetic behavior by solvent-mediated, thermally-induced phase transitions in a trimorphic MOF-based magnetic sponge system. Journal of the American Chemical Society 2013;135:4040–4050. DOI: https://doi.org/10.1021/ja312347p
[65] Landee CP, Turnbull MM. Recent developments in low-dimensional copper(II) molecular magnets. European Journal of Inorganic Chemistry 2013;13:2266–2285. DOI: https://doi.org/10.1002/ejic.201390053
[66] Ye HY, Zhang Y, Noro SI, Kubo K, Yoshitake M, Liu ZQ, Cai HL, Fu DW, Yoshikawa H, Xiong RG, Nakamura T. Molecule-displacive ferroelectricity in organic supramolecular solids. Scientific Reports 2013;3:2249. DOI: https://doi.org/10.1038/srep02249
[67] Alexandropoulos DI, Cunha-Silva L, Pham L, Bekiari V, Christou G, Stamatatos TC. Tetranuclear lanthanide(III) complexes with a zigzag topology from the use of pyridine-2,6-dimethanol: Synthetic, structural, spectroscopic, magnetic and photoluminescence studies. Inorganic Chemistry 2014;53:3220–3228. DOI: https://doi.org/10.1021/ic5001045
[68] Winter SM, Mailman A, Oakley RT, Thirunavukkuarasu K, Hill S, Graf DE, Tozer SW, Tse JS, Mito M, Yamaguchi H. Electronic and magnetic structure of neutral radical FBBO. Physical Review B 2014;89:214403. DOI: https://doi.org/10.1103/PhysRevB.89.214403
[69] Giannopoulos DP, Thuijs A, Wernsdorfer W, Pilkington M, Christou G, Stamatatos TC. Supramolecular chains of high nuclearity Mn(III) barrel-like single-molecule magnets. Chemical Communications 2014;50:779–783. DOI: 10.1039/C3CC47094F
[70] Pineda EM, Chilton NF, Marx R, Dörfel M, Sells DO, Neugebauer P, Jiang D, Collision D, van Slageren J, McInnes EJL, Winpenny REP. Direct measurement of dysprosium(III) dysprosium (III) interactions in a single-molecule magnet. Nature Communications 2014;5:5243. DOI: https://doi.org/10.1038/ncomms6243
[71] Quintero PA, Rajan D, Peprah MK, Brinzari TV, Fishman RS, Talham DR, Meisel MW. Pressure-induced enhancement of the magnetic anisotropy in $Mn(N(CN)_2)_2$. Physical Review B 2015;91:014439. DOI: https://doi.org/10.1103/PhysRevB.91.014439
[72] Brambleby J, Goddard PA, Johnson RD, Liu J, Kaminski D, Ardavan A, Steele AJ, Lancaster T, Manuel P, Baker PJ, Singleton J, Schwalbe SG, Spurgeon PM, Tran HE, Peterson PK, Corbey JF, Manson JL. The magnetic ground state of two isostructural polymeric quantum magnets, [Cu (HF_2)(pyrazine) $_2$]SbF_6 and [Co(HF_2)(pyrazine) $_2$]SbF_6, investigated with neutron powder diffraction. Physical Review B 2015;92:134406. DOI: https://doi.org/10.1103/PhysRevB.92.134406
[73] Gomez-Aguirre LC, Pato-Doldán B, Mira J, Castro-Garcia S, Antonia Senaris-Rodriguez M, Sánchez-Andújar M, Singleton J, Zapf V. Magnetic ordering-induced multiferroic behavior in $[CH_3NH_3][Co(HCOO)_3]$ metal-organic framework. Journal of the American Chemical Society 2016;138:1122–1125. DOI: https://doi.org/10.1021/jacs.5b11688
[74] Jain P, Stroppa A, Nabok N, Marino A, Rubano A, Paparo D, Matsubara M, Nakotte H, Feibig M, Picozzi S, Choi ES, Cheetham AK, Draxl C, Dalal NS, Zapf VS. Switchable electric polarization and ferroelectric domains in a metal-organic framework. Npj Quantum Materials 2016;1:16012. DOI: https://doi.org/10.1038/npjquantmats.2016.12
[75] Huang C, Du Y, Wu H, Xiang H, Deng K, Kan E. Discovery of intrinsic ferromagnetic ferroelectricity in transition metal halides monolayer. Physical Review Letters 2018;120:147601. DOI: 10.1128/JB.01484-06
[76] Coste SC, Pearson TJ, Freedman DE. Magnetic anisotropy in heterobimetallic complexes. Inorganic Chemistry 2019;58:11893–11902. https://doi.org/10.1021/acs.inorgchem.9b01459

[77] Dragulescu-Andrasi O A, Hietsoi A, Aoengar P, Dunk W, Stubbs V, Arroyave A, Kovnir K, Shatruk M. Dicyanometalates as building blocks for multinuclear iron(II) spin-crossover complexes. Inorganic Chemistry 2019;58:11920–11926. https://doi.org/10.1021/acs.inorgchem.9b01121

[78] Liu J, Kittaka S, Johnson RD, Lancaster T, Singleton J, Sakakibara T, Kohama Y, van Tol J, Ardavan A, Williams BH, Blundell SJ, Manson ZE, Manson JL, Goddard PA. Unconventional field-induced spin gap in an $S = 1/2$ staggered chain. Physical Review Letters 2019;122:057207. DOI: https://doi.org/10.1103/PhysRevLett.122.057207

[79] Johnson AI, Islam F, Canali CM, Pederson MR. A multiferroic molecular magnetic qubit. The Journal of Chemical Physics 2019;151:174105. https://doi.org/10.1063/1.5127956

[80] Chikara S, Gu J, Zhang X-G, Cheng H-P, Smythe N, Singleton J, Scott B, Krenkel E, Eckert J, Zapf VS. Magnetoelectric behavior via a spin state transition. Nature Communications 2019;10:4043. https://doi.org/10.1038/s41467-019-11967-3

[81] Brooks JS. Magnetic field dependent and induced ground states in organic conductors. Reports on Progress in Physics 2008;71:126501. DOI: https://doi.org/10.1088/0034-4885/71/12/126501

[82] Goddard PA, Singleton J, Sengupta P, McDonald RD, Lancaster T, Blundell SJ, Pratt FL, Cox S, Harrison N, Mason JL, Southerland HI, Schlueter JA. Experimentally determining the exchange parameters of quasi-two-dimensional Heisenberg magnets. New Journal of Physics 2008;10:083025. DOI: https://doi.org/10.1088/1367-2630/10/8/083025

[83] Xiao F, Woodward FM, Landee CP, Turnbull MM, Mielke C, Harrison N, Lancaster T, Blundell SJ, Baker PJ, Babkevich P, Pratt FL. Observation of 2D XY behavior observed in quasi-2D quantum Heisenberg antiferromagnets. Physical Review B 2009;79:134412. DOI: https://doi.org/10.1103/PhysRevB.79.134412

[84] Banks MG, Kremer RK, Hoch C, Simon A, Ouladdiaf B, Broto J-M, Rakoto H, Lee C, Whangbo M-H. Magnetic ordering in the frustrated Heisenberg chain system cupric chloride $CuCl_2$. Physical Review B 2009;80:024404. DOI: https://doi.org/10.1103/PhysRevB.80.024404

[85] Fortune NA, Hannahs ST, Yoshida Y, Sherline TE, Ono T, Tanaka H, Takano Y. Cascade of magnetic-field-induced quantum phase transitions in a spin-(1/2) triangular-lattice antiferromagnet. Physical Review Letters 2009;102:257201. DOI: 10.1103/PhysRevLett.102.257201

[86] Musfeldt JL, Vergara LI, Brinzari TV, Tung LC, Wang YJ, Schlueter JA, Manson JL, Sun C, Whangbo M-H. Magnetoelastic interactions in a two-dimensional quantum antiferromagnet. Physical Review Letters 2009;103:157401. DOI: 10.1103/PhysRevLett.103.157401

[87] Zvyagin SA, Cizmár E, Ozerov M, Wosnitza J, Feyerherm R, Manmana SR, Mila F. Field-induced gap in a quantum spin-1/2 chain in a strong magnetic field. Physical Review B 2011;83:060409(R). DOI: https://doi.org/10.1103/PhysRevB.83.060409

[88] Brinzari TV, Chen P, Sun Q-C, Liu J, Tung L-C, Wang YJ, Schlueter JA, Singleton J, Manson JL, Whangbo M-H, Litvinchuk AP, Musfeldt JL. Quantum critical transition amplifies magnetoelastic coupling in $Mn[N(CN)_2]_2$. Physical Review Letters 2013;110:237202. DOI: 10.1103/PhysRevLett.110.237202

[89] Sheu YM, Chang YM, Chang CP, Li YH, Babu KR, Guo GY, Kurumaji T, Tokura Y. Picosecond creation of switchable optomagnets from a polar antiferromagnet with giant photoinduced Kerr rotations. Physical Review X 2019;9:031038. DOI: https://doi.org/10.1103/PhysRevX.9.031038

[90] Allemand P-M, Khemani KC, Koch A, Wudl F, Holczer K, Donovan S, Grüner G, Thompson JD. Organic molecular soft ferromagnetism in a fullerene C_{60}. Science 1991;253:301–302. DOI: 10.1126/science.253.5017.301

[91] Jain P, Ramachandran V, Clark RJ, Dong Zhou H, Toby BH, Dalal NS, Kroto HW, Cheetham AK. Multiferroic behavior associated with an order-disorder hydrogen bonding Ttransition in metal-organic frameworks (MOFs) with the perovskite ABX_3 architecture. Journal of the American Chemical Society 2009;131:13625–13627. DOI: https://doi.org/10.1021/ja904156s

[92] Weng DF, Wang ZM, Gao S. Framework-structured weak ferromagnets. Chemical Society Reviews 2011;40:3157–3181. DOI: 10.1039/C0CS00093K

[93] Di Sante D, Stroppa A, Jain P, Picozzi S. Tuning the ferroelectric polarization in a multiferroic metal-organic framework. Journal of the American Chemical Society 2013;135:18126–18130. https://doi.org/10.1021/ja408283a

[94] Horiuchi S, Kumai R, Tokura Y. High-temperature and pressure-induced ferroelectricity in hydrogen-bonded supramolecular crystals of anilic acids and 2,3-di(2-pyridinyl)pyrazine. Journal of the American Chemical Society 2013;135:4492–4500. https://doi.org/10.1021/ja400318v

[95] Stroppa A, Barone P, Jain P, Perez-Mato JM, Picozzi S. Hybrid improper ferroelectricity in a multiferroic and magnetoelectric metal-organic framework. Advanced Materials 2013;25:2284–2290. https://doi.org/10.1002/adma.201204738

[96] Wei Qin M, Gong X, Chen TA, Shastry R, Sakidja G, Yuan M, Hersam C, Wuttig M, Ren S. Multiferroicity of carbon-based charge-transfer magnets. Advanced Materials 2015;27:734–739. https://doi.org/10.1002/adma.201403396

[97] Nagabhushana GP, Shivaramaiah R, Navrotsky A. Thermochemistry of multiferroic organic-inorganic hybrid perovskites $[(CH_3)_2NH_2][M(HCOO)_3]$ (M = Mn, Co, Ni, and Zn). Journal of the American Chemical Society 2015;137:10351–10356. https://doi.org/10.1021/jacs.5b06146

[98] Kinyon JS, Clar R, Dala NS, Choi ES, Zhou H. Ferroelectricity in the gapless quantum antiferromagnet NH_4CuCl_3. Physical Review B 2015;92:144103. DOI: https://doi.org/10.1103/PhysRevB.92.144103

[99] Horiuchi S, Kobayashi K, Kumai R, Minami N, Kagawa F, Tokura Y. Quantum ferroelectricity in charge-transfer complex crystals. Nature Communications 2015;6:7469. DOI: https://doi.org/10.1038/ncomms8469

[100] Xu B, Ren S. Integrated charge transfer in organic ferroelectrics for flexible multisensing materials. Small 2016;12:4501. DOI: 10.1002/smll.201600980

[101] Owczarek M, Hujsak KA, Ferris DP, Prokofjevs A, Majerz I, Szklarz P, Zhang H, Sarjeant AA, Stern CL, Jakubas R, Hong S, Dravid VP, Stoddart JF. Flexible ferroelectric organic crystals. Nature Communications 2016;7:13108. DOI: https://doi.org/10.1038/ncomms13108

[102] Shum WW, Schaller JN, Miller JS. Magnetic ground state and phase diagram, H_c (T), for magnetically ordered $[Ru_2(O_2CMe)_4]_3Cr(CN)_6]$. The Journal of Physical Chemistry C 1998;112:7936–7938. DOI: https://doi.org/10.1021/jp711835f

[103] Agust G, Cobo S, Gaspar AB, Molna G, Ould-Moussa N, Szila-Villo PA, Vieu C, Carmen-Munoz M, Real JA, Bousseksou A. Thermal and light-induced spin crossover phenomena in new 3D Hofmann-like microporous metal-organic frameworks produced as bulk materials and nanopatterned thin films. Chemistry of Materials 2008;20:6721–6732. DOI: https://doi.org/10.1021/cm8019878

[104] Zhou L, Shinde N, Hu A, Cook C, Murugesu M, Song Y. Structural tuning of energic material bis(1H-tetrazol-5-yl)amine monohydrate under pressures probed by vibrational spectroscopy and x-ray diffraction. The Journal of Physical Chemistry C 2014;118:26504–26512. DOI: https://doi.org/10.1021/jp507291m

[105] Lekin K, Phan H, Winter SM, Wong JWL, Leitch AA, Laniel D, Yong W, Secco RA, Tse JS, Desgreniers S, Dube PA, Shatruk M, Oakley RT. Heat, pressure and light-induced interconversion of bisdithiazolyl radicals and dimers. Journal of the American Chemical Society 2014;136:8050–8062. https://doi.org/10.1021/ja502753t

[106] Talham DR, Meisel MW. Thin films of coordination polymer magnets. Chemical Society Reviews 2011;40:3356–3365. DOI: 10.1039/C1CS15015D

[107] Prescimone A, Morien C, Allan D, Schlueter JA, Tozer SW, Manson JL, Parsons S, Brechin EK, Hill S. Pressure-driven orbital reorientations and coordination-sphere reconstructions in [CuF_2 (H_2O) $_2$ (pyz)]. Angewandte Chemie International Edition 2012;51:7490–7494. DOI: 10.1002/anie.201202367

[108] Jaffe A, Lin Y, Mao WL, Karunadasa HI. Pressure-induced conductivity and yellow-to-black piezochromism in a layered Cu-Cl hybrid perovskite. Journal of the American Chemical Society 2015;137:1673–1678. https://doi.org/10.1021/ja512396m

[109] Knaapila M, Guha S. Blue emitting organic semiconductors under high pressure: Status and outlook. Reports on Progress in Physics 2016;79:0066601. https://doi.org/10.1088/0034-4885/79/6/066601

[110] O'Neal KR, Lee JH, Kim M, Manson JL, Liu Z, Fishman RS, Musfeldt JL. Competing magnetostructural phases in a semiclassical system. Nature Quant. Mater 2017;2:65. DOI: https://doi.org/10.1038/s41535-017-0065-0

[111] Lengyel J, Wang X, Choi ES, Besara T, Schonemann RU, Kumar-Ramakrishna S, Holleman J, Blockmon AL, Hughey KD, Balicas L, McGill S, Musfeldt JL, Siegrist S, Dalal N, Shatruk M. Antiferroelectric phase transition in a proton transfer salt of squaric acid and Light-switchable exchange-coupled magnet. ACS Applied Electronic Materials 2,3-dimethylpyrazine. Journal of the American Chemical Society 2019;141:16279–16287. DOI: https://doi.org/10.1021/jacs.9b04473

[112] Rajan D, Cain JM, Brinzari TV, Ferreira CF, Rudawski NG, Felts AC, Meisel MW, Talham DR. Light-switchable exchange-coupled magnet. ACS Applied Electronic Materials 2019;1:2471–2475. DOI: 10.1021/acsaelm.9b00520

[113] Basov DN, Averitt RD, Hsieh D. Toward properties on demand in quantum materials. Nature Materials 2017;16:1077–1087. DOI: https://doi.org/10.1038/nmat5017

[114] Grochala W, Hoffmann R, Feng J, Ashcroft NW. The chemical imagination at work in very tight places. Angewandte Chemie International Edition 2007;46:3620–3642. DOI: 10.1002/anie.200602485

[115] Goodenough J. Magnetism and the Chemical Bond. New York, Wiley, 1963. DOI: https://doi.org/10.1016/0029-5582(64)90241-X

[116] Spaldin NA, Ramesh R. Advances in magnetoelectric multiferroics. Nature Materials 2019;18:203–212. DOI: https://doi.org/10.1038/s41563-018-0275-2

[117] Duncan HD, Dove MT, Keenb DA, Phillips AE. Local structure of the metal-organic perovskite dimethylammonium manganese(II) formate. Dalton Transactions 2016;45:4380. DOI: 10.1039/C5DT03687A

[118] Collings IE, Bykov M, Bykova E, Hanfland M, van Smaalen S, Dubrovinsky L, Dubrovinskaia N. Disorder-order transitions in the perovskite metal-organic frame-works [(CH3)2 NH2][M (HCOO)3] at high pressure. Crystal Engineering Communications 2018;20:3512–3521. https://doi.org/10.1039/C8CE00617B

[119] Stroppa A, Jain P, Barone P, Marsman M, Perez-Mato JM, Cheetham AK, Kroto HW, Picozzi S. Electric control of magnetization and interplay between orbital ordering and ferroelectricity in a multiferroic metal-organic framework. Angewandte Chemie International Edition 2011;50:5847–5850. DOI: 10.1002/anie.201101405

[120] Besara T, Jain P, Dalal NS, Kuhns PL, Reyes AP, Kroto HW, Cheetham AK. Mechanism of the order-disorder phase transition and glassy behavior in the metal-organic framwork [(CH_3) $_2NH_2$]Zn($HCOO)_3$. Proceedings of the National Academy of Sciences 2011;108:6828–6832. https://doi.org/10.1073/pnas.1102079108

[121] Wang W, Yan L-Q, Cong J-Z, Zhao Y-L, Wang F, Shen S-P, Zou T, Zhang D, Wang S-G, Han X-F, Sun Y. Magnetoelectric coupling in the paramagnetic state of a metal-organic framework. Scientific Reports 2013;3:2024. DOI: https://doi.org/10.1038/srep02024

[122] Maczka M, Gagor A, Macalik B, Pikul A, Ptak M, Hanuza J. Order-disorder transit ion and weak ferromagnetism in the perovskite metal formate frameworks of $[(CH_3)_2NH_2][M(HCOO)_3]$ and $[(CH_3)_2ND_2][M(HCOO)_3]$ (M = Ni, Mn). Inorganic Chemistry 2014;53:457–467. dx.doi.org/10.1021/ic402425n

[123] Maczka M, Almeida da Silva T, Paraguassu W, Ptak M, Hermanowicz K. Raman and IR studies of pressure- and temperature-Induced Phase transitions in $[(CH_2)_3NH_2][Zn(HCOO)_3]$. Inorganic Chemistry 2014;53:12650–12657. dx.doi.org/10.1021/ic502426x

[124] Sanchez-Andujar M, Presedo S, Yanez-Vilar S, Castro- Garcia S, Shamir J, Senaris-Rodriguez MA. Characterization of the order-disorder dielectric transition in the hybrid organic-inorganic perovskite-like formate $Mn(HCOO)_3[(CH_3)_2NH_2]$. Inorganic Chemistry 2010;49:1510–1516. DOI: https://doi.org/10.1021/ic901872g

[125] Walker HC, Duncan HD, Le MD, Keen DA, Voneshen DJ, Phillips AE. Magnetic structure and spin-wave excitations in the multiferroic magnetic metal-organic framework $(CD_3)_2ND_2[Mn(DCO_2)_3]$. Physical Review B 2017;96:094423. DOI: https://doi.org/10.1103/PhysRevB.96.094423

[126] Wang X-Y, Gan L, Zhang SW, Gao S. Perovskite-like metal formates with weak ferromagnetism and as precursors to amorphous materials. Inorganic Chemistry 2004;43:4615–4625. DOI: https://doi.org/10.1021/ic0498081

[127] Vinrod K, Deepak CS, Sharma S, Sornadurai D, Satya AT, Racindran TR, Sundar CS, Bharathi A. Magnetic behavior of the metal organic framework $[(CH_3)_2NH_2]Co(HCOO)_3$. RSC Advances 2005;5:37818. https://doi.org/10.1039/C5RA01417D

[128] Tian Y, Stroppa A, Cahi Y, Yan L, Wang S, Barone P, Picozzi S, Sun Y. Cross coupling between electric and magnetic orders in a multiferroic metal-organic framework. Scientific Reports 2015;4:6062. https://doi.org/10.1038/srep06062

[129] Tain Y, Wang W, Chai Y, Cong J, Shen S, Yan L, Wang S, Han X, Sun Y. Quantum tunneling of magnetization in a metal-organic framework. Physical Review Letters 2014;112:017202. DOI: https://doi.org/10.1103/PhysRevLett.112.017202

[130] Coleman P, Schhofield AJ. Quantum criticality. Nature 2005;433:226–229. DOI: https://doi.org/10.1038/nature03279

[131] Clune AJ, Hughey KD, Lee C, Abhyankar N, Ding X, Dalal NS, Whangbo M-H, Singleton J, Musfeldt JL. Magnetic field-temperature phase diagram of multiferroic $[(CH_3)_2NH_2]Mn(HCOO)_3$. Physical Review B 2017;96. DOI: 104424.

[132] Abhyankar N, Kweon JJ, Orio M, Bertaina S, Lee M, Choi ES, Fu R, Dalal NS. Understanding ferroelectricity in the Pb-free perovskite-like metal-organic framework $[(CH_3)_2NH_2]Zn(HCOO)_3$: Dielectric, 2D NMR, and theoretical studies. The Journal of Physical Chemistry C 2017;121:6314–6322. https://doi.org/10.1021/acs.jpcc.7b00596

[133] Hughey KD, Clune AJ, Yokosuk MO, Al-Wahish A, O'Neal KR, Fan S, Abhyankar N, Xiang H, Li Z, Singleton J, Dalal NS, Musfeldt JL. Phonon mode links ferroicities in multiferroic $[(CH_3)_2NH_2]Mn(HCOO)_3$. Physical Review B 2017;96. DOI: 180305(R).

[134] Chitnis AV, Bhatt H, Maczka M, Deo MN, Garg N. Remarkable resilience of the formate cage in a multferroic metal organic framework material: Dimethyl ammonium manganese formate. Dalton Transactions 2018;47:12993–13005. DOI: 10.1039/c8dt03080d

[135] Clune A, Harms N, O'Neal K, Hughey K, Smith K, Obeysekera D, Haddock J, Dalal N, Yang J, Liu Z, Musfeldt JL. Developing the pressure – temperature – magnetic field phase diagram of multiferroic $[(CH_3)_2NH_2]Mn(HCOO)_3$. Inorganic Chemistry 2020;59:10083–10090. DOI: https://dx.doi.org/10.1021/acs.inorgchem.0c01225

[136] Mao HK, Bell PM, Shaner JW, Steinberg DJ. Specific volume measurements of Cu, Mo, Pd, and Ag and calibration of the ruby R_1 fluorescence pressure gauge from 0.06 to 1 Mbar. Journal of Applied Physics 1976;49:3276–3283. DOI: 10.1063/1.325277
[137] Mao HK, Xu J, Bell PM. Calibration of the ruby pressure gauge to 800 kbar under quasi-hydrostatic conditions. Journal of Geophysical Research 1986;91:4673–4676. DOI: 10.1029/JB091iB05p04673
[138] Musfeldt JL, Brinzari TV, O'Neal KR, Chen P, Schlueter JA, Manson JL, Litvinchuk AP, Liu Z. Pressure-temperature phase diagram reveals spin-lattice intereactions in $Co[N(CN)_2]_2$. Inorganic Chemistry 2017;56:4950–4955. DOI: https://doi.org/10.1021/acs.inorgchem.6b03097
[139] Fateley WG, Dollish FR, McDevitt NT, Bentley FF. Infrared and Raman Selection Rules for Molecular and Lattice Vibrations: The Correlation Method. New York, Wiley-Intersience, 1972. ISBN: 9780471256205
[140] Piepho SB, Schatz PN. Group Theory in Spectroscopy with Applications to Magnetic Circular Dichroism. New York, Wiley, 1983. DOI: 10.1002/bbpc.19840881020
[141] Litvin DB. Ferroelectric space groups. Acta Crystallographica 1986;42:44–47. DOI: 10.1107/S0108767386099920
[142] Yamamoto T, Uruichi M, Yamamoto J, Yakushi K, Kawamoto A, Taniguchi H. Examination of the charge-sensitive vibrational modes in bis(ethylenedithio)tetrathiafulvalene. The Journal of Physical Chemistry. B 2005;109:15226–15235. DOI: 10.1021/jp050247o
[143] Rini M, Tobey R, Dean N, Itatani J, Tomioka Y, Tokura Y, Schoenlein RW, Cavalleri A. Ultrafast phase-transition induced by selective vibrational excitation in a magnetoresistive manganite. Nature 2007;449:72–74. DOI: https://doi.org/10.1038/nature06119
[144] Onda K, Gihara SO, Yonemitsu K, Maeshima N, Ishikawa T, Okimoto Y, Shao X, Nakano Y, Yamochi J, Saito G, Koshihara SY. Photoinduced change in the charge order pattern in the quarter-filled organic conductor $(EDO\text{-}TTF)_2$ PF_6 with a strong electron-phonon interaction. Physical Review Letters 2008;101:067403. DOI: 10.1103/PhysRevLett.101.067403
[145] Xu. XS, de Groot J, Sun Q-C, Sales BC, Mandrus D, Angst M, Litvinchuk AP, Musfeldt JL. Lattice dynamical probe of charge order and antiferroelectricity in $LuFe_2O_4$. Physical Review B 2010;82:014304. DOI: 10.1103/PhysRevB.82.014304
[146] Liu MK, Wagner M, Abreu E, Kittiwatanakul S, McLeod A, Fei Z, Goldflam M, Dai S, Fogler MM, Lu J, Wolf SA, Averitt RD, Basov DN. Anisotropic electronic state via spontaneous phase separation in strained vanadium dioxide films. Physical Review Letters 2013;111:096602. DOI: 10.1103/PhysRevLett.111.096602
[147] Basistyy R, Stanislacchuk TN, Sierenko AA, Litvinchuk AP, Kotelyanskii M, Carr GL, Lee N, Wang X, Cheong W. Infrared-active optical phonons and magnetic excitations in the hexagonal manganites $RMnO_3$ (R=Ho, Er, Tm, Yb, and Lu). Physical Review B 2014;90:024307. DOI: 10.1103/PhysRevB.90.024307
[148] Kung -H-H, Baumbach RE, Bauer ED, Thorsmolle VK, Zhang W-L, Haule K, Mydosh JA, Blumberg G. Chirality density wave of the "hidden order" phase in URu_2Si_2. Science 2015;347:1339–1342. DOI: 10.1126/science.1259729
[149] Zhou Y, Millis AJ. Charge transfer and electron-phonon coupling in monolayer FeSe on Nb-doped $SrTiO_3$. Physical Review B 2016;93:224506. DOI: 10.1103/PhysRevB.93.224506
[150] O'Neal KR, Zhou J, Cherian JG, Turnbull MM, Landee CP, Jena P, Liu Z, Musfeldt JL. Pressure-induced structural transition in copper pyrazine dinitrate and implications for quantum magnetism. Physical Review B 2016;93:104409. https://doi.org/10.1103/PhysRevB.93.104409

[151] Ackermann M, Brüning D, Lorenz T, Becker P, Bohaty L. Thermodynamic properties of the new multiferroic material $(NH_4)_2[FeCl_5(H_2O)]$. New Journal of Physics 2013;15:123001. DOI: https://doi.org/10.1088/1367-2630/15/12/123001

[152] Clune AJ, Nam J, Lee M, Hughey KD, Tian W, Fernandez-Baca JA, Fishman RS, Singleton J, Lee JH, Musfeldt JL. Magnetic field-temperature phase diagram of multiferroic $(NH_4)_2FeCl_5$. H_2O. Npj Quantum Materials. 2019;4:44. https://doi.org/10.1038/s41535-019-0180-1

[153] Mito M, Kawae T, Takeda K, Takagi S, Matsushita Y, Deguchi H, Rawson JM, Palacio F. Pressure-induced enhancement of the transition temperature of a genuine organic weak-ferromagnet up to 65 K. Polyhedron 2001;20:1509–1512. DOI: 10.1016/S0277-5387(01)00644-1

[154] Bellin C, Mafety A, Narayana C, Giura P, Rousse G, Itié JP, Polian A, Saitta AM, Shukla A. Disorder-order phase transition at high pressure in ammonium fluoride. Physical Review B 2017;96:094110. DOI: https://doi.org/10.1103/PhysRevB.96.094110

[155] Capitani F, Marini C, Caramazza S, Dore P, Pisanu A, Malavasi L, Nataf L, Baudelet F, Brubach J-B, Roy P, Postorino P. Locking of methylammonium by pressure-enhanced H-bonding in $(CH_3NH_3)PbBr_3$ hybrid perovskite. The Journal of Physical Chemistry C 2017;121:28125–28131. https://doi.org/10.1021/acs.jpcc.7b11461

[156] Hughey KD, Clune AJ, Yokosuk MO, Li J, Abhyankar N, Ding X, Dalal NS, Xiang H, Smirnov D, Singleton J, Musfeldt JL. Structure-property relations in multiferroic $[(CH_3)_2NH_2]M(HCOO)_3$ (M= Mn, Co, Ni). Inorganic Chemistry 2018;57:11569–11577. https://doi.org/10.1021/acs.inorgchem.8b01609

[157] Aoyama T, Yamauchi K, Iyama A, Picozzi S, Shimizu K, Kimura T. Giantspin-driven ferroelectric polarization in $TbMnO_3$ under high pressure. Nature Communications 2014;5:4927. https://doi.org/10.1038/ncomms5927

Alexander F. Goncharov

Simple molecules at high pressures: transitions to extended and atomic states

1 Introduction

Atoms in molecules are bound by strong covalent bonds. In contrast, molecules are held together by weak (e.g., van der Waals) bonds in the condensed state, which determine their low boiling and melting points. Molecular compounds are highly compressible and crystallize in a variety of structures determined by the molecular shape and intermolecular coupling. However, the application of pressure drastically changes the bond hierarchy as the inter- and intramolecular bonds change incongruently, whereby the inter- and intramolecular bond strengths eventually become comparable. At this point, the molecules demise, and extended or atomic states take over. These major transformations may or may not be associated with changes in the electronic structure. The overarching questions for these transformations are the related pathways and how do they diverge for different materials. Why a strongly bonded nitrogen molecule (N_2) becomes unstable at nearly 100 gigapascals (GPa) [1], while a weaker bonded diatomic oxygen (O_2) is expected to be stable up to terapascals (TPa) [2]? This diversity suggests that simple scaling relations [3, 4] are not applicable. Yet another open question concerns the high-pressure behavior of hydrogen and hydrogen-bearing molecules and novel poly- and super-hydrides [5, 6] as there is an expectation of enhancements of quantum effects under pressure [7–9], which may result in a fluxional behavior of hydrogen atoms in both molecular and nonmolecular phases.

Simple molecular compounds, which consist of light elements, are fundamental to understanding the interiors of the Earth and planets. Hydrogen, helium, oxygen, and carbon are the most, and nitrogen is the seventh most abundant element in the solar system. Therefore, they must retain a very sizable portion of planetary interiors [10–12]. However, the mode the elements (including the light element) associate in the planetary interiors likely forming a great variety of compounds remains highly uncertain. Moreover, the physical properties of the light elements and their compounds are not definitively understood at the conditions of planetary interiors. The transformations of simple molecules and their compounds to the extended or atomic states occur at quite high pressures (especially for hydrogen-rich compounds) making it difficult to perform such investigations [5, 6]. The knowledge of physics and chemistry of compounds made of light elements is critical for modeling planetary structure,

Alexander F. Goncharov, Earth and Planets Laboratory, Carnegie Institution for Science, 5251 Broad Branch Road, NW, Washington, DC 20015

https://doi.org/10.1515/9783110668612-004

composition, dynamics, and evolution [12]. Additionally, exploring electrical and thermal conductivity of deep planetary materials [13] would provide valuable insight into planetary observations such as internally generated magnetic field [14–16]. Due to the necessity to reach extreme pressure–temperature (P–T) conditions, probing simple molecules at the planetary interior conditions is enormously challenging requiring highly refined experimental and theoretical techniques.

High-pressure techniques continue to develop with an accelerated speed [17]. This includes the generation of more extreme conditions of pressure, temperature, and external fields (e.g., high strain rates) and the development of the associated specialized probes. Static and dynamic compression experiments and first-principles theoretical calculations are the major contributors into this research field. The advances in these topics are motivated by exciting discoveries reported recently and by great implications of the high-pressure research for geosciences [18], for materials research [19–21], and for understanding of matter under extreme conditions [22–24]. The previous relevant reviews on the behavior of molecules under high pressure [25–27] are greatly outdated because the latest discoveries found largely unexpected new behaviors. This is particularly related to novel chemistry rules at high pressures, which are being currently recognized [19, 28], as well as often unexpected structural complexity of both molecular and polymerized systems [29–31]. For molecular materials at high pressures, the unexpected behaviors that have been recently established are the disproportionation of familiar molecules and stabilization of molecular associations with a different composition and structure [32]. These phenomena are expected to be facilitated for molecules containing hydrogen atoms, for example, ammonia [33] and molecular hydrogen [34].

In this chapter, I provide a critical review of the investigation of behavior of simple molecular compounds under extreme conditions with the major emphasis on the pathways of transformations to the extended and monatomic states. The materials discussed here include halogens and H_2 (please see recent review papers on high-pressure behavior of hydrogen [5, 6]), N_2, O_2, CO_2, and H_2O. Another review paper on a similar subject of this author is presented in ref. [35].

2 Molecules under pressure – overview of high-pressure behavior

Under pressure, molecular compounds undergo a number of intermediate steps in their structural and electronic modifications toward forming extended and monatomic states (Figure 1). The states associated with these transformations may vary for different molecules depending on the chemical composition and molecular geometry. The common intermediate phenomena for molecular system under pressure include orientational ordering, bond competition, molecular disproportionation, and molecular dissociation followed by polymerization or formation of atomic nonmolecular states. The transitions between different regimes are largely driven by pressure-induced densification, which results in the electronic delocalization. The structural and chemical bond modifications are accompanied by a variety of electronic transformations including metallization. The extended states often form symmetric atomic configuration, but these tend to experience symmetry breaking under pressures developing distorted complex but dense structures with the dispersed bonds and charge transfer between atoms. The crystal lattice ionization often occurs via polarization of covalent bonds in extended structures representing another common mechanism of densification [22]. It should be noted that this mechanism is common not only for extended structures of molecular crystals but also for many other classes of materials including molecular crystals [32], ionic solids [36], and noble gas compounds [37]. Formations of electrides – ionic bonded materials with the valence electronic pairs that form unique anions in the crystal lattice – represent another pathway for high-pressure densification [38–40]. Ionization is also common at high temperatures because of the entropy effects, which results in stabilization of unique superionic phases [41], where light elements (e.g., hydrogen) can move freely, while heavy atoms are forming a steady lattice. On the other hand, metallic atomic states are subject to charge density instability, especially in the intermediate regime between insulating molecular and monatomic metallic structures [42].

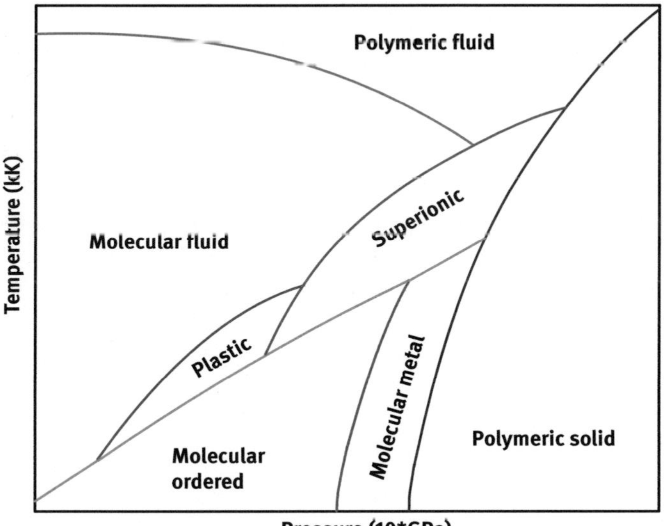

Figure 1: Conceptual phase diagram of molecular crystals at high pressures.

Upon crystallization, molecular solids typically form *plastic crystals*, where the molecules are weakly interacting permitting to retain some degrees of rotational or conformational freedom, thus making them dynamically disordered. *Orientational ordering* [27, 43] occurs as a result of increase in intermolecular interaction, so the associated orientationally ordered phases naturally favor low temperatures and high pressures. At low temperatures, the rotations are naturally diminished making even shallow minima in the potential energy accessible, which results in reduction of symmetry and formation of the ordered phases. Likewise, application of high pressure increases the intermolecular coupling and leaves smaller space for molecules to rotate again favoring ordering. Hydrogen-bearing molecules (H_2 and CH_4) possess quantum rotational degrees of freedom making this scenario even more complicated [44, 45].

The *bond competition* is common for materials with heterogeneity in chemical bonding such as molecular crystals, which are characterized by a hierarchy of interatomic interactions including intramolecular (polar) covalent and intermolecular hydrogen and van der Waals bonds [25]. In addition, intermolecular coupling can be represented by electrostatic bonds between differently charged functional groups in compounds [32] and molecules of the same kind experiencing charge transfer. Under pressure, the weakest bonds yield the most, which results in their shortening and changing in the bond character. The classic example is the behavior of hydrogen bonds, which transform to polar covalent ones compressed to sufficiently high pressures [46].

The pressure-induced molecular *dissociation* occurs due to delocalization of the valence electrons because of the molecular electronic structure destabilization due to an abrupt increase of the electronic kinetic energy, which scales with the distance faster than the electrostatic energy [22, 25]. Therefore, on compression, compounds with the lower bond order or metals become energetically preferable as their electrons occupy the states with the lower kinetic energy. However, the molecular covalent bonding patterns (Figure 2) can survive the application of high pressures in the case of the molecular geometry that matches well the crystal structure of an extended state as in the case of the ice VII–ice X transition [47]. The molecular collapse into the extended structure may or may not be associated with the formation of a metallic phase, for example, ice X with symmetric hydrogen bonds is insulating [46] as well as polymeric nitrogen [48]; however, hydrogen is definitively expected to be a metal at the transformation to a monatomic state [49, 50]. Moreover, molecular solids can metallize before transformation into the extended states via the electronic band broadening and crossing related to an increase in the intermolecular coupling [51, 52] as, for example, has been recently reported in hydrogen [53, 54].

The high-pressure (density) *ionization* at high temperatures occurs in a regime, where the entropy contribution becomes substantial and the chemical bonding is disrupted. This is very visible for hydrogen-bearing molecules in that the hydrogen atoms can become extremely mobile at intermediate temperatures below the melting line of a molecular crystal leading to the formation of superionic phases. Note that these ionized states are different from the plastic phases introduced above, as hydrogen atoms are not bonded in superionic phases. It is interesting that these superionic states have low electronic conductivity comparable to semiconductors; however, at higher temperatures they are expected to border with electronically conducting fluids. On the other hand, the fluid states at sufficiently high pressure–temperature conditions undergo a Mott-type plasma phase transition, which is expected to occur via an abrupt band gap closure [55]. Such metallization transition may or may not be accompanied by structural changes, for example, molecular dissociation [56].

The evolution in the molecular structure and bonds under pressure outlined above is a greatly simplified representation. The real sequence of transformations under pressure depends significantly on the nature of the molecule and its bonding scheme. The most thermodynamically stable structures are those, which are the most effectively packed. This is determined as a complex balance between the bonding energy (determined by electron hybridization) and electrostatic repulsion. Application of pressure results in a great increase in the electronic kinetic energy (e.g., 1 eV per 100 GPa), which is comparable to the strength of any chemical bond. Nevertheless, the molecular electronic structure plays a paramount role in the stability of the molecule at high pressure as will be discussed below. The molecular bonding configurations are readily predicted by the valence-shell electron–pair repulsion theory dictating that the bonds and non-bonding valence electron pairs (lone pairs) would be as far apart as possible (Figure 2).

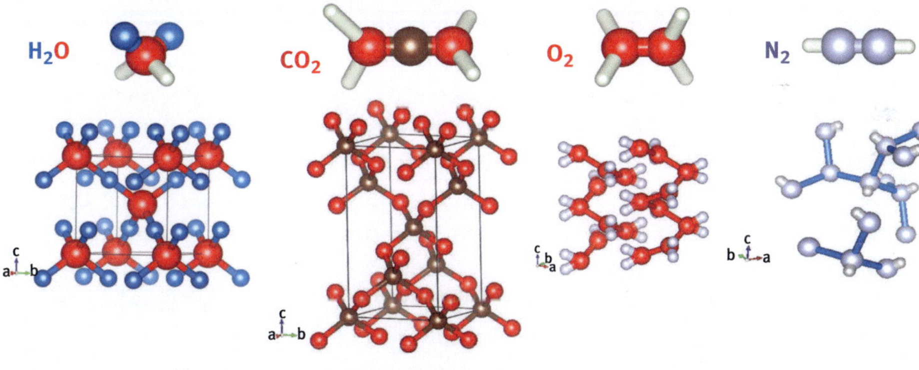

Figure 2: Molecular shapes and extended structures of simple molecular crystals: H_2O (ice X [46]), CO_2 (β-cristobalite [186, 188]), O_2 (theoretically predicted spiral-like $I4_1/acd$ structure [2]), and N_2 (cubic gauche structure, cg [201]). The cylinders show the directions of the lone pairs.

Moreover, the transition to extended (nonmolecular) states strongly depends on kinetic effects as many of these transformations involve a necessity to overcome a large potential barrier, which commonly results in great hysteretic effects. These barriers can be often overcome by choosing a specific *P–T* pathway, for example, involving the application of high temperatures facilitating the transition. The existence of high transformation barriers suggests that polymeric structures can exist as metastable phases at much lower pressures compared to those at the thermodynamic boundary, providing an opportunity for recovery of new materials (e.g., polymeric) to ambient conditions.

Until recently, the information about the structural pathways, which experience molecular crystals at high compressions, was limited because of the challenges in both experiments and theory. However, recent developments of experimental and theoretical methods enabled informative experiments and predictive theoretical calculations, which drastically improved our understanding of the behavior of simple molecules at high pressure.

3 Technical developments

Reaching very high pressures of interest for transformation of molecular systems into the extended states is very challenging because these materials are very compressible thus typically requiring a several fold compression to reach the transformation. In static experiments, only diamond-anvil cell (DAC) is capable for generating very high pressures in excess of 100 GPa. In a DAC, the sample is compressed by two opposing diamond anvils in a pressure chamber drilled in a thin gasket between diamonds (Figure 3). To reach very high compression states, one needs to thin down the gasket, which is made of very strong and thus incompressible material (e.g., Re) that does not extrude away easily because of the diamond tip deformation (cupping [57, 58]) and high friction. The conventional anvil design, which include a small culet and bevels, has been recently revised toward a toroidal anvil shape [59, 60] (Figure 3), which increased the pressure range and allowed to perform the pressure reversal in a more reliable fashion as has been demonstrated in the recent experiments in hydrogen [54]. These improvements are due to less extreme stresses experienced by the anvils if this new shape is used. Machining the diamond-anvil culets (using focused ion beam) to a desired shape represents a major leap in developing static high-pressure techniques to terapascal pressures.

Another recent major DAC technical advance is related to the development of instrumentation to reach simultaneous conditions of very high pressure and temperature. Implementation of pulsed laser heating approach combined with time domain measurements provides a new opportunity to reach higher temperature states and also to reduce the probability of unwanted diffusion and diamond failure [61–65]. Such experiments have been recently applied to a number of molecular materials to probe the regime of ionization in the liquid state at high temperatures [65–69].

Probing molecular materials requires extremely refined experimental tools. The most critical is the structural information, which presents a great challenge because the majority of materials described here are low-Z ones. Thus, application of the traditional X-ray diffraction (XRD) techniques is very difficult especially in the case of hydrogen-bearing molecules, as the scattering cross section by light atoms which is proportional to Z^2, where Z is the atomic number, is very small. This was at least partially circumvented by using single-crystal diffraction techniques [70–78]. Use of

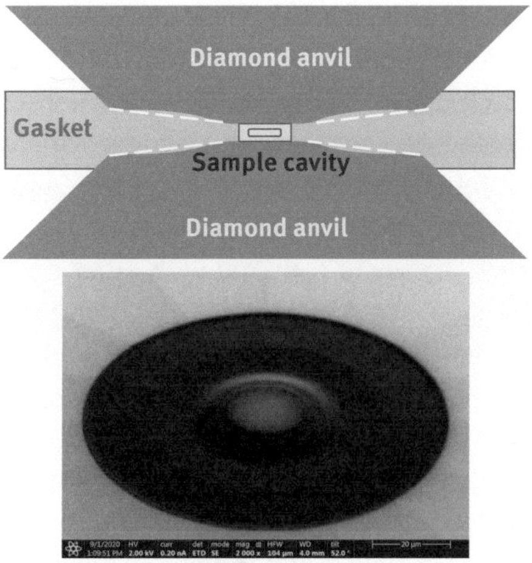

Figure 3: Diamond anvils to reach very high pressures. The sample is positioned in a hole in a gasket and is normally surrounded by a pressure medium (left panel). New toroidal-shaped anvils can be made by machining common beveled anvils (right panel presents the electronic image of one anvil); their shape is shown by white dashed lines.

neutron diffraction provides an access to the structural positions of light atoms as the scattering efficiencies are different and, generally, more sensitive to low-*Z* elements [79, 80]. However, the neutron radiation source is much larger than the X-ray synchrotron one, thus making its usage for small DAC sample much less efficient. Nevertheless, recent development of the Paris-Edinburgh cells and DAC with large conically supported diamond anvils made it possible to extend structural research of low-*Z* materials up to the 100 GPa pressure range [81, 82].

Probing the electronic properties of molecular materials at extreme conditions, which are often intimately connected to the transitions into the extended states, represents another great challenge. Optical spectroscopy techniques (discussed below), which are commonly used in this case, have significant limitations in terms of the spectral range that can be accessed because diamonds are not transparent in the ultraviolet (UV) spectral range, where the electronic band gap of the common molecular crystals are at low pressures; moreover, the band gap of stressed diamonds decreases with pressure moving to the visible at multimegabar pressures [83]. X-ray spectroscopy techniques in a DAC have been shown to be very useful in probing the electronic properties of the materials under extremes [84]. Recently, the electronic band gap structure of compressed hydrogen [17, 85] and helium (He) has been measured up to 90 and 17 GPa, respectively, in the spectral range of 20–40 eV using inelastic X-ray scattering [86].

Yet another major development in DAC technology is a recently emerging technique of X-ray free-electron laser heating combined with almost instantaneous XRD snapshots using the same fs pulses [87]. This technique in combination with pulsed laser heating holds a promise to become very instrumental in capturing fast chemical reactions and phase transformations of molecular solids and fluids at extreme *P–T* conditions.

Optical spectroscopy techniques in DAC, which can be easily combined with other techniques (e.g., electrical and X-ray) and microscopy, are among the most widely used in the field because they provide direct information about elastic, vibrational, and electronic properties under extreme *P–T* conditions. Moreover, Raman spectroscopy of the stressed diamond edge [88, 89], which is easy to combine with other optical spectroscopy tools, has become one of the most frequently used techniques at extreme conditions (especially at very high pressures [53, 54]) because it is fast, simple, and reliable even though it can be a subject of a number of systematic uncertainties [90] related to an indirect nature of these measurements; indeed, it probes the state of the anvils confining the sample and not the sample itself. The successful use of Raman and Brillouin spectroscopy at high pressures has been greatly enhanced by combining these techniques with laser heating in a DAC [91–93], thus opening new avenues for *in situ* measurements of phase and electronic transformations, melting, and chemical reactivity of molecular crystals at high temperatures [65, 68, 94–99]. Raman spectroscopy probe [100] is a fast and very informative tool to identify the phase stability and phase change as well as chemical transformations, especially in the case of molecular compounds. The intramolecular fundamental vibrations (often called vibrons for diatomic molecules) serve as the fingerprints of stable molecular (or chemical) state, and, furthermore, conveniently inform about the orientational ordering and presence of different crystallographic sites via vibrational (Davydov) [101–103] and crystal field splitting [31, 34, 104]. On the other hand, the Raman spectrum of intermolecular vibrations (often called phonons) governed by the Raman selection rules is intimately related to the crystal lattice structure. The intermolecular vibrations also include free (called roton in H_2 and H_2-bearing compounds) and restricted (librons) rotational motions, the presence of which in the spectra provide information about the lattice symmetry and also the state of the molecular rotational motion. The overall spectrum of the optical vibrational excitation provides the direct information about the vibrational dynamics of the materials, shedding light on the proximity to the phase and chemical transformation, and changes in the electronic structure including ionization. Infrared (IR) spectroscopy also probes the vibrational excitations providing a complementary information as this technique is sensitive to the dipole moment oscillations unlike Raman, which probes the polarizability. Synchrotron IR spectroscopy in a DAC has been an important tool for investigation of molecular solids at high pressures [54, 105]. Moreover, IR spectroscopy combined with a wide range of near-IR–visible–UV optical spectroscopy tools probes the electronic properties of materials via investigating how the electronic excitations couple to radiation [103, 106]. On the other hand, Brillouin spectroscopy probes very low frequency vibrational excitations, corresponding to acoustic modes near the Brillouin zone center, thus assessing the elastic properties of materials. Application of Brillouin spectroscopy to molecular compounds at very high pressures is very challenging because of the necessity to have a large angular optical access and sufficient scattering sample volume [107–109]; however, recently Brillouin spectroscopy has been used in Mbar pressure range [109, 110]. Also, a new laser optical technique, picosecond laser acoustics, has been developed and become frequently used to measure elasticity up to very high pressures beyond 100 GPa [111, 112]. This technique allowed to extend the pressure range of investigation of the elastic properties of molecular solids [113, 114].

4 Halogens and hydrogen

Halogens and hydrogen form a series of molecules, H_2, F_2, Cl_2, Br_2, and I_2, which count five out of seven (O_2 and N_2 should be added to complete this list) known stable diatomic molecules that consist of the same element. Hydrogen formally belongs to the first group, however, given that it is short by one electron in the partially filled 1*s* shell, it can be positioned in the seventh group together with the halogens, which are also short by one electron in their partially occupied electronic *p* shell. The difference is in the chemical behavior where hydrogen often loses one electron in forming chemical bonds, while the halogens do not, and, instead attract one electron to fill the *p* shell. The lightest molecules H_2, F_2, and Cl_2 are gases at ambient conditions, while Br_2 is a liquid and I_2 is a solid. The molecules in this series are single bonded making it possible to investigate their systematics depending on the atomic weight and the electronic shell number, which is partially occupied. The phenomena of interest include molecular dissociation and metallization, which appear to be similar in these materials, at least for the most accessible members Cl_2, Br_2, and I_2, but the relevant pressure range varies by more than an order of magnitude. Accordingly, the heaviest molecules are the most studied, while the lightest (H_2, F_2, and Cl_2), where the transitions are shifted well above 100 GPa, are much more challenging for detailed definitive investigations. In H_2, metallization is expected (and recently reported [53, 54, 115]) above 400 GPa, which is on the verge of current static compressions capability; however, H_2 is very different electronically and, strictly speaking, does not belong to the halogen sequence as will be discussed below. On the other hand, F_2, which does belong, is very toxic and chemically reactive and because of this it is very little experimentally studied, thus making it a good object for predictions based on simple scaling arguments [4, 116] and theoretical calculations (see below).

Experiments show that Cl_2, Br_2, and I_2 demonstrate universal trends under compression experiencing a series of very similar physical and chemical transformations [4, 117]. This universal behavior of the halogen progression can be rationalized by monitoring the properties as a function of a scaled volume parameter $\tilde{v} = V/r_{s^3}$, where V is the volume per atom and r_s is the intramolecular bond length, thus scaling the difference in the specific volumes and intramolecular bond lengths. At low pressures, halogens crystallize in an orthorhombic *Cmca* structure, where molecules are associated into the layers of zigzag chains (Figure 4). Under compression, the difference between inter- and intramolecular interatomic distances becomes smaller, and the bond competition occurs. Counterintuitively, extended X-ray absorption fine structure spectroscopy show that the bond lengths in Cl_2, Br_2, and I_2 initially increase indicating weakening of intramolecular bonds [118, 119], which is consistent with the Raman spectroscopy observations [120–122]. Interestingly, at higher pressures the bond lengths start dropping demonstrating a cusp. The existence of this cusp is connected with other phenomena including a theoretically proposed structural phase transition to another molecular phase (coexisting with *Cmca* structure) with a space group *C*2/*m* (a subgroup of *Cmca*), where the bond lengths of two molecules in the unit cell become slightly different [123, 124]. The existence of this molecular phase is supported by XRD measurements and observations of concomitant splitting of some Raman bands as well as appearance of a new low-frequency Raman mode [122–124]. However, it has been alternatively proposed that these XRD and Raman observations above 13 GPa can be explained by a complex four-dimensional incommensurately modulated structure [125]. The intermolecular distances decrease very fast before the cusp transition, which can be seen from a very high initial compressibility of the halogen solids and a rapid increase of the Raman intermolecular mode frequency with pressure. Above the cusp transition, where the intramolecular distance splits and starts decreasing with pressure, the character of the intermolecular interactions change with the second neighbor intermolecular interactions contributing into this balance more significantly hereby hindering the molecular rotations [124]. It has been empirically found that the cusp transition occurs when the scaled volume parameter $\tilde{v} = 1.55 +/- 0.02$ for Cl_2, Br_2, and I_2 (Figure 5) [118].

Compressing a *Cmca*–*C*2/*m* phase mixture well above the cusp transition results in a shortening of all bonds and metallization in molecular state. The molecular state in this regime is characterized by enhanced intermolecular interactions, and the intramolecular and nearest intermolecular interatomic distances become much closer to each other, for example, 2.75 Å versus 3.2 Å in I_2 at 14 GPa [126], where it metallizes [52]. It appears that Cl_2, Br_2, and I_2 metallize at different pressures but it has been inferred [118] that this occurs at the compression levels corresponding to the same scaled volume $\tilde{v} = 1.37$ (Figure 5). This relation has been derived from another phenomenological approach called the Herzfeld criterion [127, 128], which determines that a substance would be a metal when its density ρ becomes larger than the M/R ratio, where M is the molar weight and R is the molar refractivity, $R = (4/3)\pi\alpha$, where α is the molar polarizability. Application of this criterion to halogen molecules demonstrates that metallization occurs at the same scaled volume as the molecular dissociation was reported $\tilde{v} = 1.29$ in Br_2 and I_2 [4, 117]. However, more recent data suggest that metallization via the band gap close occurs at a smaller compression level corresponding to $R\rho/M = 0.9$ and $\tilde{v} = 1.37$. This appears to be matching well the experiments including the latest data on Cl_2 reporting metallization at 200 GPa and molecular dissociation at 258 GPa [122] (Figure 5).

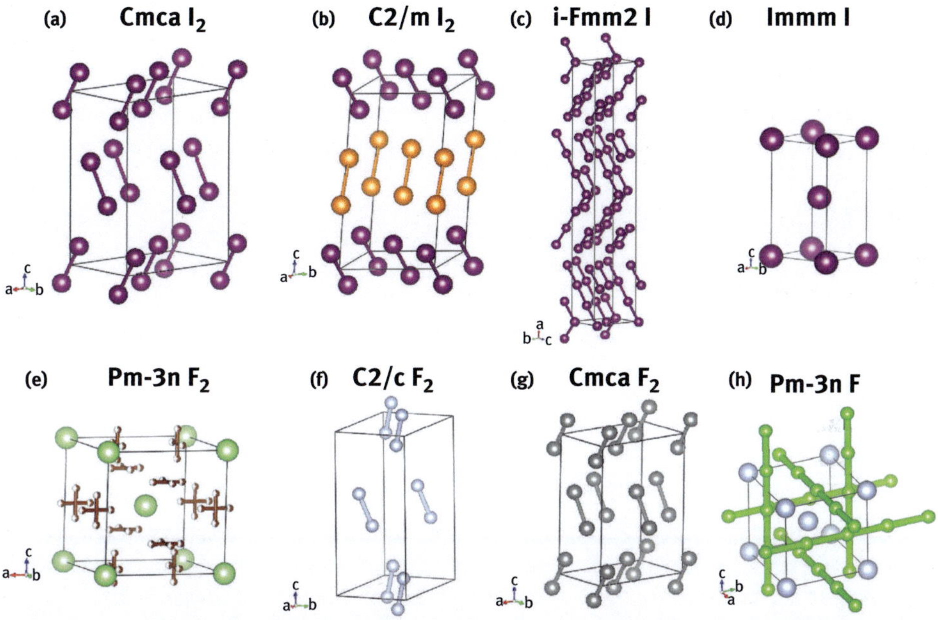

Figure 4: Crystal structures of halogens forming molecular crystals and extended structures at high pressures. (a) *Cmca* I_2 below 24 GPa (phase I); (b) *C2/m* I_2 between 12.5 and 24 GPa (phase I′); (c) *i-Fmm*2 incommensurate modulated iodine between 23 and 30 GPa (phase V); (d) *Immm* monatomic iodine between 25 and 43 GPa (phase II); (e) *Pm*-3*n* F_2 between 2 and 10 GPa at 300 K (phase β); (f) *C2/m* F_2 at 0.06 GPa at 10 K and above 10 GPa at 300 K (phase α); (g) *Cmca* F_2 above 66 GPa (theory); (h) *Pm*–3*n* F_2 above 2.9 TPa (theory).

Before transforming into a monatomic states at higher pressures, Cl_2, Br_2, and I_2 demonstrate an interesting intermediate molecular-atomic state with varying interatomic distances revealed in XRD [42, 122] and Raman investigations [121]. The intermediate state has a face-centered orthorhombic structure, which demonstrates an incommensurate modulation, where the atomic positions and molecular formation/decomposition likely fluctuate [42]. Theoretical calculations performed in Br_2 show that the modulated structure is more thermodynamically and mechanically stable [129]. Raman spectroscopy demonstrates the presence of a low-frequency Raman band in this modulated phase, assigned to an amplitude mode – collective vibration which preserves the symmetry of the modulation wave [121]. This modulated phase transforms to a body-centered orthorhombic cell, *Immm*, where no Raman activity is expected. This phase can be considered as atomic in that the nearest neighbor interatomic distances converge [118]. It is interesting that even though there are large pressure ranges of the phase coexistence, the purely molecular and monoatomic phases do not coexist [122]; it is a modulated *Fmm*2 structure theoretically predicted in chlorine [130] and bromine [131] (cf. [132]), which serves as an intermediate phase between molecular and nonmolecular states.

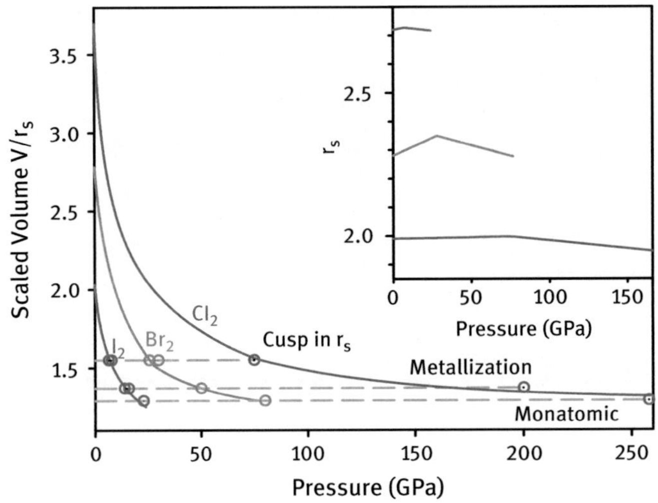

Figure 5: Compression curves of diatomic halogens. The volumes are scaled to the *P*-dependent intramolecular distances r_S. The inset shows r_S as a function of pressure. Horizontal dashed lines correspond to characteristic scaled volumes, where halogens experience a change in the compression behavior (cusp transition), metallization, and molecular full dissociation.

The above-described behavior is rather universal for halogens. However, whether this high-pressure behavior can be applied to fluorine (F) and astatine (At) needs to be seen as no relevant experimental data are available. For F_2, the available experimental data [133, 134], which are limited to ambient and close to ambient pressure, suggest that the phase sequence is different than in other halogens. At room temperature, fluorine solidifies in β-F_2, which has a cubic *Pm*-3*n* structure and is isostructural to γ-O_2 and δ-N_2. However, a monoclinic *C*2/*m* or *C*2/*c* α-F_2 becomes stable below 45.6 K at ambient pressure. Based on the Raman investigations of ref. [134], this low-temperature phase is expected to become stable at room temperature above 8–10 GPa. However, an orthorhombic *Cmca* structure, which is an established low-pressure molecular phase of Cl_2, Br_2, and I_2, is theoretically predicted to become stable above 8 GPa [135] and above 70 GPa [132] according to different theoretical calculations. The latter work also predicts a different scenario for F_2 at higher pressures compared to Cl_2, Br_2, and I_2. F_2 in *Cmca* structure is expected to be stable up to 2,500 GPa, and above this pressure it is predicted to transform to a metallic *P*4$_2$/*mmc* structure, which further transforms to a more symmetric *Pm*-3*n* structure at 3,000 GPa (Figure 4). These structures have 3/4 of the atoms arranged in three orthogonal indefinite chains, while the remaining 1/4 occupies the voids between the chains. Even more complex behavior has been theoretically predicted more recently, where *Cmca* structure is stable to 2,750 GPa and then transforms to *P*6/*mcc* structure, which consists of F_2 molecules and F polymer chains, and then into a *Pm*-3*n* structure (as in ref. [132]), which consists of F polymeric chains and F atoms [136] (Figure 4).

Astatine (At) is a heavy atom, so it is worth to compare this element's behavior to iodine. The binding energy of diatomic molecules decreases toward heavy halogens as the atoms become more massive. In the absence of experimental data, we rely on theory for At [137]. The results are markedly dependent on the level of theory, specifically on how relativistic effects are treated. Depending on whether to include spin–orbit effects into account, At forms *Cmca* molecular or *I4/mmm* atomic structure at ambient conditions suggesting At is rather a monatomic metal in the condensed state.

Having reviewed the high-pressure behavior of the whole halogen group, we can compare them with the available data for hydrogen to track the similarities and predict higher pressure behavior, which is challenging for experimental methods. Hydrogen has a much larger reduced volume than halogens because the intramolecular distance is much shorter. Thus, metallization and molecular dissociation are expected at much higher pressures; thus, the useful structural comparison should be performed with high-pressure-ordered molecular phases of hydrogen, namely phases III, IV, and V [34, 138, 139]. These phases can be only observed above 150 GPa, making their structural determination very challenging. At such pressures, XRD measured from very atom-light hydrogen crystals can only detect the most intense peaks, thus making the results ambiguous. The most recent XRD study detected three major peaks of a hexagonal close-packed (*hcp*) lattice (101, 002, and 101), which is characteristic of the plastic low-pressure phase I [73]. However, other weaker XRD peaks are not detected (or noticed), and the structural analysis does not include determination of the symmetry and the peak intensity analysis [140]. Fortunately, vibrational spectroscopy methods are sensitive to structural modifications, which solid molecular hydrogen experiences under compression. These techniques can detect orientational ordering via appearance of new lattice and fundamental (intramolecular) modes as such transitions are normally related to an increase in the unit cell dimensions and thus folding of the Brillouin zone, which makes zone boundary phonons active in Raman and IR processes [101, 141]. Vibrational spectroscopy is also sensitive to very small changes in the intramolecular distance and molecular coordination with respect to other molecules. This is reflected in a number of intramolecular vibrations (commonly called vibron in hydrogen), their frequencies, and pressure dependencies. While Raman and IR spectroscopy cannot determine the structure, it provides a clear diagnostics of the phase transitions, chemical bonding, and molecular environment.

Phase III of hydrogen is of particular interest because it is stable in a very wide pressure range (150–425 GPa) and it transforms to a metallic state by a band gap closure [54]. Raman and IR data demonstrate molecular orientational ordering of phase III [103, 141]. While compressing phase III, a strong absorption edge shifts to lower energies reaching the IR vibron mode energy above 380 GPa. At 425 GPa, the sample becomes totally absorptive above 1,000 cm^{-1} (10 μm), suggesting an abrupt transition into a metallic state [54]. In the absence of the reliable XRD data, the structure of the phases involved in this transition must be deduced from theory.

Concerning an insulating (semiconducting at high pressures) phase III, theoretical studies determined that this phase is a monoclinic *C*2/*c* with 24 atoms in the unit cell [142] or hexagonal $P6_122$ with 36 atoms in the unit cell [143], both structures are very similar but a small distortion and a difference in the stacking sequence of the layers. The layered structures are typical for high-pressure phases of hydrogen as well as halogens. Each layer consists of three inequivalent molecules sitting nearly in the plane and whose centers form a (distorted) hexagonal lattice. It is interesting that all molecular (or fluxional phases with elongated molecules referred to as "atomic" in ref. [34]) high-pressure phases of hydrogen (III, IV, and V) have a similar three-molecule structural motif [143, 144]. Phase III is expected to transform to an orthorhombic *Cmca* structure with 12 atoms in the unit cell [145], which again has the same major structural motive, but this molecular phase is metallic.

The structural building blocks of hydrogen are drastically different compared to halogens, which are also layered but the layers consist of herringbone-like packed molecules (Figure 6). Notably, the symmetry modifications with pressure are similar in hydrogen and halogens, such as lower pressure *C*2/*c* structure transforms into a *Cmca* structure, which is metallic or becomes metallic under pressure. However, structurally and chemically the molecules associate differently making problematic a direct comparison and thus predictions for hydrogen based on a possible analogy with halogens. Indeed, it would be very tempting to interpret a soft raising in intensity L_4 mode in H_2 (D_2, HD) as the amplitude mode of phase V (inferred being metallic in ref. [139]). However, this proposal needs experimental verification as the Raman data for phase V can be explained by a subtle structural modification of phase IV [144], which has a small band gap. It is interesting to note that the presence of dissimilar layers in phases IV and V results in an opening of a small band gap [144, 146], while *Cmca* structure with identical layers is metallic [145] and represents the best candidate for a metallic phase reported experimentally in ref. [54] based on optical IR investigations.

Furthermore, metallization of hydrogen in molecular phase has also been reported based on electrical conductivity measurements [53]. The results of this work, which show that hydrogen becomes a very narrow gap semiconductor or semimetal above 360 GPa, can be reconciled with optical investigation, where a direct band gap closure is reported at 425 GPa [54]. Recent theoretical calculations suggest that metallization is driven by quantum fluctuations, which results in a large reduction of the band gap between 1 and 3.5 eV [147–149]. By accounting for the quantum effects, the bond length increase of the H_2 molecules, inferred from Raman and IR spectroscopy, is reproduced as well as metallization in a molecular state due to the band gap closure. In this molecular metallic state, the density of the electronic states near the Fermi level remains low thus explaining a lack of metallic reflectivity, which is expected to occur at higher pressures at the transformation to a monatomic state [150–152]. However, the results reported by Dias and Silvera [115], which claim a transition into an atomic phase at 495 GPa based on the observation of shiny hydrogen

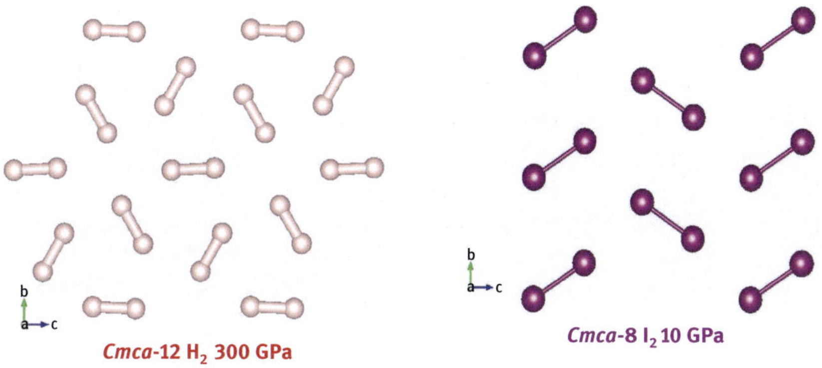

Figure 6: The arrangements of molecules in layered *Cmca* structures of H_2 and I_2.

stand out and require further confirmation. This work was criticized for a number of technical reasons such as pressure metrology, lack of the sample presence test in the DAC, inconsistency of the energy loss calibration through the diamond anvils with the documented previously diamond absorption, continuity of the results with lower pressure properties, inconsistency between the sample visual observations and the reflectance spectra, and others [153–156].

5 Nitrogen, carbon dioxide, and oxygen

5.1 Carbon dioxide

Carbon dioxide is a linear molecule with the double C–O bonds. Solid molecular CO_2 demonstrates the exemplary transition to an extended state at high pressures. Because of the unsaturated intramolecular bond scheme and lack of hydrogen, CO_2 polymerizes at much lower pressures than other simple molecules making it also a great object to study possible intermediate molecular-polymeric configurations and polymorphism of polymeric phases. The high-pressure behavior of CO_2 molecule is of great interest because the constituting elements are very abundant, thus making it one of potentially the most important materials in the interiors of planets such as Venus, the outer planets and their satellite [157], and also in the Earth's mantle [97, 158–160], where one would expect possible parallels in the behavior of carbonates with silicates, which dominate the Earth's mantle. On the other hand, these analogies call for high-pressure investigations of CO_2 in search for stronger analogs of quartz (e.g., for ultrafast transducers) or even of six-coordinated silica – stishovite [161]. Thus, the high-pressure investigations of CO_2 provide important critical information for fundamental and materials research as well as geophysics and planetary science.

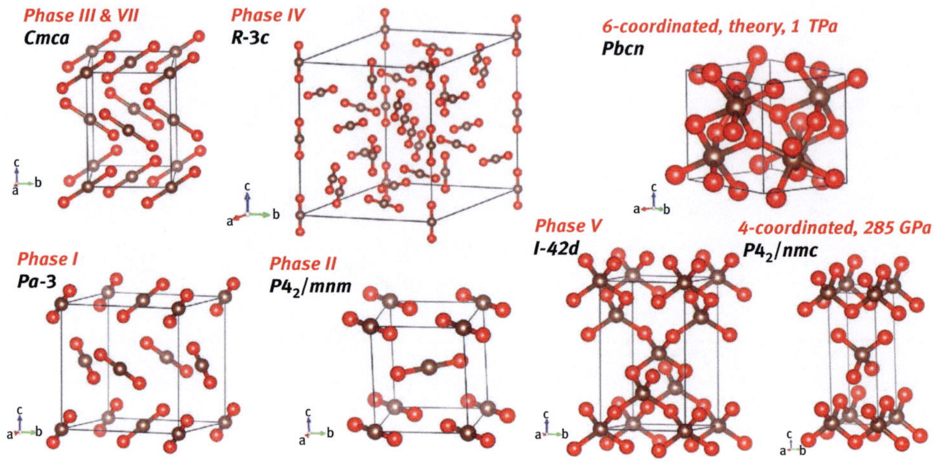

Figure 7: Structures of high-pressure phases of CO_2.

At high pressure, a familiar dry ice (phase I) is a cubic *Pa*-3 ordered phase [162], which is isostructural to other linear molecular crystals α-N_2 [163] and ortho-H_2 and para-D_2 [164] at low temperatures (Figure 7). In this structure, where the molecules are lined up along the main diagonal, the electric quadrupole–quadruple anisotropic interactions (EQQ) are minimized. At higher pressures, the EQQ interactions are overweighed by the repulsive interactions, which result in more efficient packing of the molecules leading to different kinds of orientational ordering. Compressing phase I at room temperature transforms it to an orthorhombic *Cmca* structure (phase III), where the molecules occupy the positions in the same crystallographic plane forming a herringbone structure (Figure 7), similar to halogens (see the previous chapter). Interestingly, this phase can be only entered by compression of phase I (Figure 8). In accordance with this, theoretical calculations show that this phase is metastable, and another structure, called phase II, with $P4_2/mnm$ structure is stable, while *Cmca* phase III appears to be entropy stabilized due to low-frequency shear acoustic phonons [165]. A tetragonal phase IV was determined to be molecular with the molecules in the *xy*-planes rotated by 90° in the alternating planes [166, 167]. At higher temperature, yet another structure has been found based on vibrational spectroscopy data, initially described as a *Pbnc* structure with elongated C–O bonds and strongly interacting molecules in a bent configuration [168]. These results were interpreted as manifestation of a gradual loss of molecular character via the formation of phases with intermediate bonding schemes. Later, the results for phase IV have been challenged by direct single-crystal diffraction measurements, which showed rather complex but purely molecular structure with a large unit cell (36 molecules), which does not reveal elongated or bent C–O bonds [169]. Finally, the domain of stability of *Cmca* structure, which was found as metastable phase III at room temperature, was discovered near the melting line [170] when compressing phase I above 640 K or solidifying melt above 12 GPa. At these conditions, stress-free single crystals (cf. phase III) can be grown, thus allowing to determine the structure of this phase VII using single-crystal XRD. It remains unclear whether phases III and VII are truly isostructural.

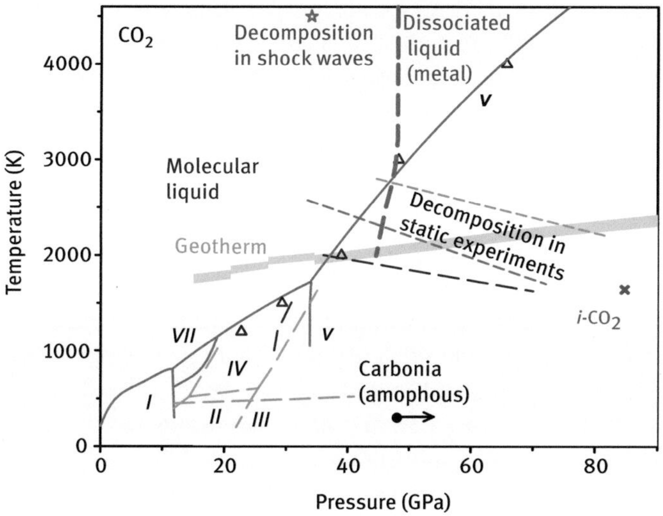

Figure 8: Phase diagram of CO_2 at high P–T conditions. Red lines are drawn based on experimental data presented in refs. [97, 169] below a triple point between phases IV, V, and liquid at 34 GPa; the melting line at higher pressures is drawn to the theoretical calculations of ref. [174], presented as open blue triangles. Theoretically computed phase lines between the solid phases are shown by thin blue [174] and cyan [175] dashed lines. A solid filled black circle with an arrow corresponds to a transformation to an amorphous a-CO_2 carbonia reported in ref. [171]. A dashed gray dashed line is a kinetic phase line between phases II and III [194]. Dashed dark blue, pink, and dark yellow lines are the proposed CO_2 decomposition lines from refs. [97, 158, 178], respectively. A thick green dashed line is a computed liquid–liquid transformation from ref. [183]. An x cross corresponds to the conditions of formation of ionic carbonate i-CO_2 from ref. [179]. An open pink star corresponds to dissociation of CO_2 in shock compression experiments [180]. The Earth's geotherm is shown by a thick gray line [318].

The molecular phases of CO_2 tend to polymerize at high pressures; however, the transition is sluggish at low temperatures as the system needs to overcome the potential barrier between molecular and network (polymeric) phases. Indeed, compressing phase III along isotherms below 680 K results in the formation of an amorphous state, carbonia, which is structurally homologous to other group IV dioxide glasses such as a-SiO_2 and a-GeO_2 [171] (Figure 8). Carbonia demonstrates disappearance of all molecular signatures in vibrational spectra and progressive formation of single C–O bonds. According to ab initio calculations, this kinetically arrested metastable intermediate state contains carbon in a mixed three- and fourfold coordination and does not contain octahedrally coordinated sites, which makes it qualitatively different with respect to heavier dioxide glasses [172]. Heating molecular phases slightly above 1,000 K at pressures as low as 33 GPa results in formation of a polymeric phase V (Figures 7 and 8) [97, 173]. While heated, metastable phase III transforms to stable phases II and IV (<30 GPa), and then to phase V.However, above 40 GPa, where the kinetic barrier becomes lower, phase III directly transforms to phase V. The phase boundary between the molecular and polymeric CO_2 is expected to have a positive slope. This can be judged based on the Clausius–Clapeyron relation $(dP/dT) = (\Delta s/\Delta v)$, where Δs and Δv are the specific entropy and volume change in the phase transition, respectively. The volume decrease of the transition is large (~20% at 39 GPa for the II–V transition [167]), while the vibrational frequencies are expected to experience a net reduction due to a change in the C–O bond order. The experimentally inferred phase line between molecular and polymeric CO_2 [97, 173] agrees well with the results of free energy theoretical calculations for the IV–V transition line [174, 175]. The calculated [174] melting line of molecular CO_2 also agrees well with the experiments [97, 170], while for polymeric phase the experiment [97] and theory [174] agree in that the melting line increases the slope compared to molecular phases (Figure 8). Polymeric phase V can be brought down below 6 GPa at room temperature and has been controversially claimed to be recoverable to ambient pressure at 191 K [176, 177].

Laser heating experiments [97, 158, 178] reported decomposition of CO_2 to carbon and molecular oxygen at various *P*–*T* conditions above 35 GPa, which was suggested to be a factor preventing to detect the melting line in experiments to higher pressures [97]. A possible decomposition of CO_2 at extreme *P*–*T* conditions relevant to the Earth's interiors is an important issue potentially impacting the deep carbon cycle [158, 160, 179] and affect redox state of the lower mantle [97]. The *P*–*T* conditions of proposed decomposition widely vary in static [97, 158, 178] and dynamic [180] high-pressure experiments (Figure 8) likely representing differences in experimental conditions and a variation in the detectivity limits of dissociation depending on the experimental technique (e.g., XRD vs Raman spectroscopy). Indeed, some experiments do not show the decomposition even at much more extreme static *P*–*T* conditions [179, 181]. Theoretical calculations show that decomposition is energetically unfavorable [174, 182, 183] to at least 200 GPa and 10,000 K. This discrepancy is likely related to an increased chemical reactivity due to catalytic effects related to heating of laser absorbers (e.g., Ir in experiments [97]) as proposed in ref. [181]. However, recent molecular dynamics (MD) calculations [184] show that at 60 GPa CO_2–V melts into a dissociated metallic state, where O_2 molecules can form giving another explanation to experimental observations of ref. [97]. Moreover, these calculations show that the decomposition of CO_2 can start as low as 2,000 K.

Finally, we discuss polymorphism of polymeric CO_2 modifications. Initially, it has been proposed that polymeric phase V is similar to that of tridymite (orthorhombic $P2_12_12_1$) – a high-temperature polymorph of silica [167, 185]. Later, these results have been challenged as three groups reported much simpler tetragonal β-cristobalite structure (*I*-42*d*) (Figure 7) [186–188] based on high-quality XRD patterns obtained on the directly laser-heated samples using a CO_2 laser at 10.6 μm wavelength. In agreement with Raman spectroscopy observations, phase V forms irrespectively of the molecular precursors (cf. [167]) and in a whole explored pressure range of 40–120 GPa (see also [181]). These results agree very well with the theoretical structure predictions [182, 189–191]. These theoretical predictions and experimental data up to 120 GPa [181, 187] challenge the existence of an "ionic" *i*-CO_2 [179], which is similar to aragonite $CaCO_3$ in a disproportionated NO–NO_3 phase [32, 192] (Figure 8), where triply coordinated planner CO_3^- and linear CO^+ ions form an extended 2D-layered structure. Instead, theoretical calculations predict the stability of tetrahedrally coordinated structures in a wide pressure range of 30–900 GPa [191]. The β-cristobalite polymeric structure (*I*-42*d*) of CO_2 is stable to at least 200 GPa, and at higher pressures a layered $P4_2/nmc$ is expected to become more stable [191, 193] (Figure 7), where the fourfold carbon coordination persists. It is interesting that both these phases demonstrate very similar oxygen face-centered cubic (*fcc*) like packing of oxygen atoms, while the carbon atoms residing in tetrahedral sites are different [193]. The sixfold coordinated carbon is predicted to become stable above 900 GPa [191, 193], which is still experimentally extremely challenging to reach, and the most stable structures are expected to be an orthorhombic *Pbcn* and cubic pyrite *Pa*-3 (Figure 7). A stishovite structure ($P4_2/mnm$), which is the most stable for six-coordinated silica, is not energetically viable for carbon compounds. These results cast doubt on the previous reports about six-coordinated carbon in phase VI of CO_2 above 50 GPa [194].

The available results show that carbon dioxide is an archetypal material to study the transformation of molecular to extended states. In the molecular states, CO_2 demonstrates remarkable phase polymorphism with a variety of differently oriented structures and hysteretic behavior, while polymorphism of polymeric phases is limited and so far was not sufficiently explored. High-temperature experiments in laser-heated DAC pose substantial technical difficulties as solid CO_2 remains insulating at high pressures and does not couple directly to near-IR laser radiation. Single-crystal XRD studies are very limited because of difficulties in the sample preparation, while the results of powder XRD are often unreliable because of the sample texture and parasitic XRD signal due to the presence of unwanted laser absorber and gasket materials. Vibrational spectroscopy data are also often confusing because the difference in the spectra due to various carbon coordination is not obvious [193]. This creates a wealth of contradicting reports, which require independent verification. More work is needed to establish the high-pressure polymeric phases and to understand their structure and properties. Also, confirming the recovery of polymeric CO_2 at low temperature [176, 177] would perhaps make the first extended state solid, which is stable at ambient pressure.

5.2 Nitrogen

Molecular nitrogen is a uniquely triply bonded diatomic molecule, which has the strength that is the second after the triple C–O bond. Nitrogen is the only group V element forming a diatomic molecule; other heavier elements are polymeric at ambient conditions except the second light element phosphorus, which can form a metastable P_4 molecular modification [195] but normally is also polymeric. The destabilization of the triple bond in N_2 molecule under pressure is expected to be a clear case of the competition of the kinetic and potential energies which results in molecular dissociation and polymerization at moderate pressures. Indeed, shock experiments detected such transition at around 30 GPa and 6,000 K, which was interpreted as molecular dissociation [196], while double-shock experiments [197, 198] reaching pressures up to 100 GPa, likely entered polymeric liquid state, and, finally, reverberating shock compression revealed a metallic (polymeric) fluid state above 120 GPa [199, 200]. Polymerization of molecular nitrogen in a solid state has been predicted theoretically at rather moderate pressures of 50 GPa [201–203]. However, realization of this state in static high-pressure experiments met considerable challenges because the experimental pressure conditions of polymerization appear to be much higher. The phase diagram of nitrogen is extremely rich for the molecular and nonmolecular parts (Figure 9); there are numerous P–T-induced phase transformations and many of them are kinetically controlled [204]. Thus, establishing the stable phase boundaries is not straightforward and requires concerted efforts of theory and experiment.

The phase diagram of nitrogen is extremely complicated in the region of molecular phase's stability (Figure 9). Some 12 molecular phases have been reported. However, many theoretically predicted phases have not been observed (see [29, 205] and references therein), making it possible that these will be discovered in the future. Similar to light halogens and hydrogens, plastic orientationally disordered phases – *hcp*-like β-N_2 [206] similar to phase I of H_2 (even though N_2 molecules may have restricted rotation motion pattern) and cubic δ-N_2 with molecules occupying two distinct crystallographic positions with different patterns of rotational motion: sphere- and disk-like [207, 208] (isostructural to δ-F_2 [209] and γ-O_2 [210]) – interface molecular fluid and orientationally ordered phases, which crystallize at higher pressures and lower temperatures. The low-temperature N_2 phases include ordered cubic α-N_2 [163] (below 36 K at $P \leq 0.6$ GPa) and ordered tetragonal γ-N_2 [206] (at low T and $0.6\ \text{GPa} < P < 2.3$ GPa). The former *Pa*-3 α-N_2, which is isostructural to CO_2-I and para-D_2, has a typical quadrupole ordered structure, while the latter γ-N_2 transformed from this one by compression yields more compact molecular packing due to molecular rotation to the *ab* planes. This results in a tetragonal $P4_2/mnm$ structure similar to that of phase II of CO_2 (see above). This transition illustrates the *intermolecular bond competition* scheme, which governs the phase transformations in the intermediate pressure range. At room temperature, the δ-N_2 phase, which crystallizes at 5 GPa, and those that follow at higher pressures represent molecular

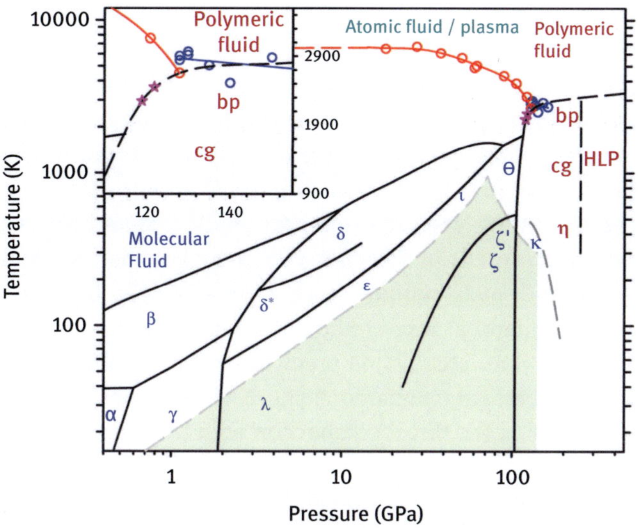

Figure 9: Phase diagram of nitrogen. The phase boundaries and phase occurrence domains are from refs. [1, 29, 68, 74, 95, 216, 219–221, 224–226, 229, 232, 238, 239, 319]. The light green area corresponds to the stability field of λ-N_2 [226]. The inset shows an enlarged view at high pressures and high temperatures. The symbols correspond to the high-temperature transformations to absorptive and conducting states (blue and red circles, respectively) [68] and crystallization of fluid nitrogen into a cg-N on compression (pink stars) [229].

packing and ordering, which are characteristic for strengthening of EQQ intermolecular interactions. This δ-phase has a cubic *Pm-3n* structure, where N_2 molecules occupy two distinct crystallographic sites with different patterns of rotational disorder, which can be represented as sphere- and disk-like [207, 208]. At 11 GPa, the vibrational spectra show an anomaly, which was related to a partial orientational ordering [104, 211] and results in a tetragonal distortion [208, 212] (space group $P4_2/ncm$) in a phase called δ^*-N_2. At 17 GPa, the structure of δ^*-N_2 further distorts transforming into a rhombohedral orientationally ordered *R-3c* ε-N_2 [207]. This sequence of the lattice transformations is consistent with a stepwise orientational ordering involving different crystallographic sites [211] (see also [213, 214]). Upon further compression, even more distorted phases appear as evidenced by changes in the Raman spectra [215, 216] and a splitting of Bragg reflections in powder diffraction patterns [217, 218], interpreted as transformations to an orthorhombic ζ-N_2 at 62 GPa and a monoclinic κ-N_2 at 110 GPa [218]. Because the XRD powder patterns are textured (have preferred orientations), which results in unreliable intensities preventing to perform the full Rietveld structural refinement, the structures of these phases could not be determined. A molecular κ-N_2 transforms to an amorphous nonmolecular phase η-N along a line with a negative *P*–*T* slope (at 150 GPa at room temperature) as evidenced by vibrational spectroscopy [219–222] and XRD [218].

Compared to other diatomic molecules, N_2 demonstrates a remarkable complexity of the phase diagram in the P–T range of stability of orientationally ordered molecular phases at high pressures and low to room temperatures. Experiments and theory do not well correspond in this regime. Indeed, theory predicts that a $P4_12_12_1$ structure is the most stable above 10 up to 56 GPa, where it is expected to transform to a nonmolecular nitrogen in a cubic gauche (cg) structure [205]. However, the experiments show a number of other than predicted orientationally ordered phases ($P4_12_12$ phase has never been documented) in this pressure range (see below), and these phases demonstrate stability (or metastability) to up to 160 GPa. On the other hand, none of the observed experiments at room-temperature molecular phases in the reversible sequence $\delta \leftrightarrow \varepsilon \leftrightarrow \zeta \leftrightarrow \kappa$ [216, 218, 223] is predicted to be a stable phase according to the first-principles theoretical calculations [29, 205]. However, δ and ε phases are partially disordered, making the direct comparison with theoretical calculations difficult.

Surprisingly, new ordered phases of nitrogen have been found as a result of various P–T excursions above 50 GPa [224]. The Raman experiments in resistively heated DAC at high temperatures up to 850 K discovered two new phases, ι and θ, at 70 and 95 GPa, respectively (see also [225]). These new phases, which have very distinct Raman spectra (characterized by only one strong Raman vibron mode suggesting that all molecules occupy the equivalent crystallographic positions unlike ε-N_2) and XRD patterns, can be quenched to room and low temperature at high pressures, and, moreover, ι and θ phases are even stable upon the pressure release down to 23 and 30 GPa, respectively, where they transform back to ε-N_2. Yet another orientationally ordered molecular phase has been synthesized while compressing α- and β-N_2 at temperatures below 90 K [226] (see also [227]). This new phase, called λ, can be detected at as low as 2 GPa and it remains stable over an exceptionally wide P–T range up to 140 GPa at low and room temperatures and below a positively sloped phase line, where it irreversibly transforms to ε-N_2 (Figure 9). This transformation line is some 100–200 K below the $\delta \leftrightarrow \varepsilon$ phase line; the latter one ends in the triple point with the fluid [95] and it demarcates partially ordered and plastic phases. At higher pressures λ-N_2 transforms to θ-N_2 at high temperatures, while it transforms to an amorphous nonmolecular nitrogen η above 140 GPa at room temperature, very much alike to the behavior of κ-N_2 at the similar conditions [219–221]. In the limit of low temperatures and low pressures, λ-N_2 shows its stability down to 0.3 GPa at 10 K, so it is likely quenchable to ambient pressures. These data show that this phase is stable (or metastable) over a very wide P–T phase space, where the ε, ζ, κ, ι, γ, and θ molecular phases (and polymeric phases, see below) of nitrogen can also be stable or metastable, thus making it so difficult to establish the stability fields in the phase diagram of nitrogen.

The direct structural characterization of high-pressure phases of nitrogen has been greatly progressed recently. The structure of the close-to-ambient pressure molecular phases α, β, and γ has been determined on rather large samples [163, 206].

The structure of ε-N_2 at 7.8 GPa [207] has been determined in a Bridgman anvil apparatus in a radial geometry. The structure of δ-nitrogen at 4.9 GPa has been solved using single-crystal XRD [228]. To determine the structure of a partially orientationally ordered δ^*-nitrogen above 11 GPa, the sample was temperature annealed in order to improve the quality of powder XRD patterns [208]. Subsequent single-crystal XRD measurements in the DAC confirmed the structures the δ and δ^* phases, and, additionally, determined the orientations of the partially disordered molecules [212]. However, at pressures above approximately 20 GPa, the structural determination becomes challenging. Nitrogen scatters X-ray weakly and the samples are very small at high pressures. On the other hand, the Compton scattering from diamonds is rather strong, thus, detecting a powder diffraction signal is difficult. Moreover, the samples typically develop a strong texture (preferred orientation), making structural investigations problematic as the use of a conventional Rietveld refinement technique becomes very difficult. These difficulties can be circumvented by either using a single-crystal XRD technique or by actively using the results of the theoretical structural predictions. The latter technique has been helpful in solving the structure of λ-N_2 at 34 GPa [226], which appears to be monoclinic $P2_1/c$ with two molecules in the unit cell – one of the most energetically favorable theoretically predicted ordered N_2 phases [205]. The structure of orientationally ordered ι-N_2 at 56 GPa [29] has been solved using single-crystal XRD on the sample, which was grown from ε-N_2 at 750 K. The structure of this monoclinic phase with a unit cell with a volume of 765 $Å^3$, which contains 48 molecules, is very complex. It cannot be solved by powder diffraction technique. Moreover, theoretical structure predictions cannot find this structure as the unit cell is too large to consider for routine structure searches. DFT (density functional theory) calculations find this structure energetically competitive: it is more stable than ε-N_2 above 20 GPa at room temperature, while another predicted but not realized structure $P4_12_12$ [205] is more energetically favorable. It is likely that this latter phase corresponds to yet poorly XRD characterized θ-N_2 crystallized at 95 GPa [224], then making the experiment and theory to reasonably agree concerning the structure and stability of molecular phases.

However, the agreement between experiments and theory is poor concerning the equilibrium transition pressure between molecular and nonmolecular (polymeric) nitrogen phases. While theoretical calculations predict the transformation to occur at nearly 50 GPa [201], the experiments suggest much higher pressures, at least above 100 GPa. Indeed, molecular phases show a tenacious stability at high pressures up to 150 GPa at 300 K [219] and 190 GPa at 80 K [220]. Compressing beyond this line, which extends down to 120 GPa and above 500 K, molecular nitrogen transforms into an amorphous semiconducting nonmolecular state η characterized by a wide range of metastability [219–221]. A polymeric cg nitrogen (cg-N) crystallizes above 110 GPa when heated above 2,000 K [1] and does not form at lower pressures. The phase transition dynamics is fast at these conditions and it has not been firmly established whether cg-N forms directly or via an intermediate phase. The presented results better match crystallization of cg-N from an amorphous η-state. The location of the equilibrium molecular-to-nonmolecular transition just above 100 GPa (Figure 9) has been recently confirmed by another experiment, where cg-N was synthesized by applying pressure at high temperatures [229]. The extension of the (meta)stability region to higher pressures is due to the fact that the transition to nonmolecular phases is kinetically hindered because of the energy barrier, which needs to be overcome upon the synthesis. Heating to temperatures, which are above the devitrification transition [95], represents the most natural strategy; however, photochemistry has been also shown to facilitate the transformation [230]. The kinetic hindrance of the transformation is also supported by the fact that a pulsed μs laser heating of κ-N_2 well beyond 2,000 K does not yield polymeric cg-N (or other polymeric phases) even after exposing the sample to several thousands of laser pulses, which only results in a partial transformation to an amorphous η-phase. On the other hand, polymeric phases remain (meta)stable on unloading to quite low pressures [231, 232] at room and low temperatures, and, moreover, they persist when laser heated to moderate temperatures (<2,500 K) down to 71 GPa [233, 234]. Given such a large range of metastability, the determination of the equilibrium transition pressure from experiments remains somewhat uncertain. The available observations also underscore the importance of taking into account of the temperature conditions of observations when comparing theory and experiments as low values of the calculated transition pressures refer to 0 K, while the majority of the phase transformation observations are at room and high temperatures. Pickard and Needs [205] and Turnbull et al. [29] note that a very large correction in enthalpy of the most stable molecular phases (which as we show here correspond to observations fairly well) is needed (0.4 eV per atom) to make the theory and experiments agree. Such correction appears to be much larger than the uncertainties due to various theoretical approximations. Theoretical calculations, which include the effects of temperature via accounting for entropy terms, seem to be needed.

Recent experimental developments of the high-pressure techniques and associated tools enabled great progress in understanding the polymorphism of polymeric nitrogen. Previously only theoretical calculations have been able to address this issue as direct structural studies above 170 GPa were very challenging. Theoretical calculations also progressed over last few decades yielding more and more reliable results. Early calculations [202, 203] largely explored the structures of polymeric phosphorus (P) and arsenic (As) from the same group in the periodic table as the most plausible candidates. However, nitrogen, which does not have *p*-electrons in its ionic core, stands out, which results in a uniquely stable cg-N structure [201]. With the advent of theoretical structure search, numerous calculations predicted a wealth of polymeric nitrogen structures [48, 205, 235, 236]. A layered structure of black phosphorus nitrogen (bp-N) was thought to be a candidate for higher pressure polymeric nitrogen >200 GPa [235]. However, another unique structure, a layered orthorhombic *Pba*2 polymeric nitrogen, was predicted to take over cg-N above 188 GPa and be stable up to 320 GPa [48]. Tomasino et al. [232] reported the synthesis of a new polymeric nitrogen, characterized by a very strong and distinct Raman spectrum, which they called LP-N. They indexed their XRD patterns using the *Pba*2 structure and assigned extra XRD peaks to another predicted metastable polymeric *C*2/*c* phase [237]. Given the observed coexistence of both polymeric phases, it is very surprising that cg-N and LP-N modifications have very different densities deviating by about 8.2% at 112 GPa [232]. If true, this would likely result in a very abrupt pressure boundary between these two phases. However, the experiments often show a mixture of η, cg, and LP-N phases after the laser heating in a wide pressure range of 126–175 GPa. Laser heating experiments [233] using a metallic heat absorber (coupler), which provide more controllable heating conditions, reported that above 150 GPa only LP-N (which appears to be bp-N, see further) is crystallized, while cg-N yields at lower pressures. These results suggest that LP-N is stable at higher pressures and higher temperatures than cg-N. Recently, the structural characterization of polymeric nitrogen phases have been greatly improved. Single-crystal XRD measurements of cg-N and LP-N have been applied to determine their structures definitively [74, 234, 238]. It appears that LP-N reported in ref. [232] has a black phosphorus structure. This structure has a density, which is only slightly larger (1% according to ref. [238]) than that of cg-N at the same pressure.

Polymeric bp-N is stable up to 180 GPa but at higher pressures laser heating did not yield any crystalline phase; it was proposed that this is because of the very close proximity in enthalpy of several phases in this pressure range [239]. Theoretical calculations suggest that a *Pba*2 structure is the most stable one up to 263 GPa, while above this pressure a unique N_{10} cage-like diamondoid structure was predicted [236]. However, experiments at 240 GPa revealed a new polymeric nitrogen (called HLP-N) [239] after laser heating up to 3,300 K. It has been identified as the previously theoretically predicted *P*42*bc* single-bonded layered polymeric nitrogen [236] based on the excellent agreement of powder XRD patterns and Raman spectra. The predicted in the same paper cage-like diamondoid structure, which can be considered as representing yet another level of densification based on the most efficient packing of pseudoatom-like ultrashort bonded N_{10} units, is still waiting for future discovery.

All the reported so far polymeric crystalline phases are insulating based on visual observations. This is in contrast to an amorphous η-N demonstrating semiconducting behavior and gradual band gap reduction with a projection at 280 GPa band gap closure [221]. In contrast, fluid nitrogen transforms into a polymeric metallic phase above 125 GPa and 2,500 K [68] (see also theoretical calculations of ref. [200]). At higher pressures, in the solid state, theory predicts a metallic *Cmca* structure, a structural analog of the halogens, to become stable at 2.3 TPa [240], followed by the transformation into a metallic *P*4/*nbm* structure consisting of partially charged $N_2^{\delta+}$ pairs and $N_5^{\delta-}$ tetrahedra, which is stable in the range of 2.5–6.8 TPa.

5.3 Oxygen

Unlike N_2 and halogens (e.g., F_2), diatomic molecular oxygen (O_2) possesses a non-zero electronic spin $S = 1$ in the ground electronic state. This fact makes oxygen a very special case of a magnetic molecular crystal, where magnetic and van der Waals interactions compete and the balance between them strongly depends on pressure and temperature. The presence of the molecular magnetic moments results in the magnetically ordered low-temperature phases. Application of pressure results in a stronger intermolecular bonding, electronic delocalization, and collapse of magnetic states. The behavior of oxygen at high pressure is a subject of two review papers [241, 242].

Three molecular oxygen phases exist at ambient pressure: α, β, and γ. A low temperature α-O_2 has an antiferromagnetically ordered monoclinic *C2/m* structure; β-O_2 phase, which is stable up to 15 GPa and 590 K, has a rhombohedral *R*-3*m* lattice with the magnetically short-range ordered molecules (Figure 10); γ-O_2, which is stable in a narrow sleeve extending almost to room temperature at 5 GPa (Figure 11) has a cubic *Pm*-3*n* structure isostructural to δ-N_2 and β-F_2 and it is paramagnetic. At high pressures and low temperature, a magnetically ordered δ-O_2 becomes stable, which has an orthorhombic *Fmmm* structure and a magnetic order varying with temperature from ferromagnetic to a mixed ferro- and antiferromagnetic and finally to an antiferromagnetic [243, 244]. The magnetic moments are oriented perpendicular to the molecular axis, which results in collinear orientation of the molecular axis in all the above structures but not in γ-O_2. The latter phase is the one which best qualifies to be considered as the plastic phase analogous to other diatomic molecules such as H_2 and N_2.

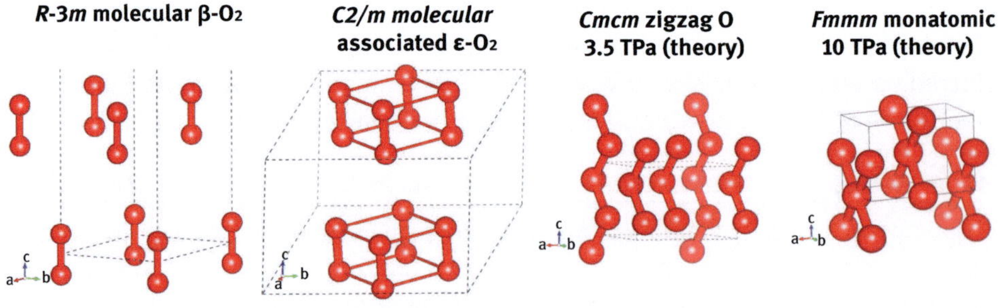

Figure 10: Selected structures of oxygen.

Even though the magnetic moments collapse at high pressures, the high-pressure phases preserve a near-collinear orientation of the molecular axis with the molecules forming close-packed layers. The structure of the high-pressure ($P > 8$ GPa) ε-O_2, which is believed to be nonmagnetic, represents an interesting case of molecular association, where four molecules form a square prism (Figure 10) [245, 246]. Another theoretically proposed herring bone chain structure with a *Cmcm* structure is very competitive in energy but it has not been observed [247, 248]. The O_8 clusters of ε-O_2 are stabilized via donor–acceptor resonance bonding scheme through the donation of a paired π^* electron from one O_2 molecule into the empty π^* orbital of the other molecule [249, 250]. This electronic transfer characterizes the bonding scheme of the O_8 clusters as a closed-shell system. Optical spectroscopy observations [251] and X-ray Raman scattering measurements [252] demonstrate a semiconducting nature of ε-O_2 up to approximately 100 GPa, which can be understood in terms of band gap opened due to a Peierls-like structural distortion.

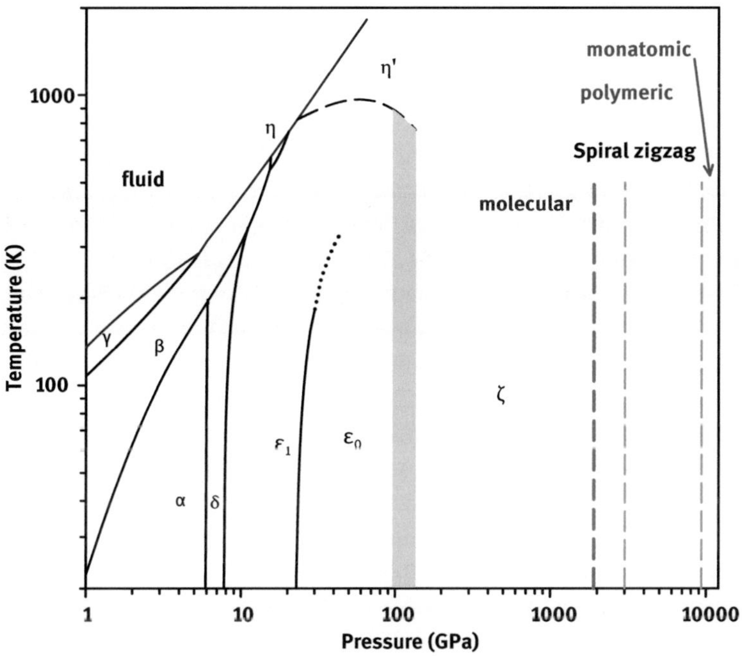

Figure 11: Phase diagram of oxygen. The phase boundaries and phase occurrence domains are from refs. [51, 241, 243, 244, 255–258, 260, 320]. Solid blue lines represent a melting curve, and solid black lines are the established thermodynamic phase lines. A dashed line between ε- and η-O_2 is from refs. [258, 259]. A dotted black line is a proposed crossover of the first-order phase transition between a short-range antiferromagnetic ε_1-O_2 and nonmagnetic ε_0-O_2 [253]. Dashed vertical lines (a red one is a boundary of molecular phases) at high pressures represent the theoretically calculated phase transitions to polymeric and monatomic phases [2].

A simple one-step collapse of the magnetic moments at the δ–ε transition has been recently questioned by first-principles theoretical calculations [253], which propose that the magnetic moments survive up to approximately 20 GPa, where the system possesses short-range antiferromagnetic correlations. Accordingly, the stability field of ε-O_2 is suggested to be split into two regions, characterized by the presence of the magnetic moments (ε_1) and their total loss (ε_0) (Figure 11). The transition (which is expected to be of the first order) line is expected to be almost pressure independent and terminate in a critical point (at about 200 K). This proposal is supported by the observations of anomalies in the pressure-dependent vibrational spectra [254, 255] above the δ–ε transition up to 20 GPa. However, more solid experimental evidences are due.

Phase stability of high-temperature phases of oxygen near the melting line is also highly affected by the presence of the magnetic moments. β-O_2 is a unique phase for molecular crystals with an *fcc* type of molecular dumbbell packing and frustrated magnetic moments. This phase is stable up to 15 GPa, where the magnetic moments are expected to collapse, thus giving the stage to a different phase. Raman spectroscopy observations [256] revealed such a phase called η-O_2 above 15 GPa and 560 K with the vibrational properties similar to β-O_2. The structure of this phase has been determined by single-crystal XRD [257]. It appears that it has an *hcp*-like stacking of the molecules aligned along the *c*-axis. The structure of the molecular centers is identical to phase I of H_2 and β-N_2; however, η-O_2 is fully ordered. This conclusion seems to contradict to a very broad low-frequency Raman feature, which suggests a disordered structure [256]. At higher pressures (>20 GPa), the stability field of η-O_2 terminates, and ε-O_2 becomes the phase which melts. However, the experiments at these *P*–*T* conditions are challenging because of oxygen reactivity in resistively heated DAC and difficulties in controlling temperature in laser-heated DAC experiments [258], so performing more experiments would be beneficial for better understanding of phase diagram at these conditions. The laser heating experiments employing powder XRD and Raman spectroscopy above 44 GPa [258] revealed a transformation of ε-O_2 into another phase dubbed as η′-O_2, which appears to be isostructural to η-O_2. The ε–η′ transition line at nearly 1,000 K, which extends up to 100 GPa, is almost temperature independent (see also [259]), suggesting that the transition is entropy driven. The structural and vibrational properties of η′-O_2 and η-O_2 are very similar and they can be represented by the same pressure dependencies. The low-frequency Raman spectra of η′-O_2 show a very broad Raman peak at about 400 cm^{-1} at 67 GPa extending to the elastic line. First-principles theoretical calculations of the electronic, structural, and thermodynamic properties [259] confirmed the structure and stability of η′-O_2 and revealed that this phase is very close to dynamic instability and it is stabilized via anharmonic effects. The soft mode corresponds to a Brillouin zone boundary transverse optical phonon, which becomes the Raman-active optical phonon in the Brillouin zone center. This qualitatively explains the observed very broad low-frequency Raman response in ordered η′-O_2 and η-O_2 [256, 258]. It is interesting that the theory predicts η′-O_2 to be metallic; the stability field of the high-pressure metallic ζ-O_2 described below is adjacent.

A structural transition from ε-O_2 to another phase ζ-O_2 detected at 96 GPa using powder and single-crystal XRD at room temperature was interpreted as isostructural with a discontinuous change in the lattice parameters [260, 261]. Raman investigations demonstrated a substantial change in the lattice vibrations but the molecular vibron clearly persists manifesting the molecular character of metallic ζ-O_2 [51]. Moreover, ζ-O_2 has been shown to be superconducting below 0.6 K [262]. The structure and properties of ζ-O_2 have been examined theoretically [247, 250, 263] and experimentally [261]. Theoretical results suggest that the ε–ζ transition is indeed isostructural and initiated by the pressure-induced deformation of the O_8 square prism, which modifies the electronic band structure. It should be noted that the connection between the band gap closure and structural change in DFT calculations was not straightforward, which has been improved using the GW approximation(the approximation is that the expansion of the self-energy in terms of the single particle Green's function *G* and the screened Coulomb interaction *W* can be truncated after the first term). However, the discrepancy in the transition pressure remained larger than usual. In the most recent experiment [261] reporting single-crystal XRD characterization of ζ-O_2, the structural refinement could not be obtained, while the reported Raman spectrum was similar to the one of ε-O_2 of ref. [51].

According to the theoretical calculations, the stability field of molecular ζ-O_2 extends to at least 250 GPa [247], which was later extended up to 1.5 TPa [2]. At this pressure, oxygen is predicted to form a square-spiral-like $I4_1/acd$ structure [2] similar to other heavier group VI elements, such as sulfur and selenium [264] (Figures 10 and 11). At about 3 TPa, this structure transforms into a zigzag chain-like phase with *Cmcm* structure, which is packed more efficiently [2]. However, in both these structures, there is no bonding between the chains, so very dense structures are not formed. This qualitatively explains why molecular structures are so stable and extend their stability to very high pressures unlike the situation in other diatomic molecules including even H_2. The molecular bond lengths are 1.15 Å and 1.1 Å in polymerized $I4_1/acd$ at 2 TPa and *Cmcm* at 3.5 TPa, respectively, which is larger than would be in a molecular phase (1.03 Å) at 2 TPa. Each oxygen forms two single bonds in a bent configuration to accommodate two lone electronic pairs. This configuration greatly limits the forms of dense oxygen which may occur. The electronic structures of these new polymeric oxygen phases greatly differ: while a spiral-like $I4_1/acd$ is a wide gap semiconductor at 2 TPa, a zigzag chain *Cmcm* is a weak metal at 3.5 TPa. Under further compression, this structure collapses into a monatomic metallic *Fmmm* structure at 9.3 TPa by merging the chains in each layer [2].

6 Water

The water molecule (H_2O) is a highly polar asymmetrical one, where the oxygen atom has two lone pairs that repel the oxygen–hydrogen bonded pairs forcing them to bend. The associated water molecules in fluid water or solid ices form bonds between neighboring hydrogen and oxygen atoms of adjacent water molecules, which are called hydrogen bonds. In water ice, the crystalline lattice is bonded via these hydrogen bonds, revealing an immense variety of different configurations at low pressures, which results in a great complexity in the phase diagram of water below 2 GPa with some 17 crystalline and 3 amorphous structures that have been documented [265–267]. At higher pressures, the second coordination shell collapses, thus reducing the structural complexity, which results in stability of denser ice structures. In ice VII, which can be considered as their prototype, oxygen atoms form a body-centered cubic (*bcc*) lattice, which nearly perfectly accommodates the optimal linear hydrogen bonds and creates two interpenetrating hydrogen bonded networks. The hydrogen bonds in ices, which are relatively weak compared to the strong covalent intramolecular bonds, are greatly affected by temperature. At sufficiently high temperature, hydrogen atoms are mobile leading to disordered phases, which are entropy driven. These hydrogen disordered phases, however, obey the ice rule: each tetrahedrally coordinated oxygen atom is covalently bonded to two hydrogen atoms forming a molecule, while it is bonded via a hydrogen bond to two hydrogen atoms from other water molecules. Naturally, the disordered ice phases (I*h*, III, IV, V, VI, VII, and XII) have (meta)stability domains adjacent to the melting line [268]. It is remarkable that the majority of hydrogen bonded and ordered ice phases have a disordered counterpart with ice II being a well-recognized exception [269–271].

In dense ices VII and VIII, the lattice compressions above 2 GPa are accompanied by a shift of the average hydrogen position toward the oxygen atom of the neighboring hydrogen bonded molecule documented via neutron diffraction and vibrational spectroscopy [46, 272–274]. Ice VII is a statically orientationally disordered phase with an averaged *bcc* oxygen lattice. Compressing ice VII causes hydrogen atoms to become more mobile changing the character of disorder from a static to a dynamic. Theoretical calculations predict a regime where the hydrogen diffusion is suppressed but molecular disorientations are allowed [275–277], which should result in the formation of plastic phases at higher temperatures with respect to the stability domain of ice VII. Such phases were predicted to occur above 6 GPa and crystallize in *bcc* and *fcc* oxygen lattices but so far these phases have not been experimentally confirmed. It is interesting that such a plastic phase does exist in solid ammonia (where hydrogen bonding is weaker) over a broad *P–T* regime [278].

At higher pressures, the hydrogen diffusion becomes a dominant mechanism for disorder [277], and this is a regime where superionic phases are expected [41, 279–285]. In superionic ice (SI), the oxygen atoms continue forming an ordered lattice (e.g., *bcc*) while the water molecules are expected to be dissociated due to a very rapid proton diffusion. In this regime, the ice rule is violated, and water consists of a number of very short-lived charged species including H_2O, H^+, O^{2-}, H_3O^+, and OH^- in a dynamic equilibrium, and their compositions are being the pressure (density) dependent [277, 280]. In SI, the charged species form a dynamic polymeric ionic state. As in the case of plastic phases, the melting line of SI is expected to increase the slope in *P–T* coordinates due to an increased entropy of SI phase. Therefore, such observations can be interpreted as manifestations of the SI as the direct probing of hydrogen dynamics is challenging [284, 286].

A wealth of experimental technique has been applied to probe SI in DAC. These studies have naturally overlapped with a study of the melting line as SI is expected to be the phase which melts. The results of the melt line investigations have been very controversial. The discrepancy between different investigations is largely dependent on the technique that was used to detect melting. Early XRD studies [287, 288] showed a monotonic increase of the melting line with pressure likely because of inability to detect the signal from possible SI. However, visual observations [289, 290], radiative temperature anomalies vs laser power [291], and Raman [284, 292] and Brillouin spectroscopy [96] reported changes in the melting line slope albeit at different pressures between 20 and 47 GPa. The impedance spectroscopy technique measurements in DAC [293] suggested that superionic states are located below the melt line and unrelated to any phase changes and changes in the melting line slope. Dynamic laser-driven compression experiments in precompressed water have been used in two regimes: shock wave and ramp compressions probing two different *P*–*T* trajectories. In shock wave experiments, the optical spectroscopy measurements detected an increase in optical conductivity which was interpreted as manifestation of superionicity [294]. In the ramp compression experiment, an abrupt transition has been detected in concomitant XRD measurements which were interpreted as a transition to *fcc* SI [295]. Recently, due to the development of synchrotron XRD and fast optical spectroscopy techniques, more consistent results have been reported, which largely reconcile the existing data [69] (Figure 12).

Synchrotron XRD in the laser-heated DAC up to 150 GPa [69] (see also [296], where *bcc* SI phase was reported) detected two changes in the melting line slope and outlined the stability fields of two new phases with *bcc* and *fcc* oxygen lattices. The superionicity was inferred based on the increased optical conductivity measured in fast optical spectroscopy experiments, the distinct density, and the greatly decreased fusion enthalpies derived from the Clausius–Clapeyron relation. The discrepancy with the ramp compression experiment [295], which reports *fcc*-SI at much lower temperatures (Figure 12), is assigned to metastability effects as the ramp compression *P*–*T* pathways pass the stability domains of the *bcc* and *fcc* SI phases on the initial stage of compression. On the other hand, the discrepancies with the early Raman works [284, 292, 297] are due to insensitivity of Raman spectroscopy technique in distinguishing between fluid water and SI solids as both have similar vibrational signatures. It appears that above 18 GPa ice VII does not melt but instead transforms into *bcc* SI and then, at higher pressures, to *fcc*-SI. XRD and Raman experiments show that this phase line changes a slope at 47 GPa (in agreement with the previous observations) suggesting that there is a triple point and another phase line, which separates ice VII from another higher pressure solid phase. In this pressure range, the oxygen–oxygen (O–O) distances are significantly contracted so the hydrogen atom

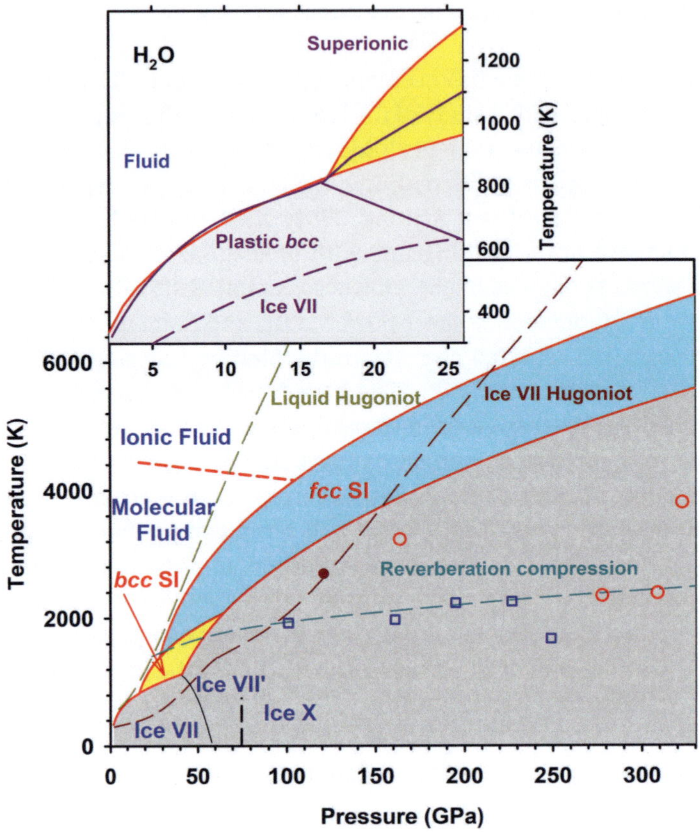

Figure 12: Phase diagram of water at high pressures and high temperatures. Red solid lines are from XRD measurements and red dashed lines are from fast optical spectroscopy in the laser-heated DAC [69]. Long dashed lines show *P–T* pathways for dynamic compression experiments [294, 295]. Symbols show the reported *P–T* conditions of normal ices (squares) and *bcc*-SI (filled dark red circle) and *fcc*-SI ices (open red circles) in these experiments. The inset shows a lower pressure–temperature part of the phase diagram, where the experimental data of ref. [69] are compared to theoretical prediction of stability fields of plastic and superionic phases [277].

becomes closer to the oxygen from the neighboring molecule, thus reducing the potential barrier between two hydrogen positions along the O–O line. Accordingly, ice VII transforms into the high-pressure phase ice VII′, where hydrogen atoms are dynamically disordered [298]. Vibrational spectroscopy data [8] demonstrate that this transition results in a substantial broadening of the vibrational excitations along the line with a negative *P–T* slope (Figure 12). Further compression of ice VII′, where hydrogens are centered dynamically but form a bimodal distribution along the O–O direction results in the formation of ice X with static symmetric hydrogen bonds [47, 299]. This occurrence manifests a transition to a nonmolecular extended phase, characterized by a breakdown of the strong covalent intramolecular bonds and formation of ionic crystal. The DAC XRD measurements indicate that oxygen *bcc* lattice persists up to at least 170 GPa [300–302], supported also by Raman (to 150 GPa) [8, 234, 274] and IR (to 220 GPa) [46, 303, 304] experiments consistently probing the vibrational dynamics. However, probing hydrogen positions, which includes addressing hydrogen bond symmetrization, remains challenging and the results often depend on the experimental technique demonstrating a spread between 30 and 100 GPa [8, 46, 82, 109, 274, 302, 305–312].

Ice X with a *bcc* oxygen lattice is expected to be stable up to 300–400 GPa where it is theoretically predicted to transform to a better packed *Pbcm* orthorhombic structure with a distorted oxygen *hcp* sublattice [313, 314]. This transition is driven by a phonon instability at the M-point of the Brillouin zone [314]. This high-pressure ice is expected to have symmetric hydrogen bonds. However, structural predictions beyond 400 GPa demonstrate further transformations of the hydrogen subsystem [49, 315, 316]. At 800 GPa, the *Pbcm* ice is predicted to transform to the *Pbca* structure, where hydrogen atoms are squeezed out sideways of the midpoint between two nearest oxygen atoms; however, they are still close to their tetrahedral sites. This nearly symmetric hydrogen bond configuration is lost in $Pmc2_1$ (800–950 GPa) and $P2_1$ (1.17 TPa) structures, where a significant rearrangement of the hydrogen bond network occurs. However, the symmetric hydrogen bond arrangement was predicted to come back again at 5.62 TPa in a layered *C2*/*m* ice, which is metallic. This prediction has been challenged in ref. [317], where it has been proposed that H_2O decomposes at nearly 5 TPa, yielding H_2O_2 and a hydrogen-rich phase preventing metallization of H_2O system.

7 Outlook

We are witnessing a great progress in our understanding of the behavior of simple molecules at extremely high pressure and a range of temperature conditions, which are highly relevant to planetary interiors. This progress is due to the major developments in experimental techniques and theoretical calculation methods. More specifically, we are now capable to know the structures of ultra-high-pressure molecular phases in a very good correspondence to theoretical predictions. Future investigations will address a number of yet open questions raised in this review. Moreover, one can envision more studies toward understanding of compositional changes in molecular crystals (e.g., in hydrides) at high pressures, and investigations of molecular mixtures. The latter ones will provide a basis for understanding of chemistry in giant planetary interiors.

References

[1] Eremets MI, Gavriliuk AG, Trojan IA, Dzivenko DA, Boehler R. Single-bonded cubic form of nitrogen. Nature Materials 2004;3:558–563.

[2] Sun J, Martinez-Canales M, Klug DD, Pickard CJ, Needs RJ. Persistence and eventual demise of oxygen molecules at terapascal pressures. Physical Review Letters 2012;108:045503.

[3] March NH. Regularities at phase transitions in molecular assemblies. Physica B: Condensed Matter 1999;265:24–30.

[4] Fujihisa H, Fujii Y, Takemura K, Shimomura O. Structural aspects of dense solid halogens under high pressure studied by x-ray diffraction – Molecular dissociation and metallization. Journal of Physics and Chemistry of Solids 1995;56:1439–1444.

[5] Goncharov A. Phase diagram of hydrogen at extreme pressures and temperatures; updated through 2019 (review article). Low Temperature Physics 2020;46:97–103.

[6] Gregoryanz E, Ji C, Dalladay-Simpson P, Li B, Howie RT, Mao H-K. Everything you always wanted to know about metallic hydrogen but were afraid to ask. Matter and Radiation at Extremes 2020;5:038101.

[7] Babaev E, Sudbø A, Ashcroft NW. A superconductor to superfluid phase transition in liquid metallic hydrogen. Nature 2004;431:666–668.

[8] Goncharov AF, Crowhurst J. Proton delocalization under extreme conditions of high pressure and temperature. Phase Transitions 2007;80:1051–1072.

[9] Guillaume CL, Gregoryanz E, Degtyareva O, et al. Cold melting and solid structures of dense lithium. Nature Physics 2011;7:211–214.

[10] Hubbard WB. Interiors of the giant planets. Science 1981;214:145–149.

[11] Guillot T. The interiors of giant planets: Models and outstanding questions. Annual Review of Earth and Planetary Sciences 2005;33:493–530.

[12] Nettelmann N, Becker A, Holst B, Redmer R. Jupiter models with improved ab initio hydrogen equation of state (H-REOS.2). The Astrophysical Journal 2012;750:52.

[13] Zhou Y, Dong Z-Y, Hsieh W-P, Goncharov AF, Chen X-J. Thermal conductivity of materials under pressure. Nature Reviews Physics 2022; 4:319–335.

[14] Stanley S, Bloxham J. Convective-region geometry as the cause of Uranus' and Neptune's unusual magnetic fields. Nature 2004;428:151–153.

[15] Connerney JEP, Benn M, Bjarno JB, et al. The Juno magnetic field investigation. Space Science Reviews 2017;213:39–138.

[16] Ness NF. The magnetic fields of Mercury, Mars, and Moon. Annual Review of Earth and Planetary Sciences 1979;7:249–288.

[17] Mao H-K, Chen X-J, Ding Y, Li B, Wang L. Solids, liquids, and gases under high pressure. Reviews of Modern Physics 2018;90:015007.

[18] Mao H-k, Hemley RJ. The high-pressure dimension in earth and planetary science. Proceedings of the National Academy of Sciences 2007;104:9114–9115.

[19] Miao M, Sun Y, Zurek E, Lin H. Chemistry under high pressure. Nature Reviews Chemistry 2020;4:508–527.

[20] Zhang L, Wang Y, Lv J, Ma Y. Materials discovery at high pressures. Nature Reviews Materials 2017;2:17005.

[21] Oganov AR, Pickard CJ, Zhu Q, Needs RJ. Structure prediction drives materials discovery. Nature Reviews Materials 2019;4:331–348.

[22] Yoo C-S. Chemistry under extreme conditions: Pressure evolution of chemical bonding and structure in dense solids. Matter and Radiation at Extremes 2020;5:018202.

[23] Ross M. Matter under extreme conditions of temperature and pressure. Reports on Progress in Physics 1985;48:1–52.

[24] Fortov VE. Matter under Extreme Conditions: Extreme States of Matter. Springer, Cham, Springer International Publishing Switzerland 2016.
[25] Hemley RJ. Effects of high pressure on molecules. Annual Review of Physical Chemistry 2000;51:763–800.
[26] Bini R, Schettino V. Molecular Crystals at High Pressure. Materials under Extreme Conditions. Imperial College Press, 2014, London 372.
[27] Hemley RJ, Dera P. Molecular crystals. Reviews in Mineralogy and Geochemistry 2000;41:335–419.
[28] Zhang W, Oganov AR, Goncharov AF, et al. Unexpected stable stoichiometries of sodium chlorides. Science 2013;342:1502–1505.
[29] Turnbull R, Hanfland M, Binns J, et al. Unusually complex phase of dense nitrogen at extreme conditions. Nature Communications 2018;9:4717.
[30] Maynard-Casely HE, Lundegaard LF, Loa I, et al. The crystal structure of methane B at 8 GPa – An α-Mn arrangement of molecules. The Journal of Chemical Physics 2014;141:234313.
[31] Bykov M, Bykova E, Pickard CJ, et al. Structural and vibrational properties of methane up to 71 GPa. Physical Review B 2021;104:184105.
[32] Somayazulu M, Madduri A, Goncharov AF, et al. Novel broken symmetry phase from N_2O at high pressures and high temperatures. Physical Review Letters 2001;87:135504.
[33] Palasyuk T, Troyan I, Eremets M, et al. Ammonia as a case study for the spontaneous ionization of a simple hydrogen-bonded compound. Nature Communications 2014;5:3460.
[34] Howie RT, Guillaume CL, Scheler T, Goncharov AF, Gregoryanz E. Mixed molecular and atomic phase of dense hydrogen. Physical Review Letters 2012;108:125501.
[35] Goncharov AF. In: Fei Y, editor. Molecular Compounds Under Extreme Conditions Dave Mao legacy 2022.
[36] Oganov AR, Chen J, Gatti C, et al. Ionic high-pressure form of elemental boron. Nature 2009;460:292-.
[37] Miao M. Noble gases in solid compounds show a rich display of chemistry with enough pressure. Frontiers in Chemistry 2020;8:570492.
[38] Ma Y, Eremets M, Oganov AR, et al. Transparent dense sodium. Nature 2009;458:182–185.
[39] Miao M-S, Hoffmann R. High pressure electrides: A predictive chemical and physical theory. Accounts of Chemical Research 2014;47:1311–1317.
[40] Dong X, Oganov AR, Goncharov AF, et al. A stable compound of helium and sodium at high pressure. Nature Chemistry 2017;9:440–445.
[41] Cavazzoni C, Chiarotti GL, Scandolo S, Tosatti E, Bernasconi M, Parrinello M. Superionic and metallic states of water and ammonia at giant planet conditions. Science 1999;283:44–46.
[42] Kenichi T, Kyoko S, Hiroshi F, Mitsuko O. Modulated structure of solid iodine during its molecular dissociation under high pressure. Nature 2003;423:971–974.
[43] Manzhelii VG, Freiman YA. Physics of Cryocrystals. New York, NY, Springer, 1997.
[44] Silvera IF. The solid molecular hydrogens in the condensed phase: Fundamentals and static properties. Reviews of Modern Physics 1980;52:393–452.
[45] Ozier I, Yi P-n, Khosla A, Ramsey NF. Direct observation of ortho-para transitions in methane. Physical Review Letters 1970;24:642–646.
[46] Goncharov AF, Struzhkin VV, Somayazulu MS, Hemley RJ, Mao HK. Compression of ice to 210 gigapascals: Infrared evidence for a symmetric hydrogen-bonded phase. Science 1996;273:218–220.
[47] Holzapfel WB. Symmetry of hydrogen-bonds in ice-VII. Journal of Chemical Physics 1972;56:712–715.
[48] Ma Y, Oganov AR, Li Z, Xie Y, Kotakoski J. Novel high pressure structures of polymeric nitrogen. Physical Review Letters 2009;102:065501.

[49] McMahon JM. Ground-state structures of ice at high pressures from ab initio random structure searching. Physical Review B 2011;84:220104.
[50] Wigner E, Huntington HB. On the possibility of a metallic modification of hydrogen. The Journal of Chemical Physics 1935;3:764–770.
[51] Goncharov AF, Gregoryanz E, Hemley RJ, Mao H-k. Molecular character of the metallic high-pressure phase of oxygen. Physical Review B 2003;68:100102.
[52] Ishikawa T, Mukai K, Tanaka Y, et al. Metallization of solid iodine in phase I: X-ray diffraction measurements, electrical resistance measurements, and ab initio calculations. High Pressure Research 2013;33:186–190.
[53] Eremets MI, Drozdov AP, Kong PP, Wang H. Semimetallic molecular hydrogen at pressure above 350 GPa. Nature Physics 2019;15:1246–1249.
[54] Loubeyre P, Occelli F, Dumas P. Synchrotron infrared spectroscopic evidence of the probable transition to metal hydrogen. Nature 2020;577:631–635.
[55] Pierleoni C, Morales MA, Rillo G, Holzmann M, Ceperley DM. Liquid–liquid phase transition in hydrogen by coupled electron–ion Monte Carlo simulations. Proceedings of the National Academy of Sciences 2016;113:4953–4957.
[56] Mazzola G, Sorella S. Distinct metallization and atomization transitions in dense liquid hydrogen. Physical Review Letters 2015;114:105701.
[57] Li B, Ji C, Yang W, et al. Diamond anvil cell behavior up to 4 Mbar. Proceedings of the National Academy of Sciences 2018;115:1713–1717.
[58] Lobanov SS, Prakapenka VB, Prescher C, et al. Pressure, stress, and strain distribution in the double-stage diamond anvil cell. Journal of Applied Physics 2015;118:035905.
[59] Dewaele A, Loubeyre P, Occelli F, Marie O, Mezouar M. Toroidal diamond anvil cell for detailed measurements under extreme static pressures. Nature Communications 2018;9:2913.
[60] Jenei Z, O'Bannon EF, Weir ST, Cynn H, Lipp MJ, Evans WJ. Single crystal toroidal diamond anvils for high pressure experiments beyond 5 megabar. Nature Communications 2018;9:3563.
[61] Goncharov AF, Beck P, Struzhkin VV, Hemley RJ, Crowhurst JC. Laser-heating diamond anvil cell studies of simple molecular systems at high pressures and temperatures. Journal of Physics and Chemistry of Solids 2008;69:2217–2222.
[62] Goncharov AF, Montoya JA, Subramanian N, et al. Laser heating in diamond anvil cells: Developments in pulsed and continuous techniques. Journal of Synchrotron Radiation 2009;16:769–772.
[63] Goncharov AF, Prakapenka VB, Struzhkin VV, Kantor I, Rivers ML, Dalton DA. X-ray diffraction in the pulsed laser heated diamond anvil cell. Review of Scientific Instruments 2010;81:113902.
[64] Aprilis G, Kantor I, Kupenko I, et al. Comparative study of the influence of pulsed and continuous wave laser heating on the mobilization of carbon and its chemical reaction with iron in a diamond anvil cell. Journal of Applied Physics 2019;125:095901.
[65] Jiang S, Holtgrewe N, Geballe ZM, et al. A spectroscopic study of the insulator – Metal transition in liquid hydrogen and deuterium. Advanced Science 2020;7:1901668.
[66] Zaghoo M, Silvera IF. Conductivity and dissociation in liquid metallic hydrogen and implications for planetary interiors. Proceedings of the National Academy of Sciences 2017;114:11873–11877.
[67] Zaghoo M, Husband RJ, Silvera IF. Striking isotope effect on the metallization phase lines of liquid hydrogen and deuterium. Physical Review B 2018;98:104102.
[68] Jiang S, Holtgrewe N, Lobanov SS, et al. Metallization and molecular dissociation of dense fluid nitrogen. Nature Communications 2018;9:2624.

[69] Prakapenka VB, Holtgrewe N, Lobanov SS, Goncharov AF. Structure and properties of two superionic ice phases. Nature Physics 2021;17:1233–1238.
[70] Mao HK, Jephcoat AP, HEMLEY RJ, et al. Synchrotron X-ray diffraction measurements of single-crystal hydrogen to 26.5 Gigapascals. Science 1988;239:1131–1134.
[71] Loubeyre P, LeToullec R, Pinceaux JP, Mao HK, Hu J, Hemley RJ. Equation of state and phase diagram of solid 4 He from single-crystal x-ray diffraction over a large P-T domain. Physical Review Letters 1993;71:2272–2275.
[72] Loubeyre P, LeToullec R, Hausermann D, et al. X-ray diffraction and equation of state of hydrogen at megabar pressures. Nature 1996;383:702–704.
[73] Ji C, Li B, Liu W, et al. Ultrahigh-pressure isostructural electronic transitions in hydrogen. Nature 2019;573:558–562.
[74] Ji C, Adeleke AA, Yang L, et al. Nitrogen in black phosphorus structure. Science Advances 2020;6:eaba9206.
[75] Bykov M, Tasca KR, Batyrev IG, et al. Dinitrogen as a universal electron acceptor in solid-state chemistry: An example of uncommon metallic compounds Na3(N2)4 and NaN2. Inorganic Chemistry 2020;59:14819–14826.
[76] Bykov M, Chariton S, Bykova E, et al. High-pressure synthesis of metal–Inorganic frameworks $Hf_4N_{20} \cdot N_2$, $WN_8 \cdot N_2$, and $Os_5N_{28} \cdot 3\ N_2$ with polymeric nitrogen linkers. Angewandte Chemie International Edition 2020;59:10321–10326.
[77] Wang Y, Glazyrin K, Roizen V, et al. Novel hydrogen clathrate hydrate. Physical Review Letters 2020;125:255702.
[78] Bykova E, Bykov M, Chariton S, et al. Structure and composition of C-S-H compounds up to 143 GPa. Physical Review B 2021;103:L140105.
[79] Ren Y, Zuo X. Synchrotron X-ray and neutron diffraction, total scattering, and small-angle scattering techniques for rechargeable battery research. Small Methods 2018;2:1800064.
[80] Goncharenko I, Loubeyre P. Neutron and X-ray diffraction study of the broken symmetry phase transition in solid deuterium. Nature 2005;435:1206–1209.
[81] Boehler R, Guthrie M, Molaison JJ, et al. Large-volume diamond cells for neutron diffraction above 90 GPa. High Pressure Research 2013;33:546–554.
[82] Guthrie M, Boehler R, Tulk CA, et al. Neutron diffraction observations of interstitial protons in dense ice. Proceedings of the National Academy of Sciences 2013;110:10552–10556.
[83] Ruoff AL, Luo H, Vohra YK. The closing diamond anvil optical window in multimegabar research. Journal of Applied Physics 1991;69:6413–6416.
[84] Shen G, Mao HK. High-pressure studies with x-rays using diamond anvil cells. Reports on Progress in Physics 2016;80:016101.
[85] Li B, Ding Y, Kim DY, et al. Probing the electronic band gap of solid hydrogen by inelastic X-ray scattering up to 90 GPa. Physical Review Letters 2021;126:036402.
[86] Mao HK, Shirley EL, Ding Y, et al. Electronic structure of crystalline ^{4}He at high pressures. Physical Review Letters 2010;105:186404.
[87] Liermann HP, Konopkova Z, Appel K, et al. Novel experimental setup for megahertz X-ray diffraction in a diamond anvil cell at the high energy density (HED) instrument of the European X-ray free-electron laser (EuXFEL). Journal of Synchrotron Radiation 2021;28:688–706.
[88] Akahama Y, Kawamura H. Pressure calibration of diamond anvil Raman gauge to 310 GPa. Journal of Applied Physics 2006;100:043516.
[89] Eremets MI, Minkov VS, Kong PP, Drozdov AP, Chariton S, Prakapenka VB. Universal diamond edge Raman scale to 0.5 terapascal: The implication to metallization of hydrogen. arXiv:220206933 [cond-matmtrl-sci] 2022.

[90] Howie RT, Gregoryanz E, Goncharov AF. Hydrogen (deuterium) vibron frequency as a pressure comparison gauge at multi-Mbar pressures. Journal of Applied Physics 2013;114:073505.
[91] Hemley RJ, Bell PM, Mao HK. Laser techniques in high-pressure geophysics. Science 1987;237:605–612.
[92] Murakami M, Asahara Y, Ohishi Y, Hirao N, Hirose K. Development of *in situ* Brillouin spectroscopy at high pressure and high temperature with synchrotron radiation and infrared laser heating system: Application to the Earth's deep interior. Physics of the Earth and Planetary Interiors 2009;174:282–291.
[93] Zhang JS, Bass JD, Zhu G. Single-crystal Brillouin spectroscopy with CO2 laser heating and variable q. Review of Scientific Instruments 2015;86:063905.
[94] Goncharov AF, Crowhurst JC. Raman spectroscopy of hot compressed hydrogen and nitrogen: Implications for the intramolecular potential. Physical Review Letters 2006;96:055504.
[95] Goncharov AF, Crowhurst JC, Struzhkin VV, Hemley RJ. Triple point on the melting curve and polymorphism of nitrogen at high pressure. Physical Review Letters 2008;101:095502.
[96] Ahart M, Karandikar A, Gramsch S, Boehler R, Hemley RJ. High P–T Brillouin scattering study of H2O melting to 26 GPa. High Pressure Research 2014;34:327–336.
[97] Litasov KD, Goncharov AF, Hemley RJ. Crossover from melting to dissociation of CO2 under pressure: Implications for the lower mantle. Earth and Planetary Science Letters 2011;309:318–323.
[98] Subramanian N, Goncharov AF, Struzhkin VV, Somayazulu M, Hemley RJ. Bonding changes in hot fluid hydrogen at megabar pressures. Proceedings of the National Academy of Sciences 2011;108:6014–6019.
[99] Ojwang JGO, McWilliams RS, Ke X, Goncharov AF. Melting and dissociation of ammonia at high pressure and high temperature. The Journal of Chemical Physics 2012;137:064507.
[100] Goncharov AF. Raman spectroscopy at high pressures. International Journal of Spectroscopy 2012;2012:617528.
[101] Goncharov AF, Eggert JH, Mazin II, Hemley RJ, Mao H-k. Raman excitations and orientational ordering in deuterium at high pressure. Physical Review B 1996;54:R15590–R3.
[102] Mao H-k, Hemley RJ. Ultrahigh-pressure transitions in solid hydrogen. Reviews of Modern Physics 1994;66:671–692.
[103] Goncharov AF, Gregoryanz E, Hemley RJ, Mao H-k. Spectroscopic studies of the vibrational and electronic properties of solid hydrogen to 285 GPa. Proceedings of the National Academy of Sciences 2001;98:14234–14237.
[104] Scheerboom MIM, Schouten JA. Anomalous behavior of the vibrational spectrum of the high-pressure delta phase of nitrogen: A second-order transition. Physical Review Letters 1993;71:2252–2255.
[105] Hemley RJ, Goncharov AF, Lu R, Struzhkin VV, Li M, Mao HK. High-pressure synchrotron infrared spectroscopy at the national synchrotron light source. Il Nuovo Cimento D 1998;20:539–551.
[106] Goncharov AF, Kong L, Mao H-k. High-pressure integrated synchrotron infrared spectroscopy system at the Shanghai Synchrotron Radiation Facility. Review of Scientific Instruments 2019;90:093905.
[107] Zha C-s, Duffy TS, Mao H-k, Hemley RJ. Elasticity of hydrogen to 24 GPa from single-crystal Brillouin scattering and synchrotron x-ray diffraction. Physical Review B 1993;48:9246–9255.
[108] Zhang JS, Hao M, Ren Z, Chen B. The extreme acoustic anisotropy and fast sound velocities of cubic high-pressure ice polymorphs at Mbar pressure. Applied Physics Letters 2019;114:191903.

[109] Shi W, Sun N, Li X, Mao Z, Liu J, Prakapenka VB. Single-Crystal Elasticity of High-Pressure Ice up to 98 GPa by Brillouin Scattering. Geophysical Research Letters, 2021; 48: e2021GL092514.
[110] Murakami M, Sinogeikin SV, Hellwig H, Bass JD, Li J. Sound velocity of $MgSiO_3$ perovskite to Mbar pressure. Earth and Planetary Science Letters 2007;256:47–54.
[111] Decremps F, Gauthier M, Ayrinhac S, et al. Picosecond acoustics method for measuring the thermodynamical properties of solids and liquids at high pressure and high temperature. Ultrasonics 2015;56:129–140.
[112] Decremps F, Antonangeli D, Gauthier M, et al. Sound velocity of iron up to 152 GPa by picosecond acoustics in diamond anvil cell. Geophysical Research Letters 2014;41:1459–1464.
[113] Goncharov AF, Gauthier M, Antonangeli D, et al. Elasticity and Poisson's ratio of hexagonal close-packed hydrogen at high pressures. Physical Review B 2017;95:214104.
[114] Kuriakose M, Raetz S, Hu QM, et al. Longitudinal sound velocities, elastic anisotropy, and phase transition of high-pressure cubic H_2O ice to 82 GPa. Physical Review B 2017;96:134122.
[115] Dias RP, Silvera IF. Observation of the Wigner-Huntington transition to metallic hydrogen. Science 2017;355:715–718.
[116] Mukose K, Fukano R, Matsuo H, Miyagi H, Yamaguchi K. First-principles studies of solid molecular halogens under pressure: Metallization pressures, Raman-active Ag modes and scaling relations. Solid State Communications 2002;123:451–456.
[117] Fujii Y, Hase K, Ohishi Y, et al. Evidence for molecular dissociation in bromine near 80 GPa. Physical Review Letters 1989;63:536–539.
[118] Shi J, Fonda E, Botti S, et al. Halogen molecular modifications at high pressure: The case of iodine. Physical Chemistry Chemical Physics 2021;23:3321–3326.
[119] San-Miguel A, Libotte H, Gauthier M, Aquilanti G, Pascarelli S, Gaspard JP. New phase transition of solid bromine under high pressure. Physical Review Letters 2007;99:015501.
[120] Johannsen PG, Holzapfel WB. Effect of pressure on Raman spectra of solid chlorine. Journal of Physics C: Solid State Physics 1983;16:L1177–L9.
[121] Kume T, Hiraoka T, Ohya Y, Sasaki S, Shimizu H. High pressure Raman study of bromine and iodine: Soft phonon in the incommensurate phase. Physical Review Letters 2005;94:065506.
[122] Dalladay-Simpson P, Binns J, Peña-Alvarez M, et al. Band gap closure, incommensurability and molecular dissociation of dense chlorine. Nature Communications 2019;10:1134.
[123] Zeng Q, He Z, San X, et al. A new phase of solid iodine with different molecular covalent bonds. Proceedings of the National Academy of Sciences 2008;105:4999–5001.
[124] Wu M, Tse JS, Pan Y. Anomalous bond length behavior and a new solid phase of bromine under pressure. Scientific Reports 2016;6:25649.
[125] Fujihisa H, Takemura K, Onoda M, Gotoh Y. Two intermediate incommensurate phases in the molecular dissociation process of solid iodine under high pressure. Physical Review Research 2021;3:033174.
[126] Takemura K, Minomura S, Shimomura O, Fujii Y, Axe JD. Structural aspects of solid iodine associated with metallization and molecular dissociation under high pressure. Physical Review B 1982;26:998–1004.
[127] Herzfeld KF. On atomic properties which make an element a metal. Physical Review 1927;29:701–705.
[128] Ross M. On the Herzfeld theory of metallization. Application to rare gases, alkali halides, and diatomic molecules. The Journal of Chemical Physics 1972;56:4651–4653.
[129] Duan D, Liu Y, Ma Y, et al. Ab initio studies of solid bromine under high pressure. Physical Review B 2007;76:104113.

[130] Li P, Gao G, Ma Y. Modulated structure and molecular dissociation of solid chlorine at high pressures. The Journal of Chemical Physics 2012;137:064502.
[131] Li P, Du X, Gao G, Sun R, Zhang L, Wang W. New modulated structures of solid bromine at high pressure. Computational Materials Science 2020;171:109205.
[132] Olson MA, Bhatia S, Larson P, Militzer B. Prediction of chlorine and fluorine crystal structures at high pressure using symmetry driven structure search with geometric constraints. The Journal of Chemical Physics 2020;153:094111.
[133] Meyer L, Barrett CS, Greer SC. Crystal structure of α-fluorine. The Journal of Chemical Physics 1968;49:1902–1907.
[134] Schiferl D, Kinkead S, Hanson RC, Pinnick DA. Raman spectra and phase diagram of fluorine at pressures up to 6 GPa and temperatures between 10 and 320 K. The Journal of Chemical Physics 1987;87:3016–3021.
[135] Lv Q, Jin X, Cui T, et al. Crystal structures and electronic properties of solid fluorine under high pressure. Chinese Physics B 2017;26:076103.
[136] Duan D, Liu Z, Lin Z, et al. Multistep dissociation of fluorine molecules under extreme compression. Physical Review Letters 2021;126:225704.
[137] Hermann A, Hoffmann R, Ashcroft NW. Condensed astatine: Monatomic and metallic. Physical Review Letters 2013;111:116404.
[138] Hemley RJ, Mao HK. Phase transition in solid molecular hydrogen at ultrahigh pressures. Physical Review Letters 1988;61:857–860.
[139] Dalladay-Simpson P, Howie RT, Gregoryanz E. Evidence for a new phase of dense hydrogen above 325 gigapascals. Nature 2016;529:63–67.
[140] Dubrovinsky L, Dubrovinskaia N, Katsnelson MI. No evidence of isostructural electronic transitions in compressed hydrogen. arXiv:191010772v1 [cond-matmtrl-sci] 2019.
[141] Goncharov AF, Hemley RJ, Mao H-k, Shu J. New high-pressure excitations in parahydrogen. Physical Review Letters 1998;80:101–104.
[142] Pickard CJ, Needs RJ. Structure of phase III of solid hydrogen. Nature Physics 2007;3:473–476.
[143] Monserrat B, Needs RJ, Gregoryanz E, Pickard CJ. Hexagonal structure of phase III of solid hydrogen. Physical Review B 2016;94:134101.
[144] Monserrat B, Drummond ND, Dalladay-Simpson P, et al. Structure and metallicity of phase V of hydrogen. Physical Review Letters 2018;120:255701.
[145] McMinis J, Clay RC, Lee D, Morales MA. Molecular to atomic phase transition in hydrogen under high pressure. Physical Review Letters 2015;114:105305.
[146] Goncharov AF, Tse JS, Wang H, et al. Bonding, structures, and band gap closure of hydrogen at high pressures. Physical Review B 2013;87:024101.
[147] Monacelli L, Errea I, Calandra M, Mauri F. Black metal hydrogen above 360 GPa driven by proton quantum fluctuations. Nature Physics 2021;17:63–67.
[148] Gorelov V, Holzmann M, Ceperley DM, Pierleoni C. Energy gap closure of crystalline molecular hydrogen with pressure. Physical Review Letters 2020;124:116401.
[149] Liao K, Shen T, Li X-Z, Alavi A, Grüneis A. Structural and electronic properties of solid molecular hydrogen from many-electron theories. Physical Review B 2021;103:054111.
[150] McMahon JM, Ceperley DM. Ground-state structures of atomic metallic hydrogen. Physical Review Letters 2011;106:165302.
[151] Azadi S, Monserrat B, Foulkes WMC, Needs RJ. Dissociation of high-pressure solid molecular hydrogen: A quantum Monte Carlo and anharmonic vibrational study. Physical Review Letters 2014;112:165501.

[152] Borinaga M, Ibañez-Azpiroz J, Bergara A, Errea I. Strong electron-phonon and band structure effects in the optical properties of high pressure metallic hydrogen. Physical Review Letters 2018;120:057402.
[153] Goncharov AF, Struzhkin VV. Comment on "Observation of the Wigner-Huntington transition to metallic hydrogen". Science 2017;357:eaam9736.
[154] Liu X-D, Dalladay-Simpson P, Howie RT, Li B, Gregoryanz E. Comment on "Observation of the Wigner-Huntington transition to metallic hydrogen". Science 2017;357:eaan2286.
[155] Eremets MI, Drozdov AP. Comments on the claimed observation of the Wigner-Huntington transition to metallic hydrogen. arXiv:170205125v1 2017.
[156] Loubeyre P, Occelli F, Dumas P. Comment on: Observation of the Wigner-Huntington transition to metallic hydrogen. arXiv:170207192v1 2017.
[157] Liu L-g. Dry ice II, a new polymorph of CO2. Nature 1983;303:508–509.
[158] Seto Y, Hamane D, Nagai T, Fujino K. Fate of carbonates within oceanic plates subducted to the lower mantle, and a possible mechanism of diamond formation. Physics and Chemistry of Minerals 2008;35:223–229.
[159] Takafuji N, Fujino K, Nagai T, Seto Y, Hamane D. Decarbonation reaction of magnesite in subducting slabs at the lower mantle. Physics and Chemistry of Minerals 2006;33:651–654.
[160] Maeda F, Ohtani E, Kamada S, Sakamaki T, Hirao N, Ohishi Y. Diamond formation in the deep lower mantle: A high-pressure reaction of MgCO3 and SiO2. Scientific Reports 2017;7:40602.
[161] Stishov SM. A new dense modification of silica. Geokhimiya 1961;10:837–839.
[162] Downs RT, Somayazulu MS. Carbon dioxide at 1.0 GPa. Acta Crystallographica Section C 1998;54:897–898.
[163] Donohue J. A refinement of the positional parameter in [alpha]-nitrogen. Acta Crystallographica 1961;14:1000–1001.
[164] Mucker KF, Talhouk S, Harris PM, White D, Erickson RA. Crystal structure of para-enriched solid deuterium below the λ transition. Physical Review Letters 1966;16:799–801.
[165] Bonev SA, Gygi F, Ogitsu T, Galli G. High-pressure molecular phases of solid carbon dioxide. Physical Review Letters 2003;91:065501.
[166] Yoo CS, Kohlmann H, Cynn H, Nicol MF, Iota V, LeBihan T. Crystal structure of pseudo-six-fold carbon dioxide phase II at high pressures and temperatures. Physical Review B 2002;65:104103.
[167] Yoo C-S, Kim M, Morgenroth W, Liermann P. Transformation and structure of silicatelike CO_2-V. Physical Review B 2013;87:214103.
[168] Yoo C-S, Iota V, Cynn H. Nonlinear Carbon Dioxide at High Pressures and Temperatures. Physical Review Letters 2001;86:444–447.
[169] Datchi F, Giordano VM, Munsch P, Saitta AM. Structure of carbon dioxide phase IV: Breakdown of the intermediate bonding state scenario. Physical Review Letters 2009;103:185701.
[170] Giordano VM, Datchi F. Molecular carbon dioxide at high pressure and high temperature. Europhysics Letters (EPL) 2007;77:46002.
[171] Santoro M, Gorelli FA, Bini R, Ruocco G, Scandolo S, Crichton WA. Amorphous silica-like carbon dioxide. Nature 2006;441:857–860.
[172] Montoya JA, Rousseau R, Santoro M, Gorelli F, Scandolo S. Mixed threefold and fourfold carbon coordination in compressed CO_2. Physical Review Letters 2008;100:163002.
[173] Santoro M, Lin J-F, Mao H-k, Hemley RJ. *In situ* high P-T Raman spectroscopy and laser heating of carbon dioxide. The Journal of Chemical Physics 2004;121:2780–2787.
[174] Teweldeberhan AM, Boates B, Bonev SA. CO_2 in the mantle: Melting and solid–solid phase boundaries. Earth and Planetary Science Letters 2013;373:228–232.

[175] Cogollo-Olivo BH, Biswas S, Scandolo S, Montoya JA. Ab initio determination of the phase diagram of CO2 at high pressures and temperatures. Physical Review Letters 2020;124:095701.
[176] Yong X, Liu H, Wu M, et al. Crystal structures and dynamical properties of dense CO_2. Proceedings of the National Academy of Sciences 2016;113:11110–11115.
[177] Datchi F, Moog M, Pietrucci F, Saitta AM. Polymeric phase V of carbon dioxide has not been recovered at ambient pressure and has a unique structure. Proceedings of the National Academy of Sciences 2017;114:E656–E7.
[178] Tschauner O, Mao H-k, Hemley RJ. New transformations of CO_2 at high pressures and temperatures. Physical Review Letters 2001;87:075701.
[179] Yoo C-S, Sengupta A, Kim M. Carbon dioxide carbonates in the earth's mantle: Implications to the deep carbon cycle. Angewandte Chemie International Edition 2011;50:11219–11222.
[180] Nellis WJ, Mitchell AC, Ree FH, et al. Equation of state of shock-compressed liquids: Carbon dioxide and air. The Journal of Chemical Physics 1991;95:5268–5272.
[181] Dziubek KF, Ende M, Scelta D, et al. Crystalline polymeric carbon dioxide stable at megabar pressures. Nature Communications 2018;9:3148.
[182] Oganov AR, Ono S, Ma Y, Glass CW, Garcia A. Novel high-pressure structures of $MgCO_3$, $CaCO_3$ and CO_2 and their role in Earth's lower mantle. Earth and Planetary Science Letters 2008;273:38–47.
[183] Boates B, Teweldeberhan AM, Bonev SA. Stability of dense liquid carbon dioxide. Proceedings of the National Academy of Sciences 2012;109:14808–14812.
[184] Wu M, Tse JS, Pan Y. Stability and properties of liquid CO_2 at high pressure and high temperature: Implications for electrical conductivities in Earth's lower mantle. Geophysical Research Letters 2015;42:5820–5827.
[185] Yoo CS, Cynn H, Gygi F, et al.. Crystal structure of carbon dioxide at high pressure: "Superhard" polymeric carbon dioxide. Physical Review Letters 1999;83:5527–5530.
[186] Datchi F, Mallick B, Salamat A, Ninet S. Structure of polymeric carbon dioxide CO_2-V. Physical Review Letters 2012;108:125701.
[187] Seto Y, Nishio-Hamane D, Nagai T, Sata N, Fujino K. Synchrotron X-ray diffraction study for crystal structure of solid carbon dioxide CO2-V. Journal of Physics: Conference Series 2010;215:012015.
[188] Santoro M, Gorelli FA, Bini R, et al. Partially collapsed cristobalite structure in the non-molecular phase V in CO_2. Proceedings of the National Academy of Sciences 2012;109:5176–5179.
[189] Dong J, Tomfohr JK, Sankey OF, Leinenweber K, Somayazulu M, McMillan PF. Investigation of hardness in tetrahedrally bonded nonmolecular CO_2 solids by density-functional theory. Physical Review B 2000;62:14685–14689.
[190] Holm B, Ahuja R, Belonoshko A, Johansson B. Theoretical investigation of high pressure phases of carbon dioxide. Physical Review Letters 2000;85:1258–1261.
[191] Lu C, Miao M, Ma Y. Structural evolution of carbon dioxide under high pressure. Journal of the American Chemical Society 2013;135:14167–14171.
[192] Yoo CS, Iota V, Cynn H, et al. Disproportionation and other transformations of N_2O at high pressures and temperatures to lower energy, denser phases. The Journal of Physical Chemistry B 2003;107:5922–5925.
[193] Lee M-S, Montoya JA, Scandolo S. Thermodynamic stability of layered structures in compressed CO_2. Physical Review B 2009;79:144102.
[194] Iota V, Yoo C-S, Klepeis J-H, Jenei Z, Evans W, Cynn H. Six-fold coordinated carbon dioxide VI. Nature Materials 2007;6:34–38.

[195] Seitz AE, Hippauf F, Kremer W, Kaskel S, Scheer M. Facile storage and release of white phosphorus and yellow arsenic. Nature Communications 2018;9:361.
[196] Nellis WJ, Holmes NC, Mitchell AC, van Thiel M. Phase transition in fluid nitrogen at high densities and temperatures. Physical Review Letters 1984;53:1661–1664.
[197] Radousky HB, Nellis WJ, Ross M, Hamilton DC, Mitchell AC. Molecular dissociation and shock-induced cooling in fluid nitrogen at high densities and temperatures. Physical Review Letters 1986;57:2419–2422.
[198] Ross M, Rogers F. Polymerization, shock cooling, and the high-pressure phase diagram of nitrogen. Physical Review B 2006;74:024103.
[199] Chau R, Mitchell AC, Minich RW, Nellis WJ. Metallization of fluid nitrogen and the Mott transition in highly compressed low-Z fluids. Physical Review Letters 2003;90:245501.
[200] Donadio D, Spanu L, Duchemin I, Gygi F, Galli G. Ab initio investigation of the melting line of nitrogen at high pressure. Physical Review B 2010;82:020102.
[201] Mailhiot C, Yang LH, McMahan AK. Polymeric nitrogen. Physical Review B 1992;46:14419–14435.
[202] McMahan AK, LeSar R. Pressure dissociation of solid nitrogen under 1 Mbar. Physical Review Letters 1985;54:1929–1932.
[203] Martin RM, Needs RJ. Theoretical study of the molecular-to-nonmolecular transformation of nitrogen at high pressures. Physical Review B 1986;34:5082–5092.
[204] Brazhkin VV, Lyapin AG. Metastable high-pressure phases of low-Z compounds: Creation of a new chemistry or a prompt for old principles?. Nature Materials 2004;3:497–500.
[205] Pickard CJ, Needs RJ. High-pressure phases of nitrogen. Physical Review Letters 2009;102:125702.
[206] Schuch AF, Mills RL. Crystal structures of the three modifications of nitrogen 14 and nitrogen 15 at high pressure. The Journal of Chemical Physics 1970;52:6000–6008.
[207] Mills RL, Olinger B, Cromer DT. Structures and phase diagrams of N_2 and CO to 13 GPa by x-ray diffraction. The Journal of Chemical Physics 1986;84:2837–2845.
[208] Hanfland M, Lorenzen M, Wassilew-Reul C, Zontone F. Structures of molecular nitrogen at high pressures. The Review of High Pressure Science and Technology 1998;7:787–789.
[209] Ivlev SI, Karttunen AJ, Hoelzel M, Conrad M, Kraus F. The crystal structures of α- and β-F_2 revisited. Chemistry – A European Journal 2019;25:3310–3317.
[210] Vegard L. Structure of solid oxygen. Nature 1935;136:720–721.
[211] Hellwig H, Daniels WB, Hemley RJ, Mao H-k, Gregoryanz E, Yu Z. Coherent anti-Stokes Raman scattering spectroscopy of solid nitrogen to 22 GPa. The Journal of Chemical Physics 2001;115:10876–10882.
[212] Stinton GW, Loa I, Lundegaard LF, McMahon MI. The crystal structures of δ and δ* nitrogen. The Journal of Chemical Physics 2009;131:104511.
[213] Goncharov A, Gregoryanz E. Chapter 8 – Solid nitrogen at extreme conditions of high pressure and temperature. In: Manaa MR., editor. Chemistry at Extreme Conditions. Amsterdam, Elsevier, 2005, 241–267.
[214] Nosé S, Klein ML. Structural transformations in solid nitrogen at high pressure. Physical Review Letters 1983;50:1207–1210.
[215] Schiferl D, Buchsbaum S, Mills RL. Phase transitions in nitrogen observed by Raman spectroscopy from 0.4 to 27.4 GPa at 15 K. The Journal of Physical Chemistry 1985;89:2324–2330.
[216] Bini R, Ulivi L, Kreutz J, Jodl HJ. High-pressure phases of solid nitrogen by Raman and infrared spectroscopy. The Journal of Chemical Physics 2000;112:8522–8529.

[217] Eremets MI, Gavriliuk AG, Serebryanaya NR, et al. Structural transformation of molecular nitrogen to a single-bonded atomic state at high pressures. The Journal of Chemical Physics 2004;121:11296–11300.
[218] Gregoryanz E, Goncharov AF, Sanloup C, Somayazulu M, Mao H-k, Hemley RJ. High P-T transformations of nitrogen to 170GPa. The Journal of Chemical Physics 2007;126:184505.
[219] Goncharov AF, Gregoryanz E, Mao H-k, Liu Z, Hemley RJ. Optical evidence for a nonmolecular phase of nitrogen above 150 GPa. Physical Review Letters 2000;85:1262–1265.
[220] Eremets M, Hemley R, Mao H-k, Gregoryanz E. Semiconducting non-molecular nitrogen up to 240 GPa and its low-pressure stability. Nature 2001;411:170–174.
[221] Gregoryanz E, Goncharov AF, Hemley RJ, Mao H-k. High-pressure amorphous nitrogen. Physical Review B 2001;64:052103.
[222] Lipp MJ, Klepeis JP, Baer BJ, et al. Transformation of molecular nitrogen to nonmolecular phases at megabar pressures by direct laser heating. Physical Review B 2007;76:014113.
[223] Goncharov AF, Gregoryanz E, Mao H-K, Hemley RJ. Vibrational dynamics of solid molecular nitrogen to megabar pressures. Low Temperature Physics 2001;27:866–869.
[224] Gregoryanz E, Goncharov AF, Hemley RJ, Mao H-k, Somayazulu M, Shen G. Raman, infrared, and x-ray evidence for new phases of nitrogen at high pressures and temperatures. Physical Review B 2002;66:224108.
[225] Tomasino D, Jenei Z, Evans W, Yoo C-S. Melting and phase transitions of nitrogen under high pressures and temperatures. The Journal of Chemical Physics 2014;140:244510.
[226] Frost M, Howie RT, Dalladay-Simpson P, Goncharov AF, Gregoryanz E. Novel high-pressure nitrogen phase formed by compression at low temperature. Physical Review B 2016;93:024113.
[227] Lorenzana HE, Lipp MJ, Evans WJ. Anomalous molecular phase of nitrogen: Implications to the phase diagram. High Pressure Research 2002;22:5–8.
[228] Cromer DT, Mills RL, Schiferi D, Schwalbe LA. The structure of N2 at 49 kbar and 299 K. Acta Crystallographica Section B 1981;37:8–11.
[229] Weck G, Datchi F, Garbarino G, et al. Melting curve and liquid structure of nitrogen probed by X-ray diffraction to 120 GPa. Physical Review Letters 2017;119:235701.
[230] Popov M. Raman and IR study of high-pressure atomic phase of nitrogen. Physics Letters A 2005;334:317–325.
[231] Eremets MI, Hemley RJ, Mao H-k, Gregoryanz E. Semiconducting non-molecular nitrogen up to 240 GPa and its low-pressure stability. Nature 2001;411:170–174.
[232] Tomasino D, Kim M, Smith J, Yoo C-S. Pressure-induced symmetry-lowering transition in dense nitrogen to layered polymeric nitrogen (LP-N) with colossal Raman intensity. Physical Review Letters 2014;113:205502.
[233] Cheng P, Yang X, Zhang X, Wang Y, Jiang S, Goncharov AF. Polymorphism of polymeric nitrogen at high pressures. The Journal of Chemical Physics 2020;152:244502.
[234] Zhang X, Wang Y, Bykov M, et al. Immiscibility in N_2–H_2O solids up to 140 GPa. The Journal of Chemical Physics 2021;154:234505.
[235] Kotakoski J, Albe K. First-principles calculations on solid nitrogen: A comparative study of high-pressure phases. Physical Review B 2008;77:144109.
[236] Wang X, Wang Y, Miao M, et al. Cagelike diamondoid nitrogen at high pressures. Physical Review Letters 2012;109:175502.
[237] Yao Y, Tse JS, Tanaka K. Metastable high-pressure single-bonded phases of nitrogen predicted via genetic algorithm. Physical Review B 2008;77:052103.
[238] Laniel D, Winkler B, Fedotenko T, et al. High-pressure polymeric nitrogen allotrope with the black phosphorus structure. Physical Review Letters 2020;124:216001.

[239] Laniel D, Geneste G, Weck G, Mezouar M, Loubeyre P. Hexagonal layered polymeric nitrogen phase synthesized near 250 GPa. Physical Review Letters 2019;122:066001.
[240] Sun J, Martinez-Canales M, Klug DD, Pickard CJ, Needs RJ. Stable all-nitrogen metallic salt at terapascal pressures. Physical Review Letters 2013;111:175502.
[241] Freiman YA, Jodl HJ. Solid oxygen. Physics Reports 2004;401:1–228.
[242] Freiman YA, Jodl HJ, Crespo Y. Solid oxygen revisited. Physics Reports 2018;743:1–55.
[243] Klotz S, Strässle T, Cornelius AL, Philippe J, Hansen T. Magnetic ordering in solid oxygen up to room temperature. Physical Review Letters 2010;104:115501.
[244] Klotz S. Magnetism in solid oxygen studied by high-pressure neutron diffraction. Journal of Low Temperature Physics 2018;192:1–18.
[245] Lundegaard LF, Weck G, McMahon MI, Desgreniers S, Loubeyre P. Observation of an O8 molecular lattice in the ε phase of solid oxygen. Nature 2006;443:201–204.
[246] Fujihisa H, Akahama Y, Kawamura H, et al. O_8 cluster structure of the epsilon phase of solid oxygen. Physical Review Letters 2006;97:085503.
[247] Ma Y, Oganov AR, Glass CW. Structure of the metallic ζ-phase of oxygen and isosymmetric nature of the ε-ζ phase transition: Ab initio simulations. Physical Review B 2007;76:064101.
[248] Neaton JB, Ashcroft NW. Low-energy linear structures in dense oxygen: Implications for the ε phase. Physical Review Letters 2002;88:205503.
[249] Meng Y, Eng PJ, Tse JS, et al. Inelastic x-ray scattering of dense solid oxygen: Evidence for intermolecular bonding. Proceedings of the National Academy of Sciences 2008;105:11640–11644.
[250] Tse JS, Yao Y, Klug DD, Desgreniers S. Bonding in the ϵ-phase of high pressure oxygen. Journal of Physics: Conference Series 2008;121:012006.
[251] Desgreniers S, Vohra YK, Ruoff AL. Optical response of very high density solid oxygen to 132 GPa. The Journal of Physical Chemistry 1990;94:1117–1122.
[252] Fukui H, Anh LT, Wada M, et al. Electronic structure of dense solid oxygen from insulator to metal investigated with X-ray Raman scattering. Proceedings of the National Academy of Sciences 2019;116:21385–21391.
[253] Crespo Y, Fabrizio M, Scandolo S, Tosatti E. Collective spin 1 singlet phase in high-pressure oxygen. Proceedings of the National Academy of Sciences 2014;111:10427–10432.
[254] Gorelli FA, Ulivi L, Santoro M, Bini R. The ε phase of solid oxygen: Evidence of an O_4 molecule lattice. Physical Review Letters 1999;83:4093–4096.
[255] Akahama Y, Kawamura H. High-pressure Raman spectroscopy of solid oxygen. Physical Review B 1996;54:R15602–R5.
[256] Santoro M, Gregoryanz E, Mao H-k, Hemley RJ. New phase diagram of oxygen at high pressures and temperatures. Physical Review Letters 2004;93:265701.
[257] Lundegaard LF, Guillaume C, McMahon MI, Gregoryanz E, Merlini M. On the structure of high-pressure high-temperature η-O2. The Journal of Chemical Physics 2009;130:164516.
[258] Goncharov AF, Subramanian N, Ravindran TR, Somayazulu M, Prakapenka VB, Hemley RJ. Polymorphism of dense, hot oxygen. The Journal of Chemical Physics 2011;135:084512.
[259] Elatresh SF, Bonev SA. Stability and metallization of solid oxygen at high pressure. Physical Chemistry Chemical Physics 2020;22:12577–12583.
[260] Akahama Y, Kawamura H, Häusermann D, Hanfland M, Shimomura O. New high-pressure structural transition of oxygen at 96 GPa associated with metallization in a molecular solid. Physical Review Letters 1995;74:4690–4693.
[261] Weck G, Desgreniers S, Loubeyre P, Mezouar M. Single-crystal structural characterization of the metallic phase of oxygen. Physical Review Letters 2009;102:255503.
[262] Shimizu K, Suhara K, Ikumo M, Eremets MI, Amaya K. Superconductivity in oxygen. Nature 1998;393:767–769.

[263] Kim DY, Lebègue S, Araújo CM, Arnaud B, Alouani M, Ahuja R. Structurally induced insulator-metal transition in solid oxygen: A quasiparticle investigation. Physical Review B 2008;77:092104.
[264] Degtyareva O, Gregoryanz E, Somayazulu M, Dera P, Mao H-k, Hemley RJ. Novel chain structures in group VI elements. Nature Materials 2005;4:152–155.
[265] Salzmann CG, Radaelli PG, Mayer E, Finney JL. Ice XV: A new thermodynamically stable phase of ice. Physical Review Letters 2009;103:105701.
[266] Falenty A, Hansen TC, Kuhs WF. Formation and properties of ice XVI obtained by emptying a type sII clathrate hydrate. Nature 2014;516:231–233.
[267] del Rosso L, Celli M, Ulivi L. New porous water ice metastable at atmospheric pressure obtained by emptying a hydrogen-filled ice. Nature Communications 2016;7:13394.
[268] Salzmann CG, Radaelli PG, Hallbrucker A, Mayer E, Finney JL. The preparation and structures of hydrogen ordered phases of ice. Science 2006;311:1758–1761.
[269] Kamb B. Crystallography of Ice. In: Whalley EJ SJ, Gold LW., editors. Physics and Chemistry of Ice. Royal Society of Canada, Ottawa. 1973, 28–41.
[270] Nakamura T, Matsumoto M, Yagasaki T, Tanaka H. Thermodynamic stability of ice II and its hydrogen-disordered counterpart: Role of zero-point energy. The Journal of Physical Chemistry B 2016;120:1843–1848.
[271] Wang Y, Glazyrin K, Roizen V, et al. Novel hydrogen clathrate hydrate. Physical Review Letters 2020;125:255702.
[272] Besson JM, Pruzan P, Klotz S, et al. Variation of interatomic distances in ice VIII to 10 GPa. Physical Review B 1994;49:12540–12550.
[273] Walrafen GE, Abebe M, Mauer FA, Block S, Piermarini GJ, Munro R. Raman and x-ray investigations of ice VII to 36.0 GPa. The Journal of Chemical Physics 1982;77:2166–2174.
[274] Goncharov AF, Struzhkin VV, Mao HK, Hemley RJ. Raman spectroscopy of dense H2O and the transition to symmetric hydrogen bonds. Physical Review Letters 1999;83:1998–2001.
[275] Himoto K, Matsumoto M, Tanaka H. Lattice- and network-structure in plastic ice. Physical Chemistry Chemical Physics 2011;13:19876–19881.
[276] Aragones JL, Vega C. Plastic crystal phases of simple water models. The Journal of Chemical Physics 2009;130:244504.
[277] Hernandez J-A, Caracas R. Proton dynamics and the phase diagram of dense water ice. The Journal of Chemical Physics 2018;148:214501.
[278] Ninet S, Datchi F, Saitta AM. Proton disorder and superionicity in hot dense ammonia ice. Physical Review Letters 2012;108:165702.
[279] Redmer R, Mattsson TR, Nettelmann N, French M. The phase diagram of water and the magnetic fields of Uranus and Neptune. Icarus 2011;211:798–803.
[280] Goldman N, Fried LE, Kuo IFW, Mundy CJ. Bonding in the superionic phase of water. Physical Review Letters 2005;94:217801.
[281] Mattsson TR, Desjarlais MP. Phase diagram and electrical conductivity of high energy-density water from density functional theory. Physical Review Letters 2006;97:017801.
[282] French M, Desjarlais MP, Redmer R. Ab initio calculation of thermodynamic potentials and entropies for superionic water. Physical Review E 2016;93:022140.
[283] Hernandez J-A, Caracas R. Superionic-superionic phase transitions in body-centered cubic H_2O ice. Physical Review Letters 2016;117:135503.
[284] Goncharov AF, Goldman N, Fried LE, et al. Dynamic ionization of water under extreme conditions. Physical Review Letters 2005;94:125508.
[285] Rozsa V, Pan D, Giberti F, Galli G. Ab initio spectroscopy and ionic conductivity of water under Earth mantle conditions. Proceedings of the National Academy of Sciences 2018;115:6952–6957.

[286] Goncharov AF, Sanloup C, Goldman N, et al. Dissociative melting of ice VII at high pressure. The Journal of Chemical Physics 2009;130:124514.
[287] Frank MR, Fei Y, Hu J. Constraining the equation of state of fluid H2O to 80 GPa using the melting curve, bulk modulus, and thermal expansivity of Ice VII11 Associate editor: D. Sverjensky. Geochimica et Cosmochimica Acta 2004;68:2781–2790.
[288] Dubrovinsky L, Dubrovinskaia N. Melting of ice VII and new high-pressure, high-temperature amorphous ice. In: Ohtani E., editor. Advances in High-Pressure Mineralogy. Geological Society of America, 2007; 421.
[289] Schwager B, Chudinovskikh L, Gavriliuk A, Boehler R. Melting curve of H_2O to 90 GPa measured in a laser-heated diamond cell. Journal of Physics: Condensed Matter 2004;16: S1177.
[290] Schwager B, Boehler R. H_2O: Another ice phase and its melting curve. High Pressure Research 2008;28:431–433.
[291] Kimura T, Kuwayama Y, Yagi T. Melting temperatures of H2O up to 72 GPa measured in a diamond anvil cell using CO2 laser heating technique. The Journal of Chemical Physics 2014;140:074501.
[292] Lin J-F, Gregoryanz E, Struzhkin VV, Somayazulu M, Mao H-k, Hemley RJ. Melting behavior of H2O at high pressures and temperatures. Geophysical Research Letters 2005;32:L11306.
[293] Sugimura E, Komabayashi T, Ohta K, Hirose K, Ohishi Y, Dubrovinsky LS. Experimental evidence of superionic conduction in H2O ice. The Journal of Chemical Physics 2012;137:194505.
[294] Millot M, Hamel S, Rygg JR, et al. Experimental evidence for superionic water ice using shock compression. Nature Physics 2018;14:297–302.
[295] Millot M, Coppari F, Rygg JR, et al. Nanosecond X-ray diffraction of shock-compressed superionic water ice. Nature 2019;569:251–255.
[296] Queyroux JA, Hernandez JA, Weck G, et al. Melting curve and isostructural solid transition in superionic ice. Physical Review Letters 2020;125:195501.
[297] Lin J-F, Militzer B, Struzhkin VV, Gregoryanz E, Hemley RJ, Mao H-k. High pressure-temperature Raman measurements of H2O melting to 22 GPa and 900 K. The Journal of Chemical Physics 2004;121:8423–8427.
[298] Benoit M, Romero AH, Marx D. Reassigning hydrogen-bond centering in dense ice. Physical Review Letters 2002;89:145501.
[299] Kamb B, Davis BL. Ice VII, the densest form of ice. Proceedings of the National Academy of Sciences 1964;52:1433–1439.
[300] Hemley RJ, Jephcoat AP, Mao HK, Zha CS, Finger LW, Cox DE. Static compression of H2O-ice to 128 GPa (1.28 Mbar). Nature 1987;330:737–740.
[301] Sugimura E, Iitaka T, Hirose K, Kawamura K, Sata N, Ohishi Y. Compression of H2O ice to 126 GPa and implications for hydrogen-bond symmetrization: Synchrotron x-ray diffraction measurements and density-functional calculations. Physical Review B 2008;77:214103.
[302] Loubeyre P, LeToullec R, Wolanin E, Hanfland M, Hausermann D. Modulated phases and proton centring in ice observed by X-ray diffraction up to 170?GPa. Nature 1999;397:503–506.
[303] Aoki K, Yamawaki H, Sakashita M, Fujihisa H. Infrared absorption study of the hydrogen-bond symmetrization in ice to 110 GPa. Physical Review B 1996;54:15673–15677.
[304] Song M, Yamawaki H, Fujihisa H, Sakashita M, Aoki K. Infrared absorption study of Fermi resonance and hydrogen-bond symmetrization of ice up to 141 GPa. Physical Review B 1999;60:12644–12650.

[305] Guthrie M, Boehler R, Molaison JJ, Haberl B, dos Santos AM, Tulk C. Structure and disorder in ice VII on the approach to hydrogen-bond symmetrization. Physical Review B 2019;99:184112.
[306] Shimizu H, Ohnishi M, Sasaki S, Ishibashi Y. Cauchy relation in dense H_2O ice VII. Physical Review Letters 1995;74:2820–2823.
[307] Polian A, Grimsditch M. New high-pressure phase of H_2O – Ice-X. Physical Review Letters 1984;52:1312–1314.
[308] Ahart M, Somayazulu M, Gramsch SA, Boehler R, Mao H-k, Hemley RJ. Brillouin scattering of H2O ice to megabar pressures. The Journal of Chemical Physics 2011;134:124517.
[309] Kuriakose M, Raetz S, Hu QM, et al. Longitudinal sound velocities, elastic anisotropy, and phase transition of high-pressure cubic H_2O ice to 82 GPa. Physical Review B 2017;96:134122.
[310] Meier T, Petitgirard S, Khandarkhaeva S, Dubrovinsky L. Observation of nuclear quantum effects and hydrogen bond symmetrisation in high pressure ice. Nature Communications 2018;9:2766.
[311] Lei J, Lim J, Kim M, Yoo C-S. Crystal structure of symmetric ice X in H2O–H2 and H2O–He under pressure. The Journal of Physical Chemistry Letters 2021;12:4707–4712.
[312] Méndez ASJ, Trybel F, Husband RJ, Steinle-Neumann G, Liermann HP, Marquardt H. Bulk modulus of H_2O across the ice VII–ice X transition measured by time-resolved x-ray diffraction in dynamic diamond anvil cell experiments. Physical Review B 2021;103:064104.
[313] Benoit M, Bernasconi M, Focher P, Parrinello M. New high-pressure phase of ice. Physical Review Letters 1996;76:2934–2936.
[314] Caracas R. Dynamical Instabilities of Ice X. Physical Review Letters 2008;101:085502.
[315] Ji M, Umemoto K, Wang C-Z, Ho K-M, Wentzcovitch RM. Ultrahigh-pressure phases of H2O ice predicted using an adaptive genetic algorithm. Physical Review B 2011;84:220105.
[316] Militzer B, Wilson HF. New phases of water ice predicted at megabar pressures. Physical Review Letters 2010;105:195701.
[317] Pickard CJ, Martinez-Canales M, Needs RJ. Decomposition and Terapascal Phases of Water Ice. Physical Review Letters 2013;110:245701.
[318] Katsura T, Yoneda A, Yamazaki D, Yoshino T, Ito E. Adiabatic temperature profile in the mantle. Physics of the Earth and Planetary Interiors 2010;183:212–218.
[319] Gregoryanz E, Goncharov AF, Sanloup C, Somayazulu M, Mao H-k, Hemley RJ. High P-T transformations of nitrogen to 170 GPa. The Journal of Chemical Physics 2007;126:184505.
[320] Benedetti LR, Antonangeli D, Farber DL, Mezouar M. An integrated method to determine melting temperatures in high-pressure laser-heating experiments. Applied Physics Letters 2008;92:141903.

H. C. Walker

Inelastic neutron scattering under pressure

1 Introduction to inelastic neutron scattering

While the structure of materials is often considered key to understanding their nature, a large number of physical properties are dependent on the dynamics of a material, including specific heat, thermal conductivity, optical and dielectric properties, and electrical resistance. Moreover, for smart and functional materials, which are often driven by a complex balance or strong coupling between competing phenomena, understanding the atomic and magnetic dynamics is essential. Inelastic neutron scattering (INS) measures the change in energy of the neutron as it scatters from a sample, giving access to a wide range of different physical phenomena from atomic diffusional or hopping motions, the rotational modes of molecules, molecular vibrations, recoil in quantum liquids, and magnetic and quantum excitations, all the way up to electronic transitions.

The most commonly used spectroscopies involve light scattering, either Raman or infrared. These are cheap and readily available in most labs and are highly sensitive. However, they are only able to measure near the Brillouin zone center and are not sensitive to all vibrational modes. Additionally, calculation of the scattered intensity is nontrivial and prone to error, and so often only the positions of observed modes can be extracted. On the other hand, INS is a highly quantitative probe, whose results are directly comparable to analytical and numerical calculations, and they do not suffer from the same optical selection rules. Neutrons are produced at reactor and spallation sources with energies ranging from μeV up to eV, and wavelengths from 10 nm to 10 pm, matching the length and energy scales in condensed matter well. This means that it is possible to access the full or even multiple Brillouin zones using INS. Furthermore, the intrinsic dipole moment of the neutron makes them interact with magnetic flux variations in the sample, such that they can be used to study magnetism, with a scattering cross section of similar magnitude to that for nuclear scattering via the strong force. The short-range nature of the strong force means that the probability for a neutron scattering from a nucleus is very low, allowing bulk penetration, and the deposition of very little energy in the sample, making

Acknowledgments: I would like to thank all the ISIS staff members involved in the development of the new TAV6 pressure cell for inelastic neutron scattering, and the ISIS Pressure and Furnace team for their excellent experimental support. Financial support from the Engineering and Physical Sciences Research Council (EPSRC: EP/S035923/1) is gratefully acknowledged.

H. C. Walker, MAPS Instrument Scientist, ISIS Neutron and Muon Source, Rutherford Appleton Laboratory, Harwell, Oxford OX11 0QX

https://doi.org/10.1515/9783110668612-005

them nondestructive. By comparison, X-rays only penetrate a few surface layers and can damage samples. The downside for neutrons is that the inelastic scattering cross section is even weaker than that for elastic neutron scattering, necessitating large samples, high neutron fluxes, and relatively long counting times.

2 INS instrumentation

Generally a neutron spectrometer is an instrument with energy analysis which measures the scattering rate as a function of momentum (**Q**) and energy (ω) transfer from a sample; the data can then be related to the scattering function $S(\mathbf{Q}, \omega)$. No single instrument design can cover the full dynamical range accessible by neutron spectroscopy: $10^{-2} < Q < 1{,}000$ nm^{-1} and $10^{-8} < \hbar\omega < 100$ eV; instead it is covered by a number of different instrument types: spin-echo, backscattering, triple-axis, time-of-flight (TOF), and Compton scattering spectrometers. Spin-echo and backscattering spectrometers are designed to achieve very high energy resolution at low energies and momenta [1], while Compton scattering spectrometers use very high energy neutrons to measure atomic momentum distributions [2]. The rest of this chapter focuses on the measurement of coherent excitations and the use of triple-axis spectrometers (TAS) and TOF spectrometers.

2.1 Triple-axis spectrometers

TAS [3] are high-flux instruments that select the scattering angle, and the incident and scattered energies, allowing counting time to be concentrated on a small important region of $S(\mathbf{Q}, \omega)$. These were the original neutron spectrometers, having been developed in the 1950s by Brockhouse and collaborators [4], but their versatility means that they are still in wide use at continuous (reactor) neutron sources. Neutrons from the source are monochromatized (E_i) by the Bragg reflection from a crystal (axis 1), then incident on the sample (axis 2), and the final energy (E_f) and direction (2θ) of the scattered neutrons are selected using a crystal analyzer (axis 3). Figure 1(a) shows a simplified schematic of a traditional TAS. TAS may be used for elastic scattering ($E_i = E_f$), to give a better signal-to-noise ratio than a standard diffractometer at a continuous source, by reducing the background; however, they are most commonly used for inelastic scattering ($E_i \neq E_f$).

In a TAS experiment, it is typical to measure either a constant energy scan or a constant **Q** scan. In the first, **Q** varies along a line while the energy transfer remains fixed, while in the second the energy transfer varies at a fixed **Q** position, with a single detector separately measuring point by point. This makes it ideal for investigating dispersive excitations, such as phonons or magnons, in a single crystal sample, as

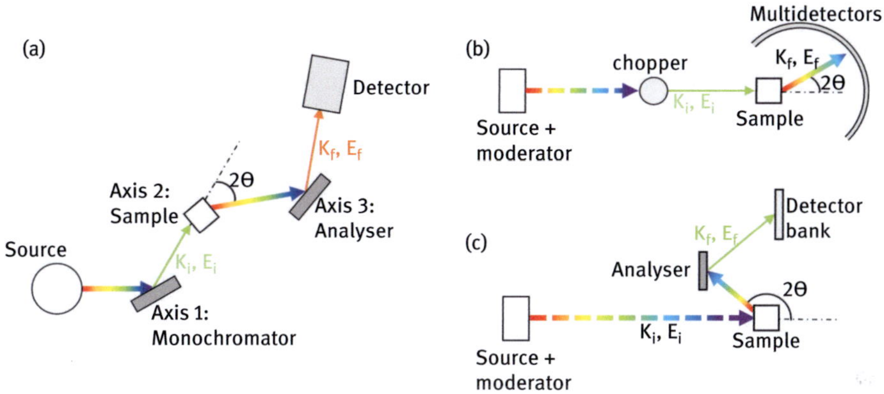

Figure 1: Simplified schematics for (a) a traditional triple-axis spectrometer, (b) a direct geometry time-of-flight spectrometer, and (c) an indirect geometry spectrometer.

measurements can concentrate on high symmetry directions to aid comparison with theory. The nonideal nature of neutron optics complicates the experiment, through beam divergence and a spread of wavelengths, leading to a distribution of momentum and energy transfers about the desired values $\mathbf{Q}_0$ and ω_0, and blurring the features in measured data compared to theory. However, this also offers a degree of freedom in designing the experiment, since extra beam divergence, leading to broader resolution, can be chosen to increase the count rate. These effects can be taken into consideration through the four-dimensional convolution:

$$I(\mathbf{Q}_0, \omega_0) = \iint R(\mathbf{Q}_0 - \mathbf{Q}, \omega - \omega_0) S(\mathbf{Q}, \omega) \mathrm{d}^3\mathbf{Q} \mathrm{d}(\hbar\omega), \tag{1}$$

where $I(\mathbf{Q}_0, \omega_0)$ is proportional to what is measured, $S(\mathbf{Q}, \omega)$ is the scattering function of the sample, and $R(\mathbf{Q}_0 - \mathbf{Q}, \omega - \omega_0)$ is the instrumental resolution function, representing the probability of detecting a neutron with $\mathbf{Q}$ and ω when the instrument is set up to measure $\mathbf{Q}_0$ and ω_0. The resolution function is a maximum at $(\mathbf{Q}_0, \omega_0)$ and decreases with distance from this point. Often the resolution function R is approximated by a Gaussian, in which case the contours of equal amplitude map out ellipsoidal surfaces, and the surface at the half maximum amplitude contour is called the resolution ellipsoid. How much features are blurred by the instrumental resolution will depend on the size, shape, and orientation of the resolution ellipsoid relative to the signal. Fortunately, this can be controlled through an optimal choice of E_i, E_f, beam divergence, the choice of crystals for the monochromator and analyzer, and the scattering sense at each of the three rotation axes. Since the measured intensity at any point in a constant energy scan is proportional to the intersection of the resolution ellipsoid with the dispersion curve, the orientation of the resolution ellipsoid affects the shape of the measured peak. This TAS property is known at focusing

TAS experiments can be performed either with fixed $\mathbf{k}_f$ or fixed $\mathbf{k}_i$, but the majority of TAS operate in fixed $\mathbf{k}_f$ mode because

- you can use the incident beam monitor to normalize so that your counts are directly proportional to $S(\mathbf{Q}, \omega)$;
- you can use a wavelength-specific filter before the analyzer to remove or at least suppress higher order scattering from the monochromators.

The main disadvantage with using fixed $\mathbf{k}_f$ mode is technical – a large beam port is required, as the whole instrument has to be moved when E_i is changed through rotating the monochromator, meaning a lot of experimental hall floor space is required.

2.2 Time-of-flight spectrometers

TOF spectrometers are a more recent development and are present at both continuous and spallation sources. In contrast to TAS, they measure wide regions of $S(\mathbf{Q}, \omega)$ simultaneously, making them ideal for surveying complex materials with multiple excitation modes and for cases where little is known in advance about the spectrum of excitations. The TOF technique maps the time of detection of a neutron to its velocity (energy), assuming that the neutrons detected originated at a single time ($t = 0$) and position. This makes them the natural spectrometer at a pulsed source, although it is also possible to create a pulsed beam at a continuous source using choppers. However, at spallation sources, there are a greater number of epithermal neutrons allowing the measurement of higher energy and momentum transfers; plus the fast neutron background decays rapidly and can be further suppressed by a background chopper, giving them an inherently lower background than at reactor sources, where the fast neutrons are present at all times.

There are two different ways to measure on a TOF spectrometer, either you fix the incident energy, and the TOF gives the final energy, or all energies are incident on the sample, and then the final energy is fixed so that the TOF gives the incident energy. These two modes of operation are known as direct and indirect geometry, respectively, depicted in Figure 1, and the design of the spectrometer dictates in which of the two modes it operates.

A typical direct geometry TOF spectrometer uses a Fermi chopper (a rotating cylinder) or a series of disk choppers, with a neutron transparent window, to select an incident energy of neutrons from a polychromatic beam. The chopper spins at a frequency equal to an integer multiple of the source pulse frequency, and, since the energy resolution is dependent on the time widths of the optical components, a faster spinning chopper is "open" for a shorter time, and therefore, has a narrower energy bandwidth but also results in a lower neutron flux. The scattered beam is then detected at a large bank of position-sensitive detectors, where the angle of the detector with respect to the sample can be converted to the momentum transferred from the neutron to the sample. The detectors are typically filled with pressurized ^{3}He gas. This makes these instruments very expensive to construct, and alternative detector designs are being developed to reduce the reliance on ^{3}He [5]. In an indirect geometry TOF spectrometer, the polychromatic beam is incident on the sample and scatters onto a crystal analyzer, where Bragg reflection selects a specific final energy to reach the detector. Typically the (002) reflection from pyrolytic graphite is used but higher orders (002*n*) will also scatter, and so often a filter is inserted between the analyzer and the detector to suppress those.

The choice of whether to use a direct or indirect geometry TOF spectrometer will depend on the goals of an experiment. With an indirect geometry instrument it is possible to reach a huge range of neutron energy loss ($\hbar\omega = [E_f, +\infty]$), whereas in direct geometry it is not possible for the energy loss to exceed the incident energy ($\hbar\omega = [-\infty, E_i]$). The energy resolution on an indirect geometry spectrometer is far superior to that on a direct geometry instrument. This is due to a combination of factors including:

- the scattering angle 2θ to the analyzer can be chosen to be close to backscattering;
- the resolution improves with the length of the primary flight path $L1$ between the moderator and the sample, which can be increased more easily and more cheaply than increasing the length of the secondary flight path $L2$ between the sample and the detector.

However, the momentum resolution is typically worse on indirect geometry instruments due to the shorter secondary flight path $L2$ and the angular area of the analyzers. In addition, kinematic restrictions mean that they cover a far smaller range of $|Q|$ at large energy transfers and generally their detectors cover a smaller solid angle. This means that molecular spectroscopy experiments looking at single atom dynamics are measured on indirect geometry instruments, while experiments looking at collective excitations, requiring momentum resolution, are performed on direct geometry instruments. The rest of this chapter focuses on coherent excitations measured under pressure, but vibrational spectroscopy measurements are also performed under pressure using indirect geometry TOF spectrometers, with a particular interest in hydrogen storage and catalysis.

3 INS high-pressure sample environment

While neutron diffraction is able to exploit the high pressures that can be achieved for small samples [6], as mentioned above, INS typically requires fairly large sample volumes. Therefore, in order to produce high pressures, larger forces are required, which are associated with rather more massive mechanical devices. Fortunately, neutrons are highly penetrating and are therefore able to reach samples having traveled through rather thick pressure vessel walls but commonly there will be more pressure vessel than sample in the beam. This will give rise to unwanted absorption, reducing neutron transmission to the detectors, but also coherent or incoherent background scattering, resulting in a reduction in the signal-to-noise ratio. However, it is possible to mitigate these problems through the right choice of pressure cell material and the optimization of vessel geometry and design for use in INS [7–10].

3.1 High-pressure gas cells

Currently, gas medium compression is one of the most popular and versatile high-pressure techniques used in neutron scattering experiments at both reactor and spallation sources [11]. The cell is permanently connected via a capillary to the compressor, allowing the pressure to be varied *in situ*, and, assuming a suitable pressure medium is used, the pressure conditions are hydrostatic. The choice of potential fluids is very limited since they need to have low neutron absorption, no hydrogen to avoid a strong incoherent background, and a high freezing pressure, even at low temperatures. This means that typically either helium or argon is used. It is clear that the material used in the construction of the pressure vessel must retain the necessary strength to endure the stresses that will occur at the intended high pressures. Safe external handling of pressurized gases limits the pressures attainable with gas pressure cells to under 0.8 GPa.

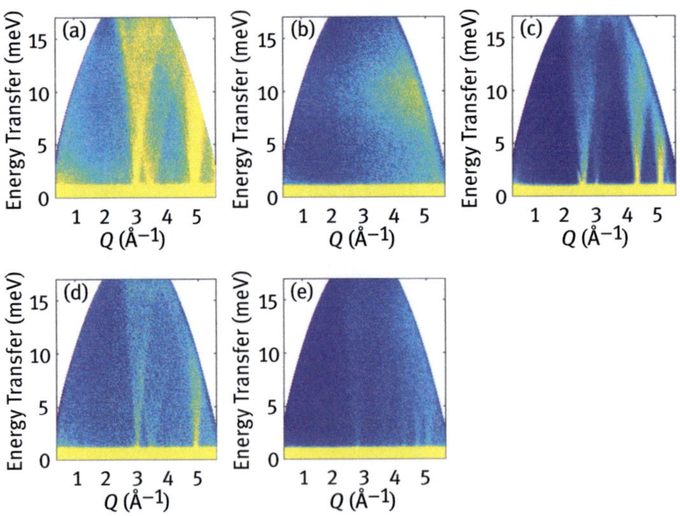

Figure 2: $S(\mathbf{Q}, \omega)$ for pressure cell materials (a) CuBe, (b) TiZr, (c) Al7075, (d) NiCrAl, and (e) TAV6 measured on LET [12] at ISIS (reproduced from [10]).The color scale represents the scattering intensity and is the same for all five plots.

Until recently, the gas cells available at neutron sources have been made out of a range of materials often used in non-neutron high-pressure measurements, such as CuBe, or which have been optimized for diffraction, such as TiZr. As part of the "Low background materials for high pressure cells used in inelastic neutron scattering experiments" project funded by SINE2020, $S(\mathbf{Q}, \omega)$ was measured for those and other materials (Figure 2). This made it clear that CuBe in particular is a very poor choice for coherent inelastic measurements, as there is a very strong highly structured phonon signal, while TiZr has a less structured diffuse high $|\mathbf{Q}|$ signal due to relatively strong incoherent scattering. Based on the combination of the results for $S(\mathbf{Q}, \omega)$ for the different cell materials, with the neutron transmission measured as a function of energy (Figure 3), a new pressure cell [10] has now been designed and built at ISIS from TAV6, a titanium alloy composed of 6 wt% Al, 4 wt% V, and 90 wt% Ti. The typical cylindrical design for the cell (Figure 4) is well suited to the direct geometry TOF spectrometers with their detectors arranged to form half a cylindrical curtain about the sample.

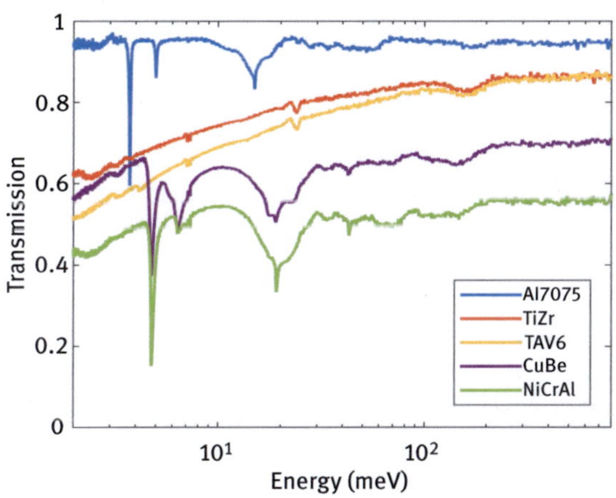

Figure 3: Neutron transmission of different pressure cell materials measured on IMAT [13] at ISIS (the data have been redrawn from [10]).

It is still clear from Figure 2(e) that there is a weak phonon signal from TAV6. For measurements of phonons from the sample within the cell, we want to minimize the cell phonon signal. Since the mass of cell material cannot be reduced below safety limits, and the cell material cannot be masked without masking the sample, one possibility is to make use of a mini-collimator [14]. This is a 1D radial collimator that the pressure cell fits inside, and which has been 3D printed and coated with Gd paint (Figure 5). The collimator vane angular spacing is optimized for the sample dimensions within the pressure cell. Unfortunately, the vanes will not only reduce the scattering from the pressure cell but will also shadow some of the scattering

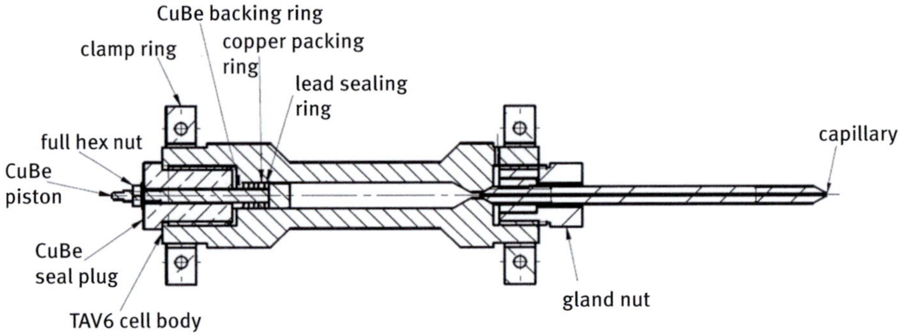

Figure 4: Schematic of 5.4 kbar TAV6 gas pressure cell developed at ISIS as a result of the "Low background materials for high pressure cells used in inelastic neutron scattering experiments" project funded by SINE2020.

Figure 5: Photographs of the 5.4 kbar ISIS TAV6 gas pressure cell and the same cell inserted into a mini-radial collimator.

from the sample. In cases where the sample is a powder, it is possible to continually oscillate the cell and collimator so as to average out the shadowing over all of reciprocal space. Such a trick is sadly not viable with single crystal samples and would require the collimator to be oscillating independently of the pressure cell.

3.2 Paris–Edinburgh presses

Paris–Edinburgh presses based on opposed anvils enable far higher pressures to be accessed and have been used in INS measurements up to 12 GPa to determine the phonon dispersion relationship in single crystals using TAS [9, 15]. However, they suffer from the sample volume being severely limited ($< 25\text{mm}^3$) and having considerably higher background due to scattering from the anvils and gaskets. Test experiments on MERLIN [16], a high flux thermal direct geometry TOF spectrometer at ISIS, using a Paris–Edinburgh press at room temperature have so far proved unsuccessful for detecting either phonons or magnetic excitations from powder samples.

3.3 Clamp cells

Occupying the intermediate pressure region are clamp cells. These are relatively simple, and hence, compact devices, but pressure can only be applied ex situ. They are typically made from similar materials to the gas cells, where the pressure limit is dictated by the yield strength of the material. For more than two decades, the main clamp cell used at neutron sources was the McWhan cell [7, 17], in which the sample capsule is in between two tungsten carbide pistons that are contained inside a biconical cylinder of either high-density sintered alumina or zirconia. The cylinder is supported radially at both ends by steel rings. In practice, clamped McWhan cells can operate up to 2.5 GPa, but the pressure has to be increased ex situ. Sadly the McWhan cells at both ILL and ISIS are currently unavailable.

An alternative large volume two-layered piston cylinder pressure cell design has been developed by Kamenev et al. [18]. The outer sleeve is made of fully hardened BERYLCO-25 alloy, and the inner cylinder is made of NiCrAl alloy. Pressures up to 1.8 GPa are generated ex situ using a hydraulic press with a removable tungsten carbide pusher, and then locked by means of a locking nut. The pressure over all operational pressures is recorded *in situ* using a manganin pressure gauge. It has been used to investigate the pressure-induced superconductivity in UGe_2 [18]. A large number of clamp cells have also been developed by Sadykov and coworkers, including an aluminum alloy cell strengthened with steel and optimized for use with the FOCUS TOF spectrometer at PSI that can operate up to 10 kbar [19].

The majority of clamp cells in INS measurements use fluorinert as the pressure transmitting medium [20]. Fluorinert liquids are hydrocarbon analogs where the hydrogen has been substituted by fluorine. This is vital because of the strong incoherent scattering cross section of hydrogen, which would substantially exceed the scattering signal from the sample of interest.

3.4 Uniaxial pressure devices

There is also strong interest from the scientific community to measure not only under hydrostatic pressure but to investigate the effects of uniaxial pressure. Due to the necessity of using larger samples in INS, it is sadly not possible to use the elegant designs with piezoelectric actuators commonly used by alternative experimental techniques [21]. Instead, large devices are typically constructed from an aluminum alloy in which single-crystal samples are mounted so that uniaxial pressure may be applied along a specific crystallographic direction with springs by means of turning screws. In the majority of cases, uniaxial pressure is being applied in this manner to detwin samples. This has proved particularly useful in INS investigations of an electronic nematic state in the high-temperature Fe-based superconductors [22]. However, the pressures involved are typically rather low, of order 0.1 kbar. Higher uniaxial pressures have been achieved using other setups, for example, two hardened steel pistons coated in Teflon in a thin-walled aluminum alloy sleeve, where the loads were generated by a Belleville spring washer system, created stresses up to 1.68 kbar [23, 24].

4 Scientific examples

4.1 Single-ion magnetic excitations

Perhaps the simplest INS experiments to perform under pressure, and hence those done earliest, are measurements of crystal field excitations. The crystalline electric field (CEF) or ligand field experienced by a magnetic ion is the potential generated by an electrostatic field from the neighboring ions and the effective field associated with the covalent bonding. The field is spatially varying and therefore acts on the orbital part of the electronic wave functions, such that the CEF will generally split the orbital degeneracy of the electrons. INS can then measure the single-ion magnetic spectrum consisting of sharp peaks centered on the transition energies between different levels, where the intensities depend on the thermal population of the initial state and the transition matrix for a magnetic dipole transition, and decrease as a function of $|Q|$ following the magnetic form factor. The energy splittings range between 1 and $1,000$ meV, with the majority of INS experiments involving rare earth elements where the CEF splits the $2J+1$-fold degenerate ground state multiplet. Fitting the observed CEF excitations allows a determination of the wave vector of the ground state, and thus places important restrictions on potential order parameters in the system.

Investigations of lanthanide CEF excitations under pressure are facilitated by the scattering cross section being proportional to $J(J+1)$ so that only relatively small sample quantities (< 1) g are required, and generally the excitations are non-dispersive (flat horizontal bands seen in $S(\mathbf{Q}, \omega)$) making them less difficult to distinguish from any background signal.

One of the first experiments to be performed in a McWhan cell was a measurement of the crystal field excitations in paramagnetic PrSb on a TAS at Brookhaven National Laboratory [25]. The cubic crystal field splits the ninefold degeneracy of the $^3H_4$ multiplet into a Γ_1 singlet, a Γ_4 triplet, a Γ_3 doublet, and a Γ_5 triplet. Interestingly, the application of pressure causes Γ_1 to Γ_4 excitation to shift down in energy, where the point-charge model would predict that with decreasing interatomic spacing a, the energy should increase as a^{-5}. This suggested the presence of a pressure-induced soft mode magnetic transition, which was indeed subsequently observed at 3 GPa [26].

Experiments are still being enthusiastically performed to measure the effects of pressure on crystal field excitations. One very recent example looks at CeCoSi, an intermetallic antiferromagnet with a very unusual $P-T$ phase diagram, due to the existence of a new magnetic ordered phase with a maximum T_C of 40 K at 1.5 GPa [27]. This is counter to the majority of Ce-based intermetallics in which hydrostatic pressure enhances the coupling between the conduction electrons and the localized Ce moments, increasing the strength of the Kondo effect, which leads to a reduced magnetic ordering temperature and shifting of the ground state closer toward a nonmagnetic heavy-fermion state. INS measurements of the CEF were performed using direct geometry TOF spectrometers IN4 at ILL for ambient pressure and LET at ISIS for high pressure in an NiCrAl clamp cell. The tetragonal CEF splits the $J = 5/2$ ground state multiplet into three doublets, and hence, one expects to see two transitions from the ground state at low temperature. At ambient pressure, upon subtraction of the phonon contribution measured in LaCoSi, two broad excitations are observed at 10.5 and 14.1 meV. Meanwhile, at 1.5 GPa, the intensity of the CEF excitations shifts or is suppressed significantly. Such a dramatic change motivates the need for single crystal neutron diffraction measurements under pressure to clarify the nature of the order parameter in the high-pressure phase.

In CEF systems with a finite exchange coupling, the elementary excitations are magnetic excitons. These are linear combinations of single-ion CEF excitations, which propagate through the crystal as a result of the magnetic coupling between the ions [28], and are therefore strictly collective rather than single-ion excitations, but the CEF is central to the physics. The excitons are dispersive, which can make them difficult to distinguish from gapped spin waves, but excitons can exist in the paramagnetic state. Praseodymium metal crystallizes in the *dhcp* structure, and the competition between crystal field, exchange, magnetoelastic, and hyperfine interactions results in nonmagnetic singlet ground states on both the hexagonal and cubic sites, but on the hexagonal sites the exchange coupling is 95% of the critical value required to induce long-range magnetic order under ambient pressure at $T = 4.2$ K.

The effect of uniaxial pressure on the excitons in single crystal praseodymium [23, 24] has been investigated on IN2 at ILL using the setup described in Section 3.4. INS measurements revealed singlet–doublet transitions and found that with increasing stress the lowest energy longitudinal exciton mode along $\Gamma - M$ exhibited an incipient soft-mode behavior at the wave vector associated with long-range sinusoidally modulated magnetic order (see Figure 6). The data presented in Figure 6(c) are in the long-range ordered phase, and the lowest energy mode is identified as a phason, an out-of-phase modulation of the lengths of the moments, with a gap. Random phase approximation calculations indicate that the phason remains at finite frequency, in contrast with a Goldstone broken-symmetry propagating mode but the magnitude of the gap is dependent on the temperature. At wave vectors close to $|Q|$, the phason mode is diffusive, and damped because of the fluctuations of the ordered moment.

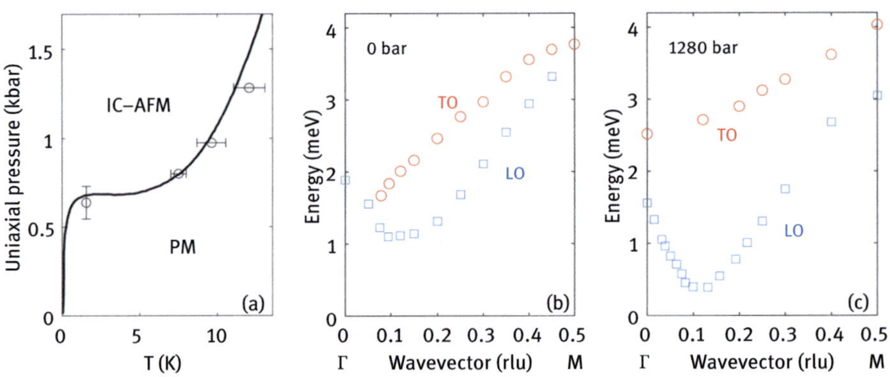

Figure 6: (a) Pressure–temperature phase diagram of praseodymium (redrawn from [29]). The solid line is a molecular field calculation for the boundary between the paramagnetic and sinusoidally modulated collinear antiferromagnetic phases, and the data points come from neutron scattering measurements. Dispersion relations for the lowest energy excitations in praseodymium at (b) zero pressure and (c) 1.28 kbar measured at $T = 5.5$ K (redrawn from [24]).

4.2 Spin waves

Going beyond single-ion magnetic excitations, the fundamental excitations of a long-range magnetic ordered state are spin waves, quantized oscillations of the magnetic moments around their average directions. Spin waves are the magnetic equivalent of the normal modes of atomic vibrations in a solid, while their energy quanta, magnons, are the magnetic quasiparticles analogous to phonons. INS is an extremely powerful tool for measuring spin waves, which allows the magnetic interactions in a system and the Hamiltonian to be determined. The majority of INS experiments looking at spin waves under pressure have been performed using TAS to exploit the higher neutron flux, but, increasingly, experiments are being performed on TOF spectrometers too.

One of the earliest measurements was performed on IN2 at ILL using an aluminum alloy clamped cell up to 4 kbar looking at the phase transition in $FeCl_2$ [30]. This revealed that the magnons simulate a two-dimensional ferromagnet with large anisotropy in both phases. The shape of the dispersion curves is the same at all pressures, and the in-plane exchange constants are independent of pressure. However, there is a sharp discontinuity in the energy gap at the phase transition, which shifts the magnons to higher energies.

One very important TAS experiment performed on TASP at SINQ under pressure was the identification of a pressure-induced quantum phase transition in the ordered phase of magnetic insulator $TlCuCl_3$ [31]. The single crystal was mounted in an aluminum/TiZr clamp cell with fluorinert as the pressure medium. At ambient pressure, quantum fluctuations prevent the formation of an ordered ground state down to $T = 0$ K in this dimer spin system, and INS reveals a gap between the singlet ground state and the triplet state. With increasing pressure, this energy gap softens until it closes at $P_c = 1.07$ kbar where antiferromagnetic ordering occurs. At 7.3 kbar, additional measurements were performed to study the spin excitations around the antiferromagnetic zone center, and compared against results for a field-induced ordered phase. This indicated a considerable increase in the spin wave stiffness in the pressure-induced ordered phase, proving that the compression of the structure leads to an increase in the effective exchange interactions. This change in the exchange interactions drives the quantum phase transition at P_c and the ordered phase can be described as the condensation of the three degenerate triplet quasiparticles into the singlet ground state.

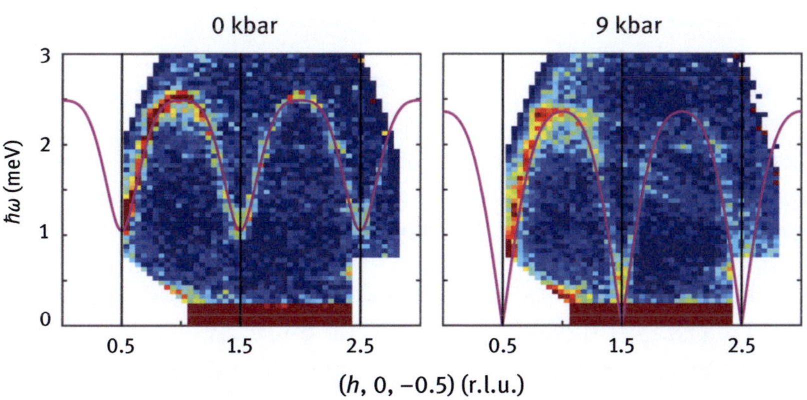

Figure 7: False-color slices through the four-dimensional inelastic neutron data sets measured in PHCC at $T = 1.5$ K, showing the dispersion of magnetic excitations in the $(h, 0, l)$ plane (reproduced from [32]). The gap in the spin waves has closed by $P = 9$ kbar.

Beautiful results have also been obtained using TOF spectrometers. Most of these experiments are performed using cold spectrometers ($E_i \lesssim 3$ meV) taking advantage of the low $Q-E$ window of $S(\mathbf{Q},\omega)$ where there is little background contamination from the cell (Figure 2), since it is below the lowest acoustic phonons. Measurements on CNCS at SNS of an $S=1/2$ frustrated gapped quantum magnet piperazinium hexachlorodicuprate in an NiCrAl pressure cell reveal that the spin singlet ground state separated by a gap to an $S=1$ triplet at ambient pressure is replaced by gapless spin waves indicative of a magnetically ordered Heisenberg spin system at 9 kbar [32] (see Figure 7), consistent with muon-spin relaxation results [33]. This contrasted with earlier results for PHCC obtained from measurements collected on SPINS, a TAS instrument, and DCS, a TOF instrument, at NIST. These showed the continued presence of a dispersive triplet excitation at higher pressures but it remained gapped up to at least 9 kbar [34]. The discrepancy between these two experiments is due to the assumption in [34] that the magnon dispersion was two-dimensional, allowing the data to be integrated over b^* to improve statistics. According to Perren et al. [32], there is also dispersion along b^* so that the integration will produce a saddle point that could be mistaken for an energy gap.

It should be noted that resonant inelastic X-ray scattering has made significant strides in recent years in the measurement of magnetic excitations, with the possibility of measuring micrometer-sized samples, and energy resolution improving year on year [35–37]. However, the requirement to measure on resonance means that, for the majority of elements, with the exception of 5*d* elements like iridium, it is difficult to perform experiments in complex sample environments [38, 39]. Therefore, it is likely that INS will remain the tool of choice for measuring magnetic excitations under pressure, particularly for quantum magnetism where the excitations are at low energy.

4.3 Lattice vibrations

The other category of collective excitations that are commonly studied using INS are lattice vibrations. Above absolute zero, the atoms in a crystal vibrate about their equilibrium positions. When the displacements are small compared with the interatomic spacings, the harmonic approximation is valid and the equilibrium position of the atoms is the minimum of a quadratic potential. The oscillations of atoms in a harmonic crystal are correlated due to the interatomic forces and are described by normal modes, states of motion in which all the atoms vibrate at the same frequency, with each individual normal mode behaving as an independent harmonic oscillator. The energy of each normal mode is therefore quantized in units of $\hbar\omega$ and is known as a phonon. INS measures the normal modes through the exchange of energy quanta leading to the creation and annihilation of phonons.

Understanding the phonons in a system can provide important information about heat transfer, the driving force behind phase transitions, the coupling mechanism in superconductors, and so on. There is therefore often strong interest in determining how pressure affects phonon modes. Measurements can be performed either on polycrystalline samples, from which the neutron-weighted phonon density of states are obtained, or on single crystals, giving the phonon dispersion curves. Powder averaging will obscure certain features of the phonons but it is typically far easier to perform polycrystalline measurements, with maximum sample mass filling the sample space, and data summed over more detectors on the TOF spectrometer to make counting times shorter.

Negative thermal expansion (NTE) materials contract on heating rather than expanding as most other materials do and have a range of applications where the coefficient of thermal expansion needs to be controlled, such as in the creation of composites for dental fillings which would not expand differently to the surrounding tooth enamel on heating causing toothache. The most studied NTE material is ZrW_2O_8, which contracts continuously between 0.3 and 1,050 K. It is proposed that the NTE behavior is associated with the transverse vibration of an oxygen atom in the W–O–Zr linkers, and so the phonons have been studied extensively with INS [40] to determine the Grüneisen parameter γ:

$$\gamma = V\left(\frac{\mathrm{d}P}{\mathrm{d}E}\right)_V \tag{2}$$

where V is volume, P is pressure, and E is energy, and the Grüneisen parameter γ_i of an individual vibrational lattice mode i of frequency ω_i is given by

$$\gamma_i = -\frac{V}{\omega_i}\frac{\partial\omega_i}{\partial V}. \tag{3}$$

Measurements were performed on the TOF spectrometer IN6 at ILL using a gas pressure cell with argon and showed a clear shift of the low-energy phonon density of states down in energy with increasing pressure up to 1.7 kbar, indicating large negative Grüneisen parameters. The temperature-dependent volume thermal expansion coefficient derived from the inelastic data is in good agreement with that obtained directly from diffraction data, indicating that the anomalous phonon behavior below 8 meV is able to account for the observed NTE behavior in ZrW_2O_8 over the whole temperature range.

Another highly relevant class of functional materials are barocalorics, which experience cooling effects at pressure-induced phase transitions, and therefore have great potential for energy-efficient refrigeration. Giant barocaloric coefficients may be attributable to highly anharmonic lattice dynamics, making a study of the phonons in barocalorics under pressure highly desirable. Such experiments have been reported using the TOF spectrometers Amateras at J-PARC and Pelican at

ACNS on NPG plastic crystals using a mesoalite clamped cell [41], and revealed that the phonons become sharper at 2.86 kbar due to reduced scattering by the orientational disorder.

For more detailed information on phonon dispersion under pressure, experiments must be performed using single crystals. These experiments are considerably more time-consuming and are further complicated by the requirement for the pressure to be sufficiently hydrostatic to avoid breaking the crystal. While the phonon–roton excitation spectrum in superfluid ^{4}He has been studied very extensively at low energy transfers [43], fortunately for those using helium in pressure cells, Gen Shirane and others have also made a comprehensive study of the lattice dynamics of solid helium, in all three different phases "hcp, bcc, and fcc" using INS on TAS [42, [44–46]. The pressure dependence was obtained by performing measurements on single crystals synthesized *in situ* on TAS at different pressures, resulting in different densities [47]. In both the hcp and fcc phases, increasing pressure was found to not only harden the phonons but also to reduce the anharmonic effects, since the He atom goes toward the bottom of the potential well as the crystal is compressed. In principle, Grüneisen parameters $\gamma_i(\mathbf{q})$ may be different for different phonon modes and may also depend both on temperature and volume. In hcp ^{4}He, the Grüneisen parameter appeared to be independent of wave vector, such that all the observed frequencies could be scaled uniformly within the experimental error; however, there is a pronounced dependence on volume (see Figure 8). By contrast, in fcc ^{4}He, the Grüneisen parameters

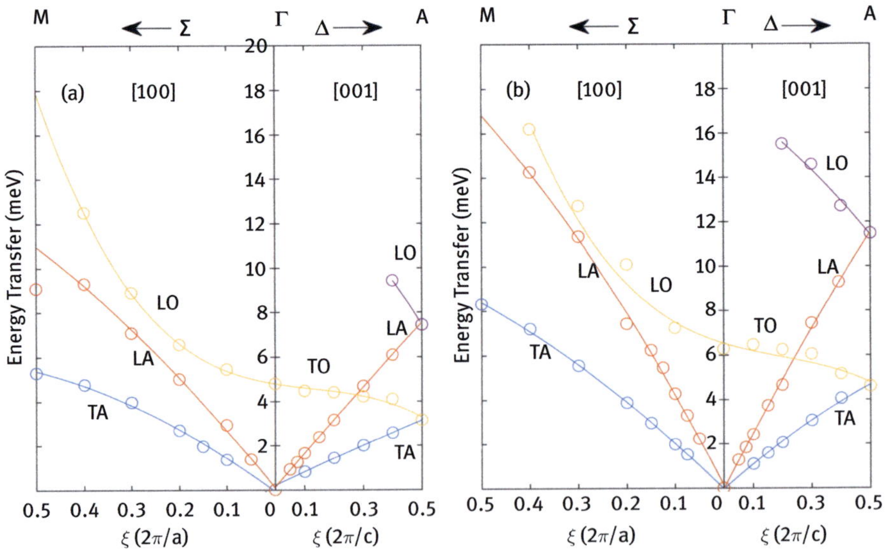

Figure 8: Phonon dispersion relations measured along [100] and [001] for hcp ^{4}He at (a) 11.61 cm^3/mol and 8.3 K, and (b) 9.41 cm^3/mol and 10 K, corresponding to P~1.4 and 3.7 kbar, respectively. Solid lines are a guide to the eye (redrawn from [42]).

appear to vary strongly as a function of wave vector for a given mode. Additionally, the Grüneisen parameters are different between longitudinal and transverse modes, albeit on a lower level than in solid Ne [47].

When measuring at low temperatures and high pressures, it may be necessary to be aware of the phonon signal arising from the solidified pressure medium, to avoid confusing this with a pressure-induced change in the sample. Figure 9 shows unpublished powder data collected on Merlin, a TOF spectrometer at ISIS [16], at $T = 10$ K and $P = 2.6$ T from the ISIS TAV6 cell described in Section 3.1, where helium was used as the pressure-transmitting medium, and demonstrates a strong dispersive excitation, which can be compared with the TAS-measured dispersion from hcp ^{4}He, scaled according to the Grüneisen parameter. This highlights the importance of doing background measurements in the pressure cell with the pressure-transmitting medium minus the sample under the same temperature and pressure conditions when the medium has solidified to avoid an incorrect identification of pressure-induced changes in the sample.

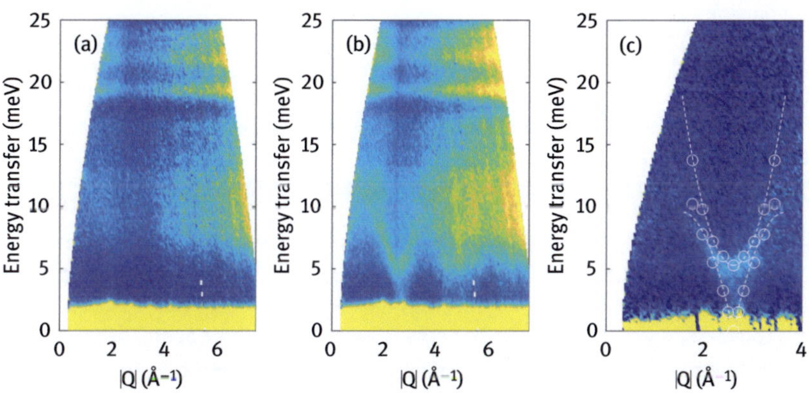

Figure 9: Measurements (unpublished) of a powder sample in the ISIS TAV6 cell on Merlin [16] at 10 K (a) at atmospheric pressure with no ^{4}He, and (b) at $P = 2.6$ kbar using ^{4}He. (c) A zoomed-in region of the results of the subtraction of (a) from (b), with a Grüneisen parameter scaled dispersion curve from [42] overlaid.

5 Future directions

With the SINE2020 project driving the development of new pressure cells specially developed for INS, and new indirect geometry TOF spectrometer designs [48, 49] to boost the incident neutron flux by several orders of magnitude, it is not unreasonable to hope that we are about to see a large growth of INS experiments performed at high pressure. In addition, with the European Spallation Source to start operating in the second half of this decade, new state-of-the-art direct geometry TOF spectrometers will be available with higher neutron fluxes than ever seen before [50], which will be a clear boon for these neutron-hungry measurements. However, as described in this chapter, TAS has proved an ideal tool for performing INS measurements under pressure, and so it is vital that we do not lose access to these instruments as the number of reactor sources continues to decline.

References

[1] Gardner JS, Ehlers G, Faraone A, Garcia-Sakai V. High-resolution neutron spectroscopy using backscattering and neutron spin-echo spectrometers in soft and hard condensed matter. Nature Reviews Physics 2020;2:103.

[2] Andreani C, Krzystyniak M, Romanelli G, Senesi R, Fernandez-Alonso F. Electron-volt neutron spectroscopy: Beyond fundamental systems. Advances in Physics 2017;66:1–73.

[3] Shirane G, Shapiro SM, Tranquada JM. Neutron Scattering with a Triple-Axis Spectrometer. Cambridge, Cambridge University Press, 2002.

[4] Brockhouse BN. Slow neutron spectroscopy and the grand atlas of the physical world. Reviews of Modern Physics 1995;67:735.

[5] Klausz M, Kanaki K, Zagyvai P, Hall-Wilton RJ. Performance evaluation of the boron coated straws detector with Geant4. Nuclear Instruments and Methods in Physics Research Section A 2019;943:162463.

[6] Klotz S. Techniques in High Pressure Neutron Scattering. Boca Raton, FL, CRC Press, Taylor & Francis Group, 2013.

[7] Bailey IF. A review of sample environments in neutron scattering. Zeitschrift Für Kristallographie 2003;218:84–95.

[8] McWhan DB. Neutron scattering at high pressure. Review Physical Applied 1984;19:715–18.

[9] Klotz S, Braden M, Besson JM. Inelastic neutron scattering to very high pressures. Hyperfine Interactions 2000;128:245–54.

[10] Kibble MG, Laliena V, Goodway CM, Lelièvre-Berna E, Kamenev KV, Klotz S, Kirichek O. Low-background materials for high pressure cells used in inelastic neutron scattering experiments. Journal of Nutrition Research 2019;21:105–16.

[11] Kirichek O, Done R, Goodway CM, Kibble MG, Evans B, Bowden ZA. Development of high pressure gas cells at ISIS. Journal of Physics: Conference 2012;340:012008.

[12] Bewley RI, Taylor JW, Bennington SM. LET, a cold neutron multi-disk chopper spectrometer at ISIS. Nuclear Instruments and Methods in Physics Research Section A 2011;637:128.

[13] Kockelmann W, Burca G, Kelleher JF, Kabra S, Zhangand S-Y, Rhodes NJ, Schooneveld EM, Sykora J, Pooley DE, Nightingale JB, Aliotta F, Ponterio RC, Salvato G, Tresoldi D, Vasi C, McPhate JB, Tremsinc AS. Status of the neutron imaging and diffraction instrument IMAT. Proceedings Physical 2015;69:71–78.

[14] Ridley CJ, Manuel P, Khalyavin D, Kirichek O, Kamenev KV. A novel compact three-dimensional laser-sintered collimator for neutron scattering. Review of Scientific Instruments 2015;86:095114.

[15] Klotz S. Phonon dispersion curves by inelastic neutron scattering to 12 GPa. Zeitschrift Für Kristallographie 2001;216:420–29.

[16] Bewley RI, Guidi T, Bennington S. MERLIN: A high count rate chopper spectrometer at ISIS. Notiziario Neutroni E Luce Di Sincrotrone 2009;14:22.

[17] McWhan DB, Bloch D, Parisot G. Apparatus for neutron diffraction at high pressure. Review of Scientific Instruments 1974;45:643.

[18] Wang W, Sokolov DA, Huxley AD. and K V Kamenev. Large volume high-pressure cell for inelastic neutron scattering. Review of Scientific Instruments 2011;82:073903.

[19] Sadykov RA, Strassle T, Podlesnyak A, Keller L, Fak B, Mesot J. High-pressure cells for study of condensed matter by diffraction and inelastic neutron scattering at low temperatures and in strong magnetic fields. Journal of Physics: Conference 2017;941:012082.

[20] Sidorov VA, Sadykov RA. Hydrostatic limits of fluorinert liquids used for neutron and transport studies at high pressure. Journal of Physics Condensed Matter 2005;17:S3005.

[21] Hicks CW, Barber ME, Edkins SD, Brodsky DO, Mackenzie AP. Piezoelectric-based apparatus for strain tuning. Review of Scientific Instruments 2014;85:065003.

[22] Song Y, Carr SV, Lu X, Zhang C, Sims ZC, Luttrell NF, Chi S, Zhao Y, Lynn JW, Dai P. Uniaxial pressure effect on structural and magnetic phase transitions in NaFeAs and its comparison with as-grown and annealed $BaFe_2As_2$. Physical Review B 2013;87:184511.

[23] McEwen KA, Stirling WG, Vettier C. Magnetic ordering in single crystal praseodymium induced by uniaxial stress. Physical Review Letters 1978;41:343.

[24] McEwen KA, Stirling WG, Vettier C. Magnetic structure and excitations in single crystal praseodymium. Journal of Magnetism and Magnetic Materials 1983;31–34:599–600.

[25] Vettier C, McWhan DB, Blount EI, Shirane G. Pressure dependence of magnetic excitations in PrSb. Physical Review Letters 1977;39:1028.

[26] McWhan DB, Vettier C, Youngblood R, Shirane G. Neutron scattering study of pressure-induced antiferromagnetism in PrSb. Physical Review B 1979;20:4612.

[27] Nikitin SE, Franco DG, Kwon J, Bewley R, Podlesnyak A, Hoser A, Koza MM, Geibel C, Stockert O. Gradual pressure-induced enhancement of magnon excitations in CeCoSi. Physical Review B 2020;101:214426.

[28] Jensen J. Physics of rare-earths: The magnetism of praseodymium. Journal de Physique Colloid 1979;40:C5–1.

[29] Jensen J, McEwen KA, Stirling WG. Magnetic excitations in the longitudinally polarized antiferromagnetic phase of praseodymium. Physical Review B 1987;35:3327.

[30] Vettier C, Yelon WB. Magnetic properties of $FeCl_2$ at high pressure. Physical Review B 1975;11:4700.

[31] Rüegg C, Furrer A, Sheptyakov D, Strässle T, Krämer KW, Güdel H-U, Mélési L. Pressure-induced quantum phase transition in the spin-liquid $TlCuCl_3$. Physical Review Letters 2004;93:257201.

[32] Perren G, Möller JS, Hüvonen D, Podlesnyak AA, Zheludev A. Spin dynamic in pressure-induced magnetically ordered phases in $(C_4H_12N-2)Cu_2Cl_6$. Physical Review B 2015;92:054413.

[33] Thede M, Mannig A, Månsson M, Hüvonen D, Khasanov R, Morenzoni E, Zheludev A. Pressure-induced quantum critical and multicritical points in a frustrated spin liquid. Physical Review Letters 2014;112:087204.

[34] Hong Y, Stock C, Cabrera I, Broholm C, Qiu Y, Leao JB, Poulton SJ, Copley JRD. Neutron scattering study of a quasi-two-dimensional spin-1/2 dimer system: Piperazinium hexachlorodicuprate under hydrostatic pressure. Physical Review B 2010;82:184424.

[35] Ament LJP, van Veenendaal M, Devereaux TP, Hill JP, van den Brink J. Resonant inelastic x-ray scattering studies of elementary excitations. Reviews of Modern Physics 2011;83:705–67.

[36] Dvorak J, Jarrige I, Bisogni V, Coburn S, Leonhardt W. Towards 10 meV resolution: The design of an ultrahigh resolution soft X-ray RIXS spectrometer. Review of Scientific Instruments 2016;87:115109.

[37] Kim J, Casa D, Said A, Krakora R, Kim BJ, Kasman E, Huang X, Gog T. Quartz-based flat-crystal resonant inelastic x-ray scattering spectrometer with sub-10 meV energy resolution. Scientific Reports 2018;8:1958.

[38] Rossi M, Henriquet C, Jacobs J, Donnerer C, Boseggia S, Al-Zein A, Fumagalli R, Yao Y, Vale JG, Hunter EC, Perry RS, Kantor I, Garbarino G, Crichton W, Monaco G, McMorrow DF, Krisch M, Sala MM. Resonant inelastic X-ray scattering of magnetic excitations under pressure. The Journal of Synchrotron Radiation 2019;26:1725–32.

[39] Kim J. Advances in high-resolution RIXS for the study of excitation spectra under high pressure. High Pressure Research 2016;36:391–403.

[40] Mittal R, Chaplot SL, Schober H, Mary TA. Origin of negative thermal expansion in cubic ZrW_2O_8 revealed by high pressure inelastic neutron scattering. Physical Review Letters 2001;86:4692.

[41] Li B, Kawakita Y, Ohira-Kawamura S, Sugahara T, Wang H, Wang J, Chen Y, Kawaguchi SI, Kawaguchi S, Ohara K, Li K, Yu D, Mole R, Hattori T, Kikuchi T, Yano S, Zhang Z, Zhang Z, Ren W, Lin S, Sakata O, Nakajima K, Zhang Z. Colossal barocaloric effects in plastic crystals. Nature 2019;567:506–10.

[42] Eckert J, Thomlinson W, Shirane G. Lattice dynamics of hcp He at high pressure. Physical Review B 1978;18:3074.

[43] Beauvois K, Dawidowski J, Fåk B, Godfrin H, Krotscheck E, Ollivier J, Sultan A. Microscopic dynamics of superfluid he: A comprehensive study by inelastic neutron scattering. Physical Review B 2018;97: 184520.

[44] Lipschultz FP, Minkiewicz VJ, Kitchens TA, Shirane G, Nathans R. Phonon dispersion measurements on a hcp He single crystal at 27 atm. Physical Review Letters 1967;19:1307.

[45] Osgood EB, Minkiewicz VJ, Kitchens TA, Shirane G. Inelastic neutron scattering from bcc He. Physical Review A 1972;5:1537.

[46] Eckert J, Thomlinson W, Shirane G. Lattice dynamics of fcc helium at high pressure. Physical Review B 1977;16:1057.

[47] Eckert J, Daniels WB, Axe JD. Phonon dispersion and mode Grüneisen parameters in neon at high density. Physical Review B 1976;14:3649.

[48] Bewley R. FARO: A new type of neutron spectrometer with flux and resolution optimized. Review of Scientific Instruments 2019;90:075106.

[49] Bewley RI. MUSHROOM: A High Count Rate Cold Indirect Geometry Spectrometer. Submitted for publication, 2020.

[50] Andersen KH, et al .The instrument suite of the European spallation source. Nuclear Instruments and Methods in Physics Research Section A 2020;957:163402.

John S. Tse, E. Ercan Alp

Application of inelastic scattering of X-rays at high pressure

1 Introduction

Spectroscopy is an experimental technique to examine the intrinsic properties of materials from analyzing light absorption and emission [1, 2]. A pedagogical example is Newton's famous experiment on the refraction of sunlight through a prism into a different color spectrum from red to violet. Now we understand the physical process involved. Sunlight is a combination of radiations of different wavelengths. When it penetrates a dielectric medium (prism), the light of different wavelengths will interact differently and become dispersed. The dispersion is dependent on the refractive index of the medium. This, in turn, is governed by the electronic structure of the material. Thus, from studying the dispersion of emerged light, one can understand the medium's nature. Nowadays, a broad spectrum of radiation(light) with different energy (color) can be generated by various techniques.

Spectroscopy using electromagnetic waves has a wide range of applications in the characterization of material properties. The energy (or frequency) of the radiation determines the specific type of interaction with matter and the application. Figure 1 shows, schematically, the electromagnetic spectrum and the corresponding interaction processes. For practical consideration, the energy of the electromagnetic waves must be commensurate with the properties to be examined. For example, molecular rotations require minimal energy and can be detected by microwave. On increasing energy, this is followed by atomic and molecular vibrations, bound state electronic excitations, photoionization of valence shell electrons and then core electrons into nuclear excitation, and Compton scattering (scattering of photon and electron) [3]. There are drawbacks of using electromagnetic radiation of unique wavelengths to probe material properties. An example is *in situ* study of the chemical reactions in a battery where the sample is confined, and long wavelength radiation cannot penetrate to study low-energy vibrations. Instead of directly investigating the process, an alternative is to perform energy loss spectroscopy [4–8].

John S. Tse, Department of Physics, University of Saskatchewan, Saskatoon, Saskatchewan, Canada S7N 5E2
E. Ercan Alp, Advanced Photon Source, Argonne National Laboratory, Lemont, IL 60439, USA

https://doi.org/10.1515/9783110668612-006

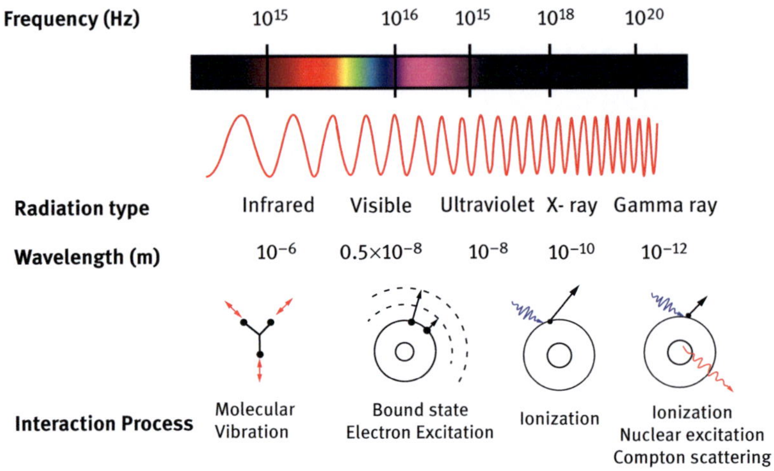

Figure 1: A picture of the electromagnetic spectrum accessible with synchrotron radiation and the relevant applications at different energy ranges.

2 Basic principle

The elementary principle of energy loss spectroscopy is straightforward [5, 8, 9]. When a particle scatters a photon (electromagnetic radiation), regardless it is an atom, an electron, or a nucleus, there may be energy exchange between the two systems. If the difference between the initial and final photon energy is zero, there is no energy exchange and the collision is elastic. On the other hand, if the difference is nonzero, the photons gain or loss energy from the particle and the process is inelastic. The energy transfer is a direct probe of the dynamic processes (i.e., nucleus, electron, or atomic) activated by the collision. Inelastic scattering with energetic photons (large momentum) prompts various interactions, such as vibrations, valence, core electron excitations, and nuclear excitations. Analysis of angular average and angular dependence of the scattered photon offers information on different aspects of the dynamic process. Angular average measurement gives information on the collective excitations, while angular-dependent measurement provides information on the single-particle dynamics. Figure 2 depicts the common inelastic scattering (inelastic x-ray scattering (IXS)) processes activated by an energetic x-ray. For example, vibrational excitons are in millielectron volt (meV), but core-level electronic excitations are in kiloelectron volt (keV). The excitation processes span an energy range of six orders of magnitude. The vast range of energy transfer poses an experimental challenge [6]. For different scattering processes, specialized monochromators and analysis optics for both incident and scattered radiation are required to assure the quality and resolution of the spectra. Since scattering is a second-order process, the intensity of the scattered light is often very weak. However, due to the high penetration power of the incidence hard x-ray, IXS may be the only viable spectroscopic technique to characterize the vibrational and electronic properties of materials that must be contained in an enclosed environment. Examples are *in situ* study of the chemical processes of a working battery and the structural and electronic changes of matter under extreme conditions. The latter topic is the focus of this chapter.

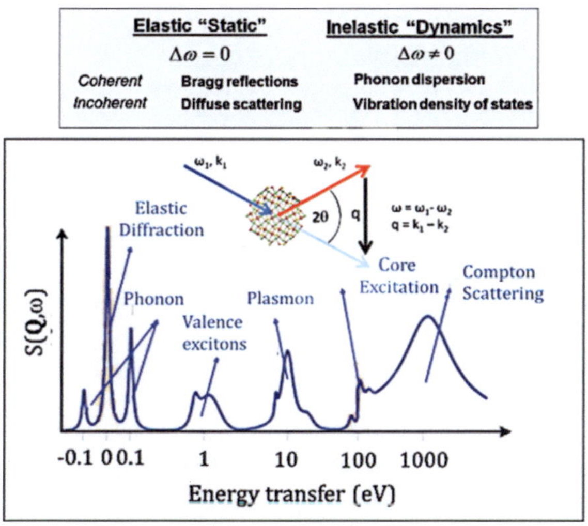

Figure 2: A schematic representation of inelastic scattering processes and the related spectroscopic techniques.

3 Application

3.1 Vibrational dynamics

The strength of the chemical bonds primarily determines the structure and properties of a material [10]. A simplistic picture of a chemical bond between a pair of atoms is illustrated in Figure 3. It is known that the stronger the binding, the higher the vibrational frequency and larger energy separation between occupied and unoccupied orbitals [11]. Therefore, studies on the vibrational and electronic spectra provide complementary information on the nature of the chemical bond. With this knowledge, other properties of the material can be deduced and elucidated. In the following sections, the application of IXS to the investigation of materials under extreme conditions, that is, high pressure and high temperature, is illustrated through selected examples.

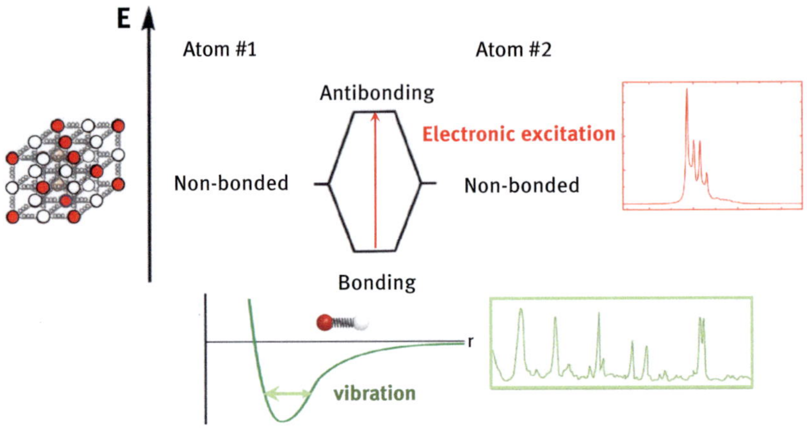

Figure 3: Simplified bonding scheme of a covalent solid showing the bonding and antibonding states and representative vibrational and electronic spectra.

3.1.1 Molecular dynamics

There are few examples on the straightforward application of energy loss spectroscopy using x-ray to study the vibrations in materials. As indicated in Figure 1, a typical vibrational spectrum spans from 10^{12} to 10^{15} Hz (1–300 meV). To resolve fine details on the vibrational modes, an energy resolution on both the incident and scattered x-rays of the order meV is required [12]. Although it is feasible to achieve this resolution with a synchrotron x-ray, the instrumentation is rather cumbersome and expensive. Nevertheless, to illustrate the principle of IXS to study vibration, the N–N stretch in liquid N_2 at 76 K and 1 bar has been recorded with a monochromatized x-ray of 13,840 eV [13]. The photons scattered were measured in a 2.6 m long spectrometer, and the energy was analyzed with a spherical silicon crystal. The spectrum collected at $q = 2.70$ Å^{-1} is shown in Figure 4. The full width at half maximum is about 5 meV (40 cm^{-1}). It should be noted that the experiment was not primarily intended to record the vibrational spectrum of N_2 but to investigate the dispersion (q-dependence) of the absolute IXS cross section for the vibrational excitation. Alternative techniques, such as Raman and infrared (IR) spectroscopy, are readily available and provide even higher resolution, perhaps are more suitable. Conventional IR and Raman techniques are also applicable at high pressure if the sample is contained in a donut-shaped gasket and compressed axially by two opposing diamonds (diamond anvil cell, DAC), which is transparent to the incident excitation radiation. However, there are a few shortcomings of these techniques that will be discussed below.

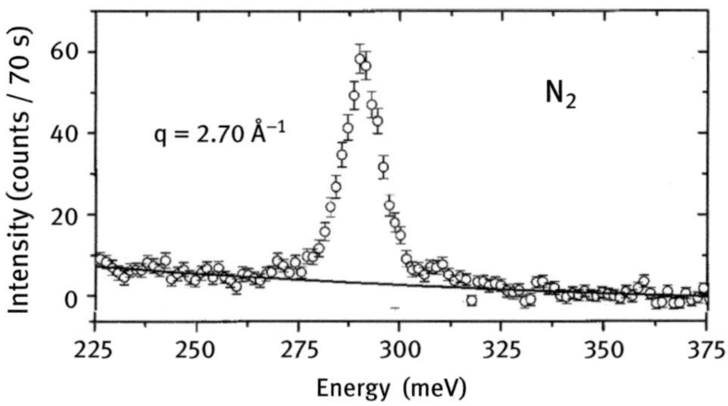

Figure 4: Inelastic scattering spectrum of N_2 showing the N–N stretch mode (taken from ref. [9]).

3.1.2 Nuclear-resonant IXS (NRIXS) and site-specific phonon density of states

Using time discrimination, IXS can provide high-resolution atom-specific vibrational spectrum for selected atoms. The method is called nuclear-resonant IXS (NRIXS). The nuclear resonant energy is selected from the white synchrotron radiation with help of combination of multi-crystal monochromator with band pass of 1 meV, and fully tunable phonon annihilation and creation spectrum. In case of $_{57}$Fe, this would be 14.4125 keV ± 150 meV. The incident x-ray beam may excite the nuclear transition from ground state with a nuclear spin state ½ to the first excited state with spin state 3/2 . The subsequent decay of 14.4 keV is accompanied by internal conversion of s-electrons to vacuum level, causing emission of 6.4 and 7.1 keV Kα, and Kβ delayed fluorescence. This signals the phonon creation or annihilation probability [14, 15] The energy required is usually in the tens of keV. Many electronic processes occur concurrently with nuclear excitation. However, the timescales for the relaxation from excited electronic and nuclear states are very different. The lifetimes for the cascade of electronic de-excitations are much shorter than the nuclear de-excitation. The timing mismatch is exploited to record the emitted radiation (nuclear fluorescence) from nuclear scattering utilizing the time structure of circulating electrons in a synchrotron. Synchrotron is a pulsed radiation source usually separated by nanosecond and duration of picosecond per pulse. Despite the low intensity, the nuclear emission spectrum can be measured from very sensitive solid-state avalanche photodiodes after a time delay to eliminate all the background contributions from the electronic relaxation (Figure 5). When the incident energy is lower than the nuclear resonant energy, there is a finite probability that the photon will annihilate a phonon, and with the extra energy it gained from this exchange, will reach the resonant energy. The signal of such a transition is the observation of "delayed" fluorescence signal at the detector. Similarly, when the incident photon energy is above the nuclear transition energy, there is a finite probability that a phonon will be created, reducing the energy of the photon to reach the nuclear resonance. The bottom rightmost spectrum in Figure 5 describes this sequence of events.

Thus, the lattice vibrations convoluted the absorption and emission spectrum in which the vibrational density of state (VDOS) can be extracted. The NRIXS has several advantages over conventional vibrational spectroscopy. First, it is atom specific. The Mossbauer effect is observed in nuclei with nuclear spin $I > 1/2$ that exhibits a substantial recoil-free probability transition to the ground state [16]. Second, unlike Raman or IR spectroscopy, there is no symmetry selection rule, and all atomic displacements along the momentum vector are allowed. The NRIXS spectrum is a projection of the specific atomic vibration from the total VDOSs. Moreover, since monochromatized high-energy x-ray can be focused to a very small spot (μm), NRIXS is very efficient to examine small samples in restrictive surroundings. NRIXS has been used successfully to study the atomic dynamics of a variety of materials. In particular, this technique allows the study of very low frequency vibrations that are not readily amenable to IR or Raman spectroscopy.

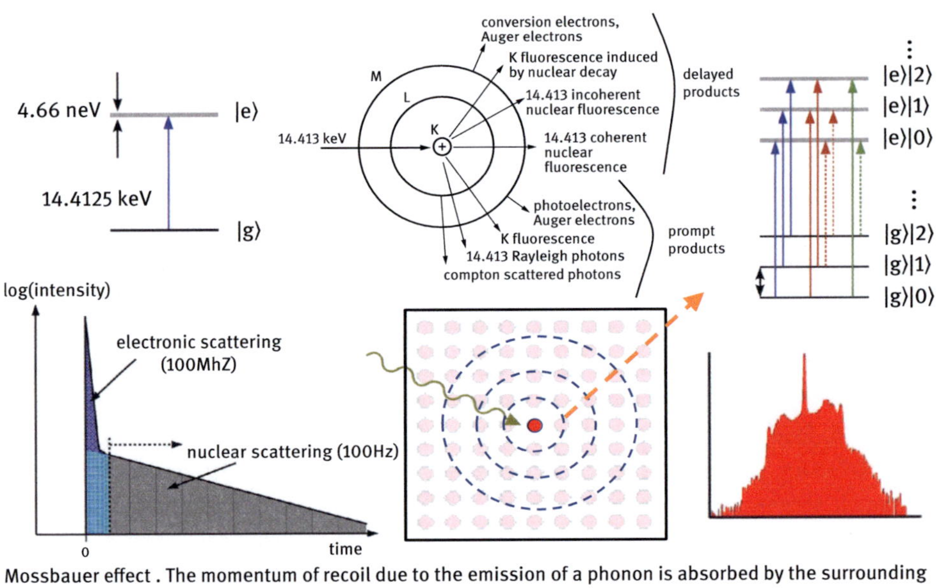

Figure 5: A picture showing the time-delayed electronic processes after the irradiation of ^{57}Fe nuclei with 14.413 keV x-ray. The nuclei transitions and the subsequent emission provide information on the vibrational density of states of the Fe atoms in the solid.

^{57}Fe is the most commonly used Mossbauer nucleus [16, 17]. The nuclear excitation energy is 14.4 keV with a 5 neV resonance width. According to the uncertainty principle, this corresponds to an excited state lifetime of 141 ns [18, 19]. When a long pulse of synchrotron radiation is incident on an Fe sample, the subsequent fluorescence from the nucleus can be discriminated after a time delay of the order of 141 ns. In the following, the advantage of this technique is highlighted in a molecular system, an Fe-containing heme protein model.

Fe is the active site in heme proteins (such as myoglobin) responsible for reversible transport of O_2, N_2, and CO. Myoglobin is a five-coordinated iron (Fe) porphyrin complex (heme) in which the axial ligand is an imidazole ring from a histidine residue of the protein. Simple molecules, such as O_2, N_2 and CO, can bind to the Fe atom as the sixth ligand in an axial position. As a result, the Fe is pulled away from the ring, triggering a protein's conformational change. The dynamics of the heme prosthetic group and the surrounding protein determine the physiological activities of hemoglobin. Insight into the Fe out-of-plane motion (doming mode) is of utmost importance. Fe is a heavy atom and only weakly binds to the nitrogen of the porphyrin ring. The doming vibration is expected to be at very low frequency and difficult to identify unequivocally. To discriminate the doming mode from the noisy background, an attempt was made to monitor the shift of an IR-active vibration by applying a small hydrostatic pressure. The experimental results on a model heme protein CO-bonded COFe(OEP)Py (carboxy-octaethyl-iron porphyrin pyridine) are reproduced in Figure 6a (HEME C-12, ^{12}C natural abundance) [20]. A weak feature exhibiting a reversible frequency shift upon compression/decompression was found at 54 cm^{-1} (6.7 meV) and was attributed to the doming mode. Evidently, the IR absorption is not well defined. The IR spectra are compared to the powder and single-crystal NRIXS spectra of the Fe(OEP)CO model heme in Figure 6b [21]. Once again, observation of frequency vibrational modes (below ~200 cm^{-1}) is challenging, but NRIXS spectra are clearly more resolved and the vibrational features are less ambiguous. In the latter case, orientation-dependent measurements on a single-crystal sample help differentiate the in-plane and out-of-plane vibrational modes. Although obscured by acoustic vibrations (see below), two features observed at 64 and 127 cm^{-1} are likely candidates for the doming modes. The

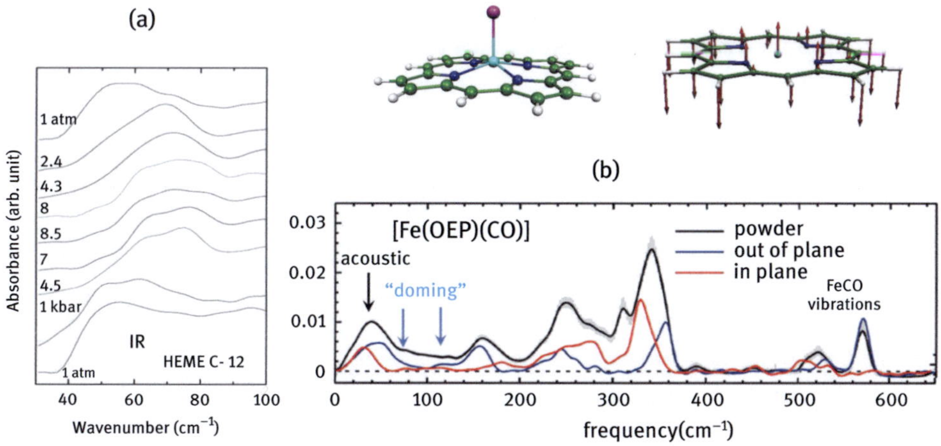

Figure 6: (a) Far infrared spectrum of compressed CO(Fe(OEP)py showing the variation of the characteristic doming vibration with pressure. (b) The IXS spectrum of CO(Fe(OEP)py clearly shows the distinctive doming mode.

internal molecular vibrations in the low-frequency region (<200 cm^{-1}) can be identified from comparison with theoretically predicted frequencies. The information leads to a better understanding of the chemical nature and strength of the chemical bond with the attached ligands and helps further understand the binding mechanism. NRIXS is an atom-specific technique. In the example illustrated here, every vibrational mode contributing to the experimental signal is directly proportional to the mean-squared displacement of the Fe atom participating in the motion [22].

Before embarking on discussing other applications on solids, we shall give a brief review of the fundamental physics of lattice dynamics. It is known that we have to supply heat (energy) in order to raise the temperature of a material. Since the total energy must be conserved, the phenomenon indicates that the external energy is converted to mechanical work. Knowing the mechanical work will give information on the interactions between the atoms forming the material. Assume that the bond holding two atoms together can be represented by a spring connecting two atoms (balls) with the spring constant governing the bond strength (Figure 3).

3.1.3 Lattice dynamics

A brief review of the elementary lattice dynamics of a crystalline solid is appropriate here. A simple model for a solid is a collection of interconnected balls and springs [23–25]. For a monatomic solid, the model can be simplified even further. One considers motions of a linear chain of identical particles (Bravais lattice can be an atom or a molecular unit) connected by springs with equal spring constant (Figure 3) in one dimension. A perturbation will set up oscillations of the particles. The solution to the Newtonian equation of motion is a sine wave-like dispersion relation for the allowed oscillator frequency (ω) with the magnitude of the wave vector (q), the direction the wave propagates (Figure 7). The exact expression and detail derivation can be found in many standard textbooks on condensed matter physics. At $q = 0$, all the atoms move in unison along the same direction. The collective motion is the translation of the unit cell, and no energy is required ($\omega = 0$). The wave vector q can only take discrete values of integral n ($q = n(2\pi/a)$, where a is the primitive cell spacing). The corresponding vibration frequencies $\omega(q)$, shown as horizontal bars, are also discrete. The available vibrational states within an energy interval are the density of states (DOS) at ω, $g(\omega)$. As shown in Figure 7, the variation of the vibration frequency with the wave vector is represented by a single sinusoidal curve and is called the acoustic branch. Furthermore, changes in the vibrational (phonon) frequency (ω) become smaller and the number of vibrational states gets denser as q approaches π/a, the Brillouin zone boundary. The DOS terminated abruptly at $q = \pi/a$, giving rise to a "peak" in $g(\omega)$. This feature is known as the van Hove singularity.

The phonon dispersion of a three-dimensional crystal behaves similarly with several additional features (Figure 8). Instead of propagating in one direction, the lattice wave has a longitudinal (along the propagation direction) and two degenerated transverse (perpendicular to the propagation) components. Furthermore, suppose that more than one type of particle is present. In that case, additional vibrations (optics mode) appear at higher frequencies and are separated from the acoustic branches by an exclusive zone (gap) at the Brillouin zone center ($q = 0$) where vibrations are forbidden (Figure 8).

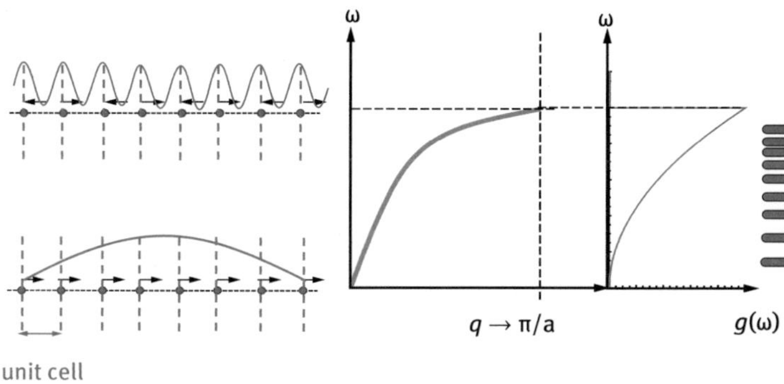

Figure 7: Theoretical results on the calculation of a model one-dimensional monatomic chain with lattice spacing a showing the frequency dispersion (ω vs k), the density of vibration states ($g(\omega)$), and the phonon wave functions at the Brillouin zone center $k = 0$ and at the zone boundary $k = \pi/a$.

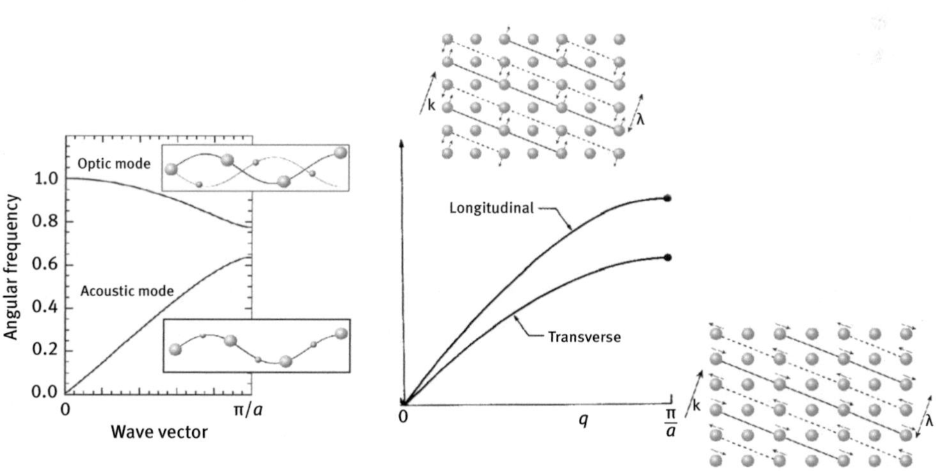

Figure 8: The longitudinal and traverse vibrations of a selected crystal plane.

Since the lattice vibrations are quantized, it is possible to derive a relationship between the DOS and the vibrations at low frequency. The lattice vibrations described above show that the atomic oscillators are coupled, forming standing waves (phonons) of different frequencies confined within the unit cell. One can count the number of standing waves per unit volume following the same procedure as the quantum mechanical particle in a box. The result shows the DOS $g(\omega)$ is a quadratic function of the frequency (ω) (Figure 9), a relationship first proposed by Debye in the theory of specific heat of solids:

$$g(\omega) = \frac{V\omega^2}{2\pi}\frac{1}{v_D} \tag{1}$$

where V is the volume of the unit cell and v_D is the average of the longitudinal (v_L) and transverse (v_T) sound velocity (Debye sound velocity):

$$\frac{3}{v_D} = \frac{1}{v_L} + \frac{2}{v_T} \tag{2}$$

Thus, the average sound velocity can be derived from a parabolic fit to the low-frequency (long wavelength limit) VDOS obtained from x-ray scattering experiment (Figure 9).

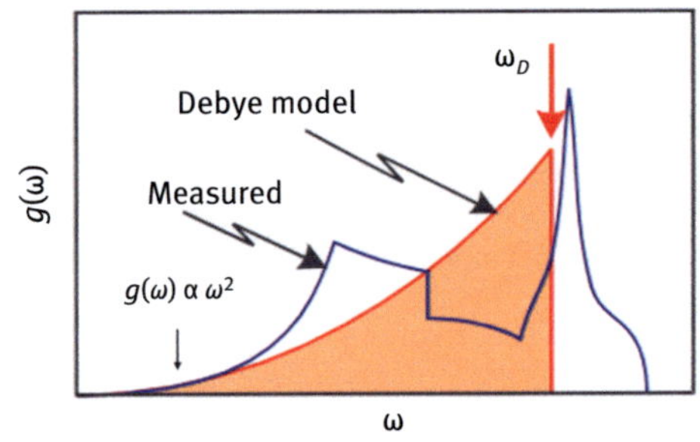

Figure 9: Comparison of the density of states $g(\omega)$ from the Debye model with a typical monatomic solid.

The average sound velocity is a significant quantity as it is related to several physical and mechanical properties of a material. For example, the bulk (K) is the derivative of the pressure–volume relationship (equation of state, EOS). The shear modulus (G) is related to the longitudinal and transverse velocities:

$$v_D = \sqrt{\frac{K}{\rho}} \tag{3}$$

$$\frac{K}{\rho} = v_L^2 - \frac{4}{3}v_T^2 \tag{4}$$

$$\frac{G}{\rho} = v_T^2 \tag{5}$$

where ρ is the density and the unit cell volume V can be determined from x-ray diffraction experiments.

IXS is a powerful probe to investigate the single particle and collective atomic dynamics in a material. Using NRIXS, the VDOSs of several specific nuclei can be measured using powder sample. On the other hand, the phonon dispersion relationship can be determined on single crystals with nonresonant scattering. The two applications and energy resolution achievable from the experiments are depicted in Figure 10.

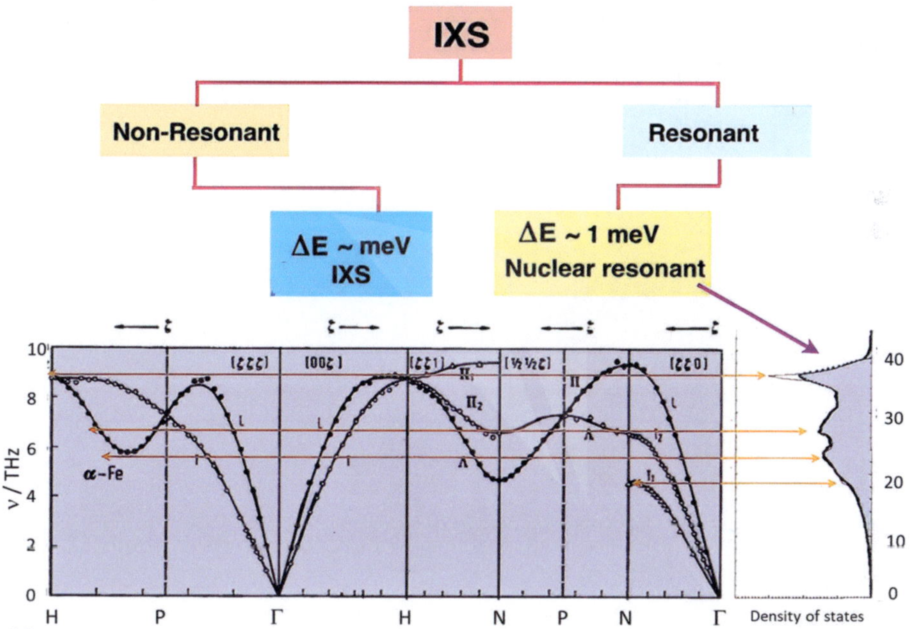

Figure 10: Schematic of nonresonant and nuclear-resonant inelastic x-ray scattering (IXS) for the determination of phonon dispersions from single crystals and site-specific vibrational density of state powder samples.

3.1.4 Applications of Fe NRIXS in earth science

Iron is the third most abundant element on the Earth and is a major component of many minerals. It was formed from shockwave nucleosynthesis from lighter hydrogen and helium in the universe. The ejection of these chemical elements by supernova explosions led to the formation of the solar nebula, a large rotating cloud of interstellar gas and dust. The contraction of the solar nebula accelerated the rotation [26]. The increased angular momentum, gravity, and inertia flattened the nebula into a protoplanetary disk. Subsequent collisions between debris created protoplanets orbiting around the nebula center. The hypothesis is that proto-Earth was formed and grown by accretion. The size increased gradually until it became so large that inertia heating melted the silicates and metals. The more massive iron then sunk into the

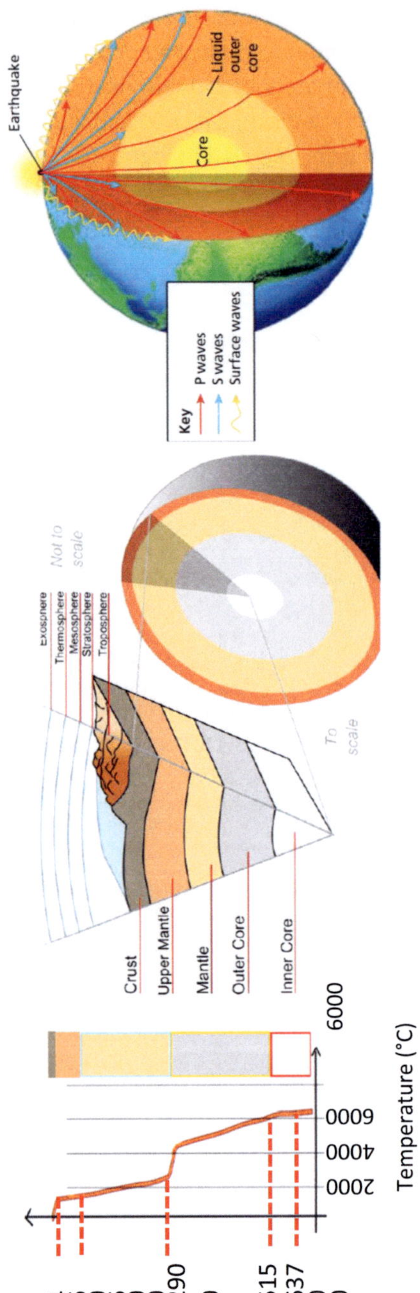

Figure 11: The different regions of inner Earth and the pressure and temperature at different depths.

interior and formed the early Earth core. Nowadays, the Earth is believed to have a solid core of majority iron surrounded by the melt. The outer core continued to evolve, eventually forming a multilayer structure composed of a mantle (inner and outer) and the crust (Figure 11) [27].

The average radius of the Earth is 6,300 km and the temperature is about 6,000 K. A vast knowledge of the Earth's structure was based on seismology, the study of seismic (sound) waves propagating within the Earth. Seismic waves are the waves caused by the sudden breaking of rock, such as an earthquake or an explosion. There are two principal types of seismic waves: body waves and surface waves. Body waves travel through the Earth's inner layers and surface waves only move along the planet's surface. The body wave travels much faster than the surface wave. The body wave is composed of two components: the *p* (primary or compressional) wave that travels along the direction (longitudinal) of propagation and the *s* (secondary or shear) wave perpendicular (transverse) to it. Since the body waves propagate through the Earth's interior, measurements of the direction, attenuation, and sound velocities provide information on the average density and elastic behavior (*vide supra*), so the composition of inner Earth can be deduced. In 1981, an international commission led by Dziewonski and Anderson gathered a large database on a variety of properties, including free oscillation center frequency measurements, surface wave dispersion observations, travel time data for many body-wave phases, and basic astronomical data (Earth's radius, mass, and moment of inertia) to construct an average one-dimensional isotropic model (preliminary reference Earth model (PREM)) of the structure of the Earth [28]. The velocities and density of the Earth from the crust to the inner core predicted by the PREM

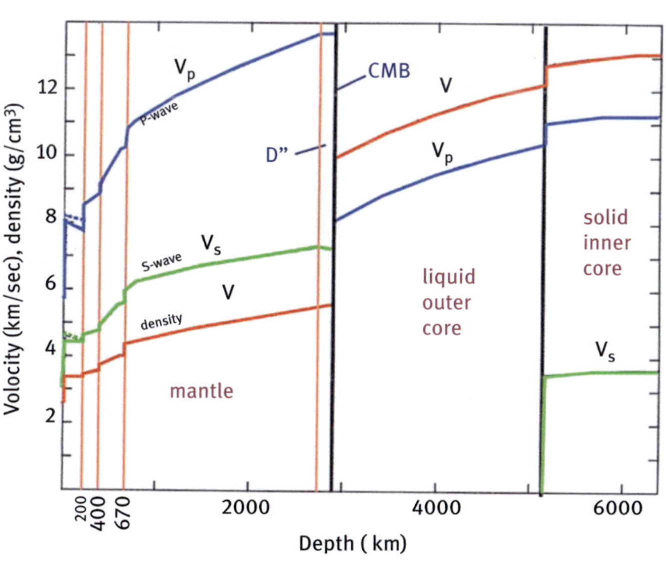

Figure 12: The preliminary reference Earth model (taken from ref. [20]).

model is shown in Figure 12. The discontinuities in the properties clearly illustrate the multilayer structure of the Earth. The PREM model has been an invaluable guide in geoscience for the past 40 years.

The Earth core does not consist solely of iron. Since nickel is also created during nucleosynthesis, it is not unreasonable to suggest that the core may also contain a small fraction of Fe–Ni alloys [29]. However, this proposition contradicts the PREM model as the average density at the core is 2–5% lower than pure Fe. The density difference strongly suggests that the Earth core must contain lighter elements. Reasonable candidates are elements, such as H, Si, and S, that are volatile and relatively abundant in the solar nebula that may be co-condensed with Fe. Since the *p*-wave velocity is the only seismological observed quantity to impose constraint on the possible compositions, it is essential to know the sound speed of the relevant light element alloys of Fe at the *P*–*T* conditions (360 GPa and 6,000 K) of the inner core. The deepest man-drill hole is only 22 km deep [30]. It is impossible to obtain firsthand knowledge of the composition and properties of the materials in the core. In the laboratory, such extreme conditions can be mimicked using a simple device, the DAC. DAC is composed of two opposing diamond anvils to press again the sample enclosed in a donut-shaped gasket. High temperature can be reached using resistance heating up to 2,000 K or laser heating up to 6,000 K. So far, pressure up to 600 GPa has been achieved [31]. The implementation of NRIXS at high-intensity third-generation synchrotrons was a breakthrough in investigating Fe and Fe-bearing minerals' phonon and electronic properties. These applications have broadened the knowledge of material properties at high pressures and temperatures. The information is fundamental for understanding the evolution and the physical conditions inside the Earth and other planetary bodies.

The first logical step is to validate the NRIXS technique at high pressure from the powder sample of pure Fe [32, 33]. As described above, a monochromatized x-ray of 14.4125 keV was used to excite the Fe nuclear transition. The recorded spectrum was converted to VDOS [34, 35]. Under ambient pressure, α-Fe is magnetic and has the body-centered cubic (bcc) structure. The VDOS of bcc Fe is compared with the result of an inelastic neutron scattering (INS) study in Figure 13a. The vibrational profiles extracted from NRIXS and INS are in excellent agreement. More significantly, the trend of the VDOS at low frequency (dotted circle) in which the sound velocity is derived is almost identical. The comparison confirms that NRIXS is an alternative/replacement to INS as the latter is only feasible up to 15 GPa. When α-Fe is heated to 910 K, it transforms to face-centered cubic (fcc) γ-Fe. Figure 13b [36] compares the VDOS of α-Fe and γ-Fe. The change in the VDOS associated with the structural transformation is clearly displayed. The signature high-energy peak in bcc Fe is shifted to a lower frequency in the fcc phase at pressures higher than 6 GPa. This observation indicates that the interaction between Fe atoms is weaker in the fcc structure.

At 13 GPa, α-Fe transforms to a hexagonal close-packed (hcp) structure (ε-Fe) and loses its magnetism [36]. At room temperature, the hcp structure persists up to

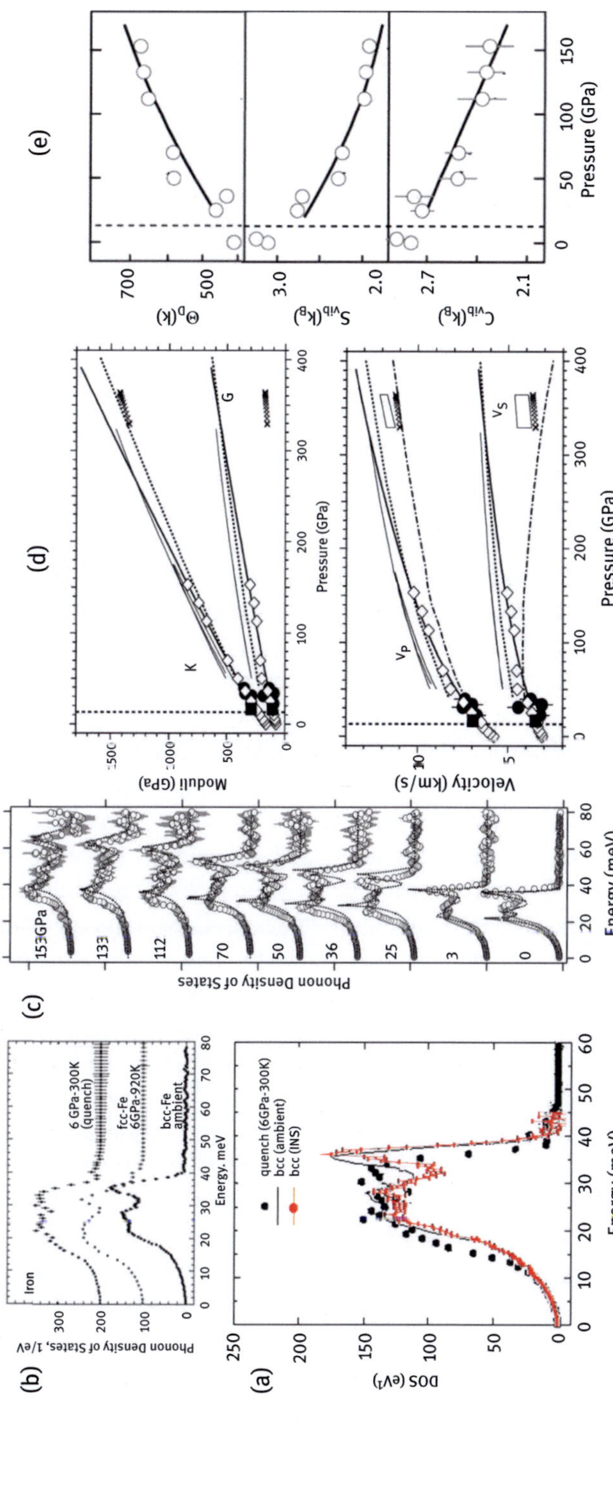

Figure 13: The Fe-projected vibrational density of states (PVDOS) extracted from ^{57}Fe nuclear-resonant inelastic scattering of (a) and (b) BCC Fe; (c) compressed hcp Fe; (d) and (e) thermodynamic parameters extracted from the analysis of PVDOS (see text) (adapted from ref. [25]).

the core pressure of 360 GPa. The VDOS of ε-Fe has been determined from the corresponding NRIXS spectra at room temperature from 25 to 153 GPa (Figure 13c), slightly above the pressure at the core–mantle boundary of 136 GPa. Since the hcp structure has two atoms per primitive cell, both the acoustic and optic branches are present (*vide supra*). The VDOS profile becomes broader, and the highest frequency band (optic mode) increases in energy with increasing pressure. The observed trend is to be expected. Intuitively, compression forces the atoms closer together, increasing the repulsion forces between atom pairs and raising the out-of-phase oscillation's vibrational frequency. Compression also helps to spread the distribution of the phonon branches leading to a broader vibrational profile. The energy resolution of the NRIXS spectra is better than 1 meV (8 cm^{-1}). The excellent quality allowed the accurate fitting of the VDOS near the long wavelength limit to eq. (1) to extract the Debye sound velocities. Employing experimental bulk modulus (K) obtained from x-ray diffraction and together with eqs. (3)–(5), the p- and s-wave velocities and the shear modulus (G) were calculated. These quantitates are in good agreement with ultrasonic data, which were available only up to 40 GPa. The extrapolated p- and s-wave velocities to the inner core region (thick solid lines in Figure 13d) are higher than with the shock-wave data (dash-dot curves) at high Hugoniot temperature (6,000 K), even with temperature corrections based on the PREM model. The higher sound velocities suggest that Earth's inner core may contain elements heavier than pure iron [37, 38]. This conclusion contradicts Birch's proposal that Earth's was too light to be composed of pure iron. The origin of the discrepancy will be addressed below. Additional thermodynamic parameters such as the Debye temperature and vibrational contribution to the heat capacity and entropy can be evaluated directly from the VDOS (Figure 12e).

The pioneering NRIXS experiment on high-pressure ε-Fe was quickly extended to other Fe-bearing minerals and alloys. Apart from many technical modifications, a major improvement is on the DAC design with a large opening allowing the emitted light to be collected at a larger solid angle. Figure 14 shows the new results obtained by Murphy et al. on pure iron [39]. At the same pressure range and temperature, it was found that the derived Debye temperatures (Figure 14a) are in agreement with previous results within the estimated errors. This observation indicates that both sets of data are consistent and reliable. In comparison, the derived compressional wave velocities (p-wave) are systematically lower than the earlier results. The difference is traced back to the EOS used to extract the bulk modulus (K). The p- and s-sound wave velocities were not determined directly from the NRIXS spectra. They were computed from eqs. (3) and (5) with the knowledge of K from separate measurement. Therefore, their values are dependence on the accuracy of K, which, in term, were calculated from the experimental EOS. In the earlier experiment, the non-hydrostatic EOS of hcp Fe was used in the calculation. Using a more appropriate hydrostatic EOS, lower values for the sound speeds were obtained. (Figure 14b). Another subtle factor is the atomic weight difference between enriched ^{57}Fe and the average natural abundance (95% ^{57}Fe). Since the vibrational frequencies are proportional to the inverse square root of the mass, a smaller mass can introduce a discrepancy of up to 2%.

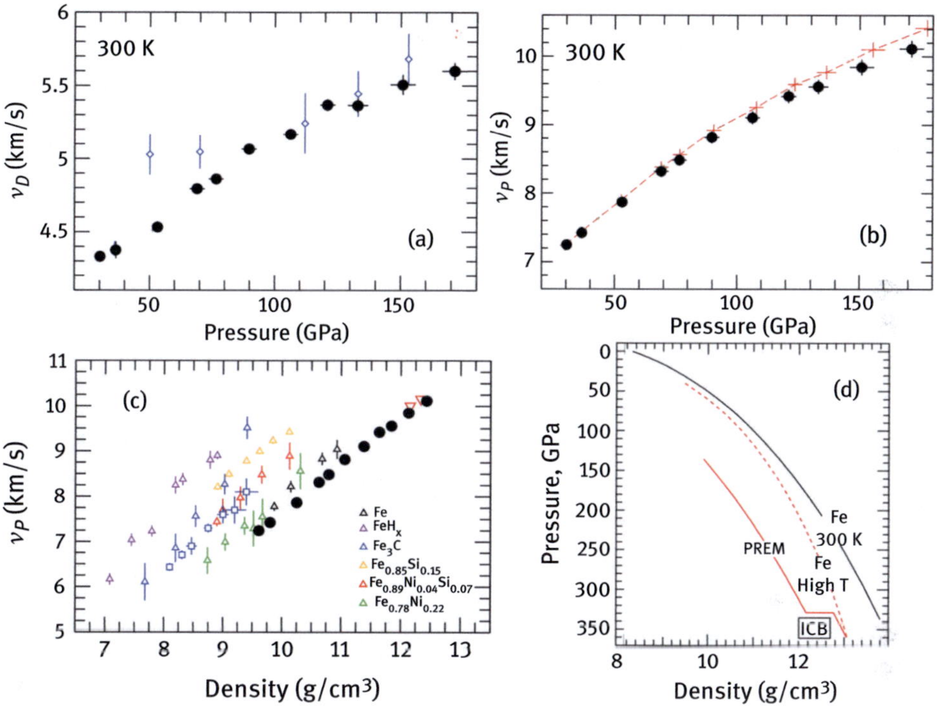

Figure 14: Sound velocities of hcp under high pressure. See text for details (taken from ref. [29]).

Birch's velocity–density relationship [40, 41] is an essential tool for constructing the Earth models and has been critically important in geophysical research. Birch's law states that there is a linear relationship relating to the sound velocity and the density with a constant term arising from the average mass of the composite system. Many experimental studies have confirmed the validity of the Birch's law. It is now believed to be universally applicable. The availability of sound velocities determined from NRIXS offers a broad *P*–*T* database for more stringent tests. Figure 14c shows plots of the velocity of the compressional (*p*) wave against the density of pure Fe and alloys with H, C, Si, and Ni with different compositions at 300 K. The linear velocity–density relationship is found to hold remarkably well for all cases over a broad range of densities corresponding to 0–280 GPa (Figure 14d).

The effect of temperature to Birch's law is not well understood. In most high-pressure studies, experiments were performed at ambient temperature of 300 K. At the Earth core, the pressure can be as high as 360 GPa and the temperature reaches 6,000 K. Is Birch's law still hold at high temperatures? Extreme heat can be generated using laser heating in a DAC (LHDAC). The sound velocities of hcp Fe up to 73 GPa and 1,700 K have been studied using LHDAC [42]. The results shown in Figure 15 are surprising. Both *p*- and *s*-waves show significant deviation from linearity at 1,700 K as compared to 300 K (Figure 15a and b). Combining with shockwave data obtained at high pressure and high temperature (Figure 15c), the deviation is even more significant at moderate density, around 10–10.5 g/cm^{-3}. However, the discrepancy seems to recover at pressure closer to the inner core. If correct, this observation may have implications on the constraints in the core as lighter elements are expected to be present. We shall return to this topic later.

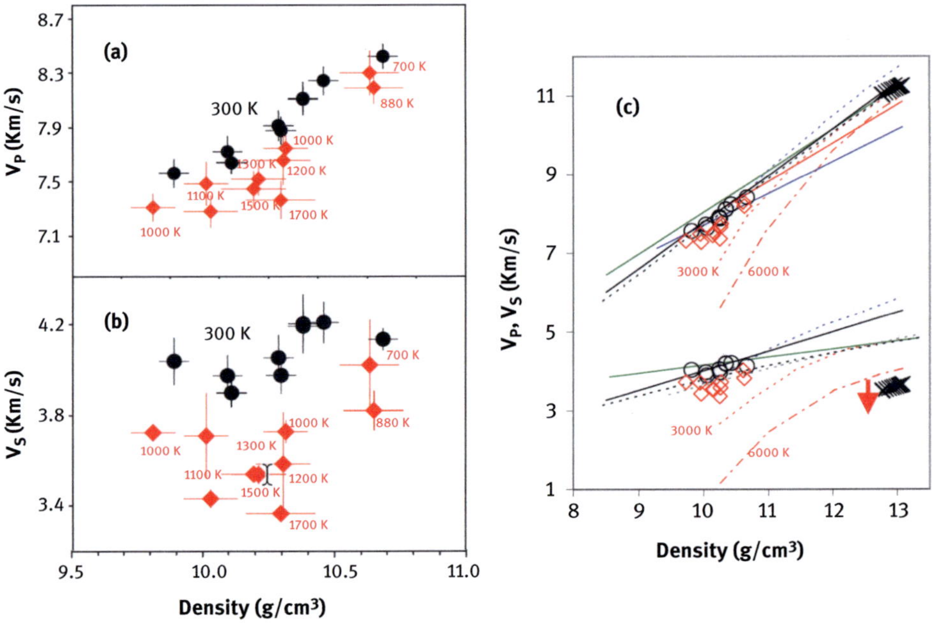

Figure 15: Pressure-dependent principal (v_P) and secondary (v_S) sound velocities of hcp Fe at (a) 200 K and (b) 300 K, and (c) compared with previous studies.

3.1.5 Sn vibrational density of states

3.1.5.1 Structural transition of Sn at high pressure

Another common nucleus of interest is tin (Sn). Sn has a very small neutron-incoherent inelastic cross section and it is not easy to determine the VDOS with INS even under ambient conditions. ^{119}Sn isotope is the second most popular nuclear resonance in the Mossbauer literature. The natural abundance of ^{119}Sn is 8.7%. The transition energy is higher than ^{57}Fe, at 23,879.4 keV, with a half-life of 17.8 ns. This corresponds to 25.7 neV resonance line width. ^{119}Sn is the second isotope observed using synchrotron radiation [43]. Therefore, elemental Sn is an excellent system to demonstrate the complementary nature of the NRIXS method.

The Sn metal at room temperature has two allotropes: β-Sn and α-Sn. The white, malleable β-Sn has a low Lamb–Mossbauer factor f_{LM} [43, 44], thus making the multi-phonon separation difficult. Sn-I (α-Sn) has an fcc structure, which is the most stable form at low temperature under normal pressure. Sn-II (β-Sn) has a body-centered tetragonal ($I4_1/amd$) structure that is stable between 287 and 505 K [45]. It transforms to another bct (bct-Sn, $I4/mmm$) structure at 9.6 GPa. Further compression leads to Sn-IV at 46.5 GPa, which has a bcc structure. Sn nucleus has a serious drawback for Mossbauer spectroscopy. The recoil fraction (f_{LM} Lamb–Mossbauer parameter) [16, 17] is very small at room temperature. It is not possible to obtain the VDOS unless the sample is cooled. A way to enhance f_{LM} is to apply pressure. This method has successfully extracted the VDOS of high-pressure Sn-III (bct-Sn) and Sn-IV [46]. The Sn VDOS at 13.1 and 37 GPa (bct, Sn-III) and 64 GPa (bcc, Sn-IV) is shown in Figure 16a. At the three pressures, dramatic broadening of the VDOS profiles and a shift in the phonon modes to higher frequency were observed. The width of the vibrational band increases to 16, 21, and 26 meV at 13.1, 27, and 64 GPa, respectively. At the same time, the high-frequency van Hove singularity peak position is blueshifted from 19 to 26 and 30 meV. The large increase of the phonon frequency with compression indicated that the Sn–Sn metallic bonds are weak at low pressure and stiffened significantly when the atoms are pressed closer together. This is the reason that the f_{LM} is very small under ambient conditions. As expected, f_{LM} increases from 10% to 50% over this pressure range (Figure 16b).

3.1.5.2 SnO

The Lamb–Mossbauer factor for Sn is expected to increase in compounds if the chemical bond is stronger between nearest neighbors. This conjecture was confirmed by the successful measurement of Sn NRIXS spectra on a textured sample of SnO with the texture axis parallel to the surface of the diamond anvils (Figure 16c) using synchrotron radiation monochromatized at the ^{119}Sn nuclear resonance energy of 22.88 keV [47]. In this case, projected Sn VDOS was obtained even at zero pressure (Figure 16d) and room temperature due to the stronger Sn–O. SnO has a layered structure formed from puckered SnO slabs stacked along the *c*-axis (Figure 16c). The Sn VDOS is sensitive to the sample's orientation with respect to the incident x-ray but only at low pressure. At ambient and 1.3 GPa, there are noticeable differences in the Sn acoustic phonon VDOS when the x-ray is incident parallel (0°) and almost perpendicular (80°) to the crystallographic *c*-axis. Compared to 0°, the intensity of the Sn -O -Sn bent modes at 80° polarized within the layers at 15 meV are enhanced. The increased density of the E_g mode is compensated by decreasing the density of the O–O breathing vibrations (A_{1g}), polarized along the *c*-axis, at 26 meV (Figure 16d). The differences, however, vanished at 6.1 GPa.

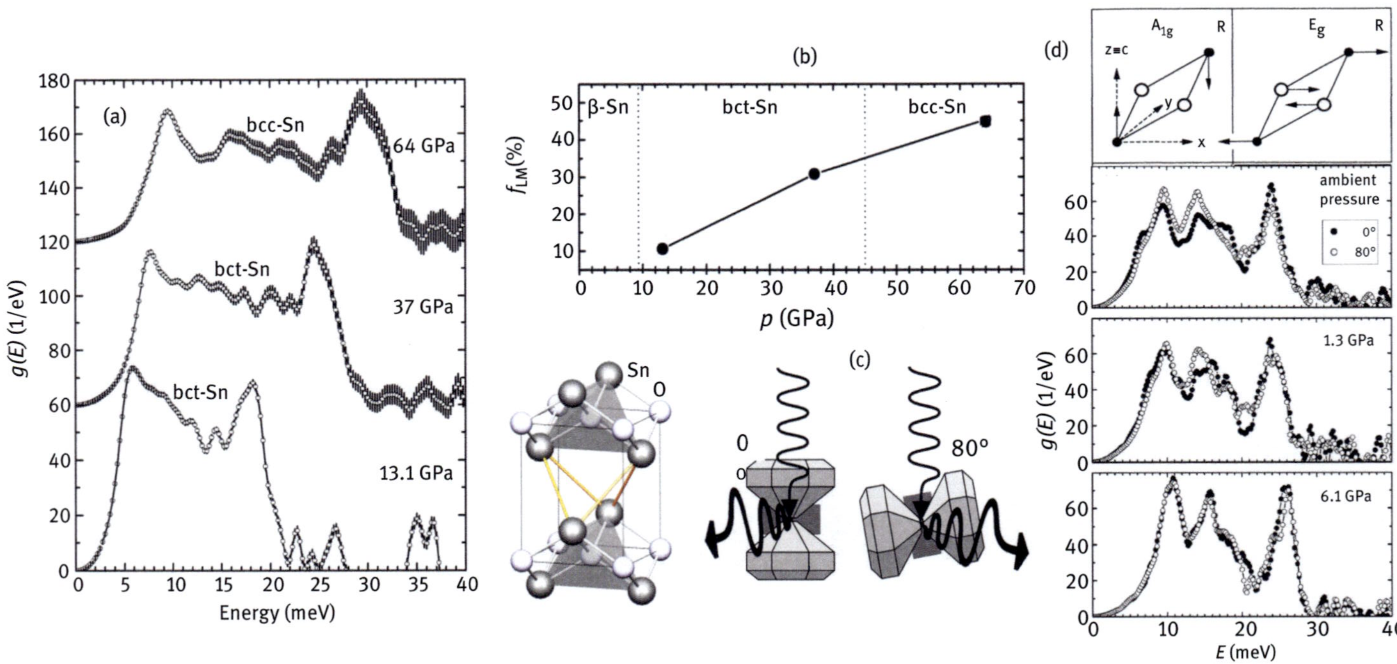

Figure 16: ^{119}Sn PVDOS of low-pressure bct-Sn and bcc-Sn at high pressure. The Lamb–Mossbauer factor at different pressures. (c) The structure of SnO and the orientation of the crystal to the incident radiation and (d) the derived angular-dependent Sn PVDOS showing the vibrational modes of different symmetries.

3.1.5.3 Pressure-induced amorphization in SnI_4

When a crystalline solid is compressed, a crystal-to-crystal transformation will usually occur at some appropriate pressure. The reason being that an ordered structure is thermodynamically more stable than a disordered solid. However, pressure-induced poly(a)morphism (PIA) is observed in some cases. It is the consequence of an abrupt structural collapse from a crystalline to an amorphous structure. The most familiar example is crystalline ice Ih [48]. When ice is compressed at temperature just below the melting line, a series of crystal-to-crystal transformations are observed. In contrast, when ice is compressed at very low temperature (100 K), a transformation to an amorphous form is observed. It is now becoming apparent that PIA is due to insufficient thermal energy to overcome the activation barrier for a crystal-to-crystal transformation at low temperatures. This phenomenon often occurs when a substantial rearrangement of atoms is required to transform from the precursor crystalline structure to the thermodynamically stable product phase. However, structures of the amorphous phases are unknown and frequently in disputes. The PIA in SnI_4 is one of the outstanding issues.

SnI_4 has a cubic structure under normal pressure–temperature conditions. When compressed, it transforms into another crystalline structure at 7.2 GPa. The onset of amorphization was observed around 15 GPa and fully established above 25 GPa. The amorphous form is maintained up to 55 GPa and then crystallizes into an fcc structure [49]. Several structures have been proposed for the amorphous phase. A Mossbauer study suggests that the molecular SnI_4 units have decomposed forming polymeric $SnI_2 + I_2$ [50]. In contrast, Raman and x-ray diffraction data support a highly distorted SnI_4 tetrahedral structure [51]. Atom-specific Sn VDOS is expected to be sensitive to the local bonding environment. For this reason, ^{119}Sn NRIXS experiment was performed [52]. The Sn VDOS from 0 to 25 GPa is shown in Figure 17a. It is significant to note that both the acoustic and Sn-I VDOS are very broad at room temperature but sharpened significantly when cooled to 65 K. The reason is that the SnI_4 tetrahedra are loosely bound and behave like free rotors at room temperature. The rotations are prohibited at low temperatures. Significant changes in the vibrational profile are observed at 8 GPa. The pressure corresponds well with the transformation to the first crystal-to-crystal transformation identified from x-ray diffraction. The distinctive Sn-I vibration at 27 meV starts to broaden with the VDOS spread toward lower frequencies. The acoustic phonons below 10 meV are also broadened and becoming almost featureless. The experimental observations suggest that (i) the Sn-I bonds are weakened but not disrupted; and (ii) the loss of distinctive collective acoustic excitations indicates that the initial cubic crystal's symmetry is reduced and becoming more disordered. Theoretical simulation of the Sn VDOS reproduced the observed trend (Figure 17b). The Sn-I vibrations are redshifted at and above 7 GPa and cover a frequency band from 18 to 22 meV. Examination of the simulated structure reveals severe distortion of the SnI_4 tetrahedra with the neighboring I atoms intruded into the first Sn-I coordination shell and serves as linkages of a polymeric network.

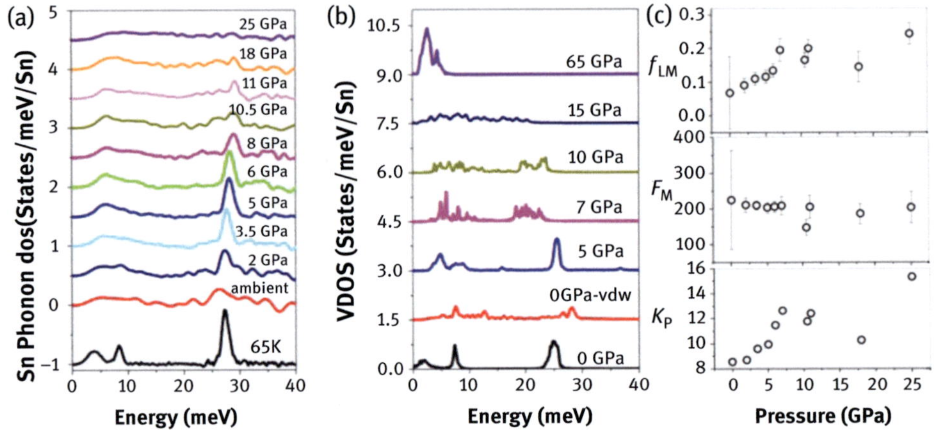

Figure 17: ^{119}Sn PVDOS SnI_4 under different pressures. (b) Sn VDOS computed from density functional molecular dynamic calculations and (c) the Lamb–Mossbauer factor at different pressures.

Relevant dynamic parameters for pressurized SnI_4 can be obtained from the analysis of the moments of the NRIXS spectra. The Mossbauer–Lamb factor (f_{LM}, recoilless fraction) is related to the mean-square displacement of the Sn atom, which shows an initial increase from 0.08 to 0.20 until ca. 8 GPa, then drops to 0.135 at 18 GPa, and then rises abruptly to 0.24 at 25 GPa. This trend can be explained that at low compression, the SnI_4 tetrahedra become more tightly packed and the Sn-I bond is stiffer, thus increasing the f_{LM} of the Sn atom. At the 7 GPa phase transition, the Sn-I bonds are weakened due to the intruded I atoms. This observation collaborates with the softening and broadening of the Sn-I VDOS due to the structural changes. Above 25 GPa, the amorphization is completed and the Sn atoms are in a progressively tighter environment as the pressure increases further.

The example discussed above illustrates that atom-specific vibrational dynamics can provide crucial information on the local bonding. Together with the first-principles calculations, it is possible to gain insight into a disordered system's local structure. The advantage of NRIXS is that the full VDOS can be determined. However, this method is limited only to some of the Mossbauer nuclei.

3.1.5.4 Superconductivity in Sn nanostructures

Although not related to high pressure, another interesting application of ^{119}Sn NRIXS is to probe the VDOS of nanostructures [53]. It is known that metastable β-Sn is a superconductor with a critical temperate (T_c) of 3.7 K. The T_c, however, can be increased by reducing the dimensionality. It was found that the T_c of nanostructured Sn is dependent on the size and morphology. From magnetic measurements (Figure 18a), Sn nano-islands with a nominal size of 40 (isl40) and 60 (isl60) nm have T_c at 4.38 and 4.17 K, respectively. For a cluster film with a nominal size of 46 nm (clus46), the T_c is 3.98 K. Electron–phonon coupling is the mechanism for the formation of Cooper pairs responsible for superconductivity. Thus, the vibrational spectrum and T_c are intimately related. To understand the cause for the enhancement of the T_c, VDOS of the samples were measured at 35 K with ^{119}Sn NRIXS.

The results for the three samples are shown in Figure 18b. In all the nanostructures, with respect to bulk tin, the high-frequency optic vibration positions are softened, and the VDOS is reduced. The reduction in the high-frequency DOS is compensated by increasing the DOSs of the low- and mid-frequency bands. This effect is most noticeable in isl40. In this sample, even the vibrational density at the top of the acoustic branch has also increased. The observation indicates stronger mixing of the optics and acoustic phonon branches. Usually, high-energy optic vibrations are localized and do not contribute much to electron–phonon coupling. The smaller nanostructure has more surface Sn atoms and weaker Sn–Sn bonds and, therefore, a lower stretching optic vibration. The softening of the optic modes leads to better mixing with the low-frequency branched and, therefore, more phonon modes can be utilized to couple with the electrons. Consequently, T_c is enhanced in the sequence isl40 > isl60 > clus46. The experiment nicely illustrates the efficiency of NRIXS in the study of the vibrational properties of nanostructures.

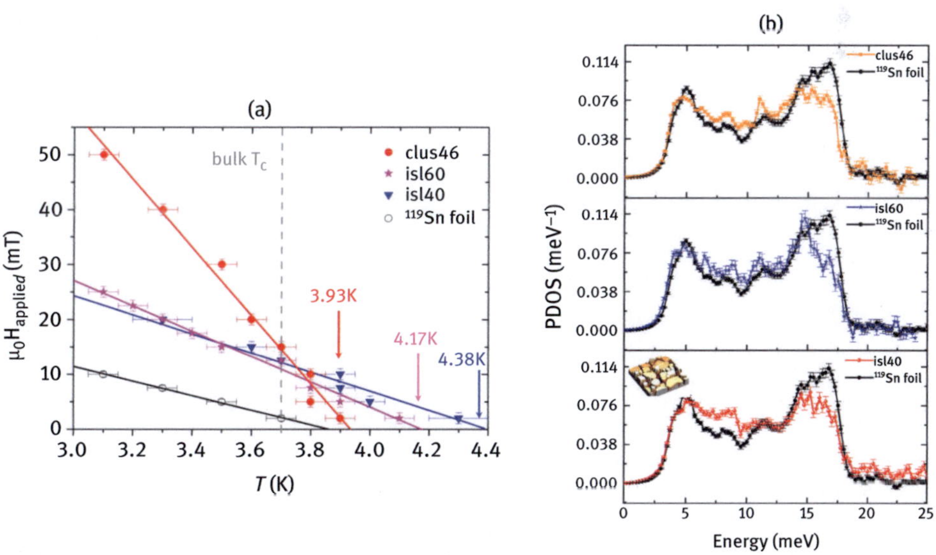

Figure 18: (a) Magnetometry data of Sn nanostructures of different sizes: isl40, is60, and clus60. (b) Orientation-dependent Sn PVDOS of the Sn nanostructures.

In passing, it is noteworthy that ^{119}Sn resonance has also been used to investigate on the problem of geophysical significance. A recent example is the study of redox and structural controls on Sn isotopic fractionations among magmas [54].

3.1.6 Phonon dispersion from IXS: high-energy-resolution IXS (HERIX)

The sound velocity of a polycrystalline sample can also be determined by IXS from the phonon dispersion measurement, that is, q-dependent frequency. A polycrystal is a collection of randomly oriented single crystallites [9]. Each crystallite scatters as a single crystal. Averaging over all the crystal orientations, only the absolute $|q|$ is defined. Due to spherical averaging, phonons in the transverse direction cancel each other in the first Brillouin zone. Only the longitudinal (compressional or p-wave) phonons can be observed. The projection of the transverse phonon modes, however, may appear at higher $|q|$ beyond the first Brillouin zone.

3.1.6.1 Sound velocity

The IXS spectrum of hcp Fe at 28 GPa recorded at $q = 4\ \mathrm{nm}^{-1}$ is shown in Figure 18a [55]. The strong peak at zero momentum transfer is the elastic line. The peak at 18 meV is the longitudinal acoustic (LA) phonon frequency of the Fe. IXS spectra at several q values were measured. Spectra at selected q values are shown in Figure 19b. A fit of the q-dependence phonon frequencies and the limiting slope evaluated at $q \to 0$ is the velocity of sound of the compressional wave. The procedure is then repeated for each pressure. In this experiment, the sound velocity was measured up to 110 GPa. A plot of the sound velocity against the specific mass demonstrates that the Birch's law is obeyed (Figure 19c). Extrapolation into the core density shows that the sound velocity is lower than the PREM model (shown as diamonds, in Figure 19c). This result is different from the earlier NRIXS study in which an opposing conclusion was reached. As noted above, the discrepancy was attributed to using the non-hydrostatic EOS to convert the Debye sound velocities to the s- and p-components. The advantage of phonon dispersion measured from IXS is that the longitudinal (p) wave is determined directly from the spectra without resorting to other data and approximations.

There has been increasing use of the q-dependent IXS method [56] to determine the velocity of Fe alloys at high pressure and high pressure to constrain the existence of light elements in the core. A recent result is the determination of the sound velocity of pure Fe up to 163 GPa and 1,600 K [57]. Once again, the Birch's law performed reliably well in the broad P–T region (Figure 20a). Furthermore, the sound velocities of several potential light candidates and Fe–Ni alloys are compared in Figure 20b. The average mass was computed assuming Vegard's law for mixed alloy and applying Birch's law for composite materials H, Si, and S. The predicted compressional waves suggested these elements are found to be suitable candidates to alloy with Fe in the core.

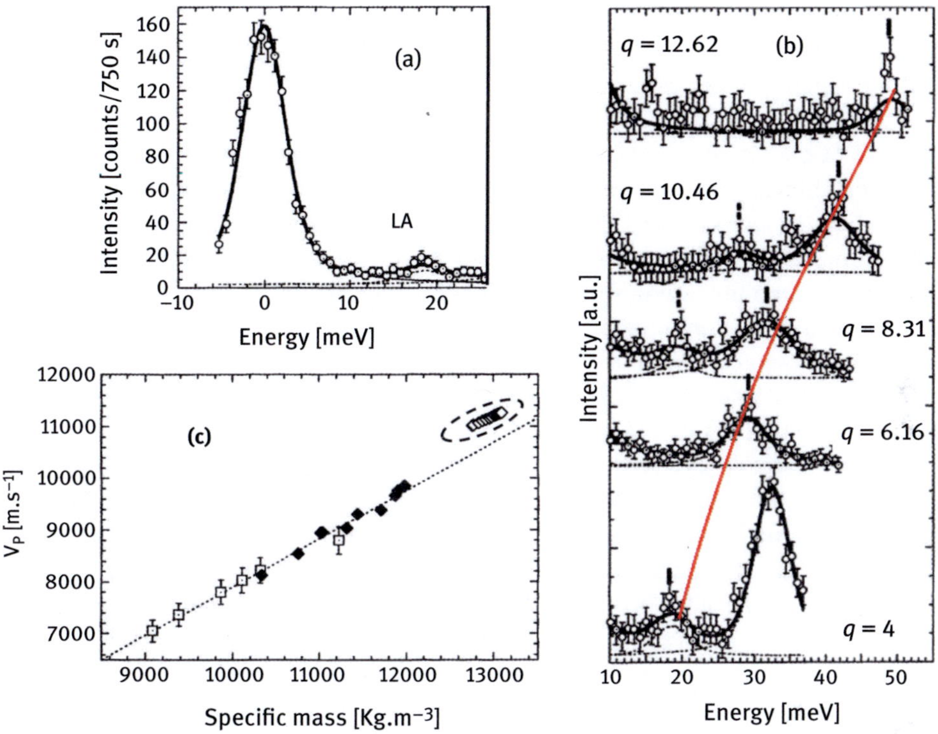

Figure 19: (a) The IXS spectrum of hcp Fe measured at 29 GPa with momentum transfer $q = 4\ \mathrm{nm}^{-1}$. (b) IXS spectra as a function of momentum transfer. (c) Principal velocity v_P (longitudinal) determined from the extrapolation of the IXS peak to $q = 0$.

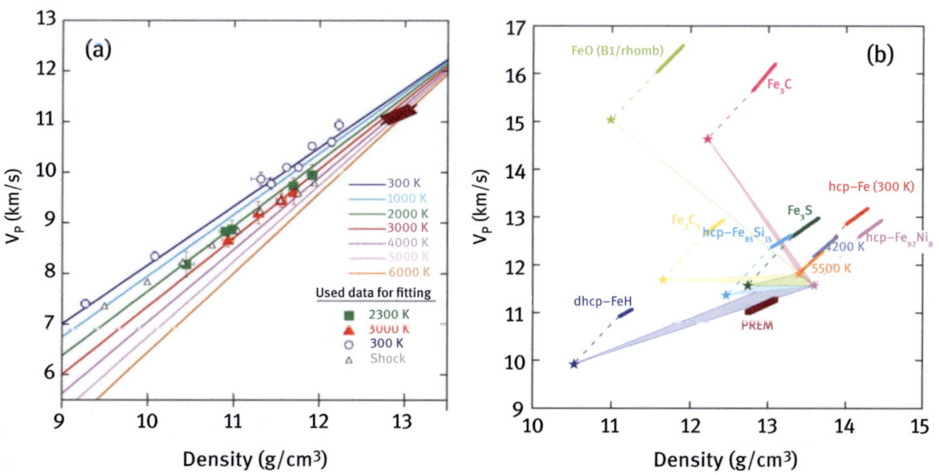

Figure 20: Comparison of experimental principal velocity v_P of (a) Fe and (b) Fe–Ni alloy with Birch's law.

3.2 Phonon dynamics

3.2.1 Phonon–phonon interactions

The IXS method is applicable to measure the phonon dispersion for all systems. However, it is difficult, if not impossible, to obtain the VDOS on polycrystalline samples. From the peak position (energy dispersion) of the q-dependent IXS spectrum, information on the lattice dynamics can be obtained [58]. Analysis of the peak profile gives details on the mechanism of the scattering process. In the following, results on the study of a phonon–phonon scattering process in an ice clathrate (methane hydrate) are presented. Clathrate hydrates are nonstoichiometric crystalline solids with different atomic and molecular gases and small molecules trapped in the cages formed by the framework of water molecules (Figure 20a) [59]. Clathrate hydrates, in particular, methane hydrate, are ubiquitous on the Earth and in planets throughout the universe. Methane hydrate is probably the largest energy reserve on the Earth. It is found in permafrost and the continental shelves. As in ice, the crystal frameworks of gas hydrates are formed from linking water molecules by hydrogen bonds. Gas hydrates have one peculiar property. The thermal conductivity of clathrate hydrates is much lower than ice and has an anomalous temperature profile resembling a glass despite the crystalline structures [60, 61]. This unusual behavior has been attributed to a process called resonant scattering, in which the localized rattling motions of the trapped guest interact strongly with the acoustic phonon branch of the ice lattice. In clathrate hydrate, the guest–host interaction is very weak, often described by the van der Waals force. Therefore, the guest motions are independent of the water lattice vibrations. The guests rattle inside the cavity and the vibrations are localized showing no energy dispersion with q. This interaction will be amplified at specific momentum transfer (q), leading to strong mixing of the lattice and the guest vibrations with the same symmetry (resonant) and enhanced the scattering of heat-carrying phonons inhibiting the thermal transport [62]. A schematic representation of this process is shown in Figure 21a. The localized guest vibrations, shown as flat lines, cross into the LA and transverse acoustic (TA) modes of the ice lattice at specific momentum transfers (q). From quantum mechanics, the guest vibrational eigenvectors and lattice vibrations are of the same symmetry. The two branches cannot cross and should avoid each other (anti-crossing rule). This process will lead to energy exchange between the two types of motions. The energy of the heat-carrying acoustic vibrations is transferred to the guests and activated the localized excitations. Effectively, this will dissipate the heat current and resulted in lower thermal conductivity. It is difficult to synthesize a methane hydrate crystal large enough for INS study. Moreover, it is not possible to access small momentum transfer by neutron scattering due to kinematic reason. Instead, IXS can be used to study a powder sample at relatively low q. To test the conjecture, phonon dispersions of a methane hydrate sample have been measured from IXS at momentum transfer (q) from 1.5 to 5 nm^{-1} (Figure 21b) [63]. The

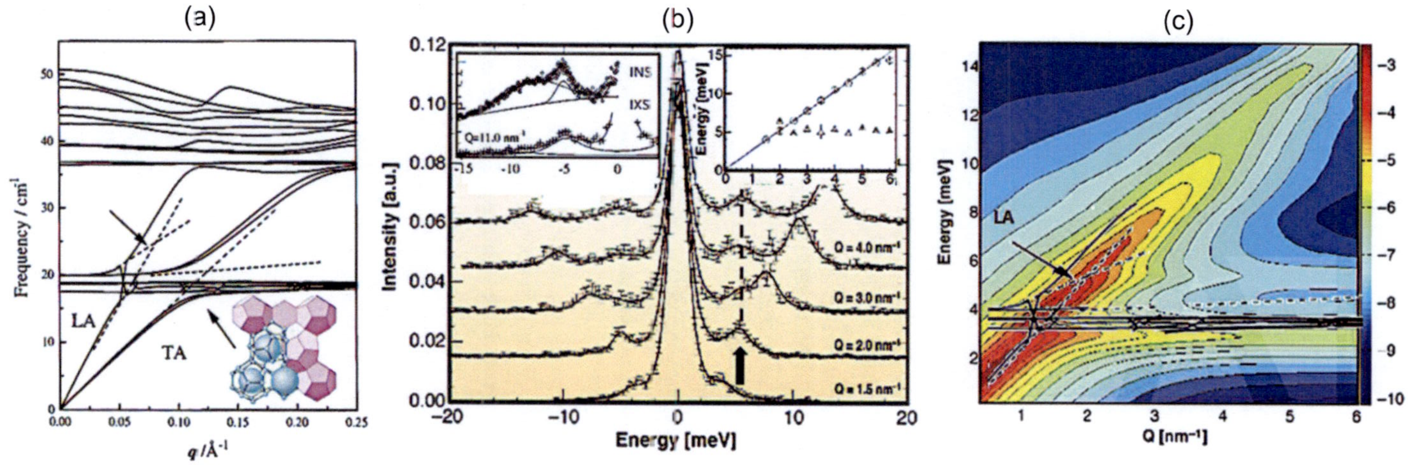

Figure 21: Comparison of theoretical (a) and experimental (c) inelastic x-ray scattering function of methane hydrate. (b) The longitudinal acoustic lattice mode and the optic guest mode can be observed.

black arrow at 5 meV in Figure 21b indicates the nondispersive (localized) translational motions of the trapped methane molecules. The higher energy peak, which changes with q, is the acoustic phonon branch of the water framework. Plot of the vibrational frequencies (ω) versus momentum transfer (q) is the phonon dispersion curve. A crossing between the localized guest phonon mode with the lattice mode is observed at $q \sim 2$ nm^{-1}. The patterns of the experimental phonon dispersion are in good agreement with theoretical calculations (Figure 21c). As alluded above (*vide infra*), only LA phonon can be observed in the IXS of a powder sample in principle. However, the TA branches can also be observed if the momentum transfer is larger than the Brillouin zone. In the inset of Figure 19b, the IXS spectra recorded at a large $q = 11$ nm^{-1} are compared with the VDOS obtained from INS. The VDOS obtained at high q in the IXS experiment agrees with the VDOS of the acoustic vibrations measured by INS. In passing, a more careful examination shows that linewidths of the localized CH_4 excitations increase from $q = 1.5$ to 3.0 nm^{-1} and then decrease. A broader linewidth near the anti-crossing at $q = 2$ nm^{-1} indicates a decrease in lifetimes of the excited vibrations, a result expected for resonant scattering.

Another test of the resonant scattering hypothesis is to apply pressure to the hydrate. By reducing the water–guest separation, the interaction becomes repulsive and no longer independent of each other. If this is correct, the resonant scattering, as described above, will no longer exist. When methane hydrate is compressed [64], the first transition occurs at 10 kbar (1 GPa), in which the initial cubic structure (MH-I) transforms into another cage structure (MH-II, hexagonal) with more smaller cavities. The second transition occurs at 20 kbar has an ice-like structure and the cages no longer exist. MH-II and MH-III were synthesized *in situ* by compressing MH-I inside a DAC. Selected IXS in coexistence with ice, high-pressure ice VII, at different wave vector transfers between 2.0 and 8.9 nm^{-1} is shown in Figure 22a and b, respectively [65]. No localized CH_4 is observed in both samples. The low-energy peak at high q (>7 nm^{-1}) is the TA mode foldback from the extended Brillouin zone (*vide infra*). Only the dispersion of the LA phonon of the high-pressure hydrate and ice VII was observed. The absence of the distinct localized mode shows that the guest is now part of the entire crystal. As described above, the tangent of the phonon dispersion extrapolated to $q \to 0$ gives the velocity of the compressional wave (v_p). The results of the analysis are shown in Figure 22c. The velocities v_p for MH-II at 17 kbar is 4,200 m/s and for MH-III at 21 kbar is 4,600 m/s. Using the experimental densities and bulk moduli, the shear wave velocity for MH-II is found to be 1,800 and 800 m/s for MH-III. The compressional wave velocity of MH-III is in substantial agreement with the extrapolated value of 4,500 m/s obtained from a recent measurement of a single crystal of methane hydrate under high-pressure Brillouin spectroscopy. The sound velocities determined may have practical applications for remote sensing. It was discovered recently that a large quantity of MH-III may be buried below the surface of Titan, a moon of Saturn, speculating that early life form may exist.

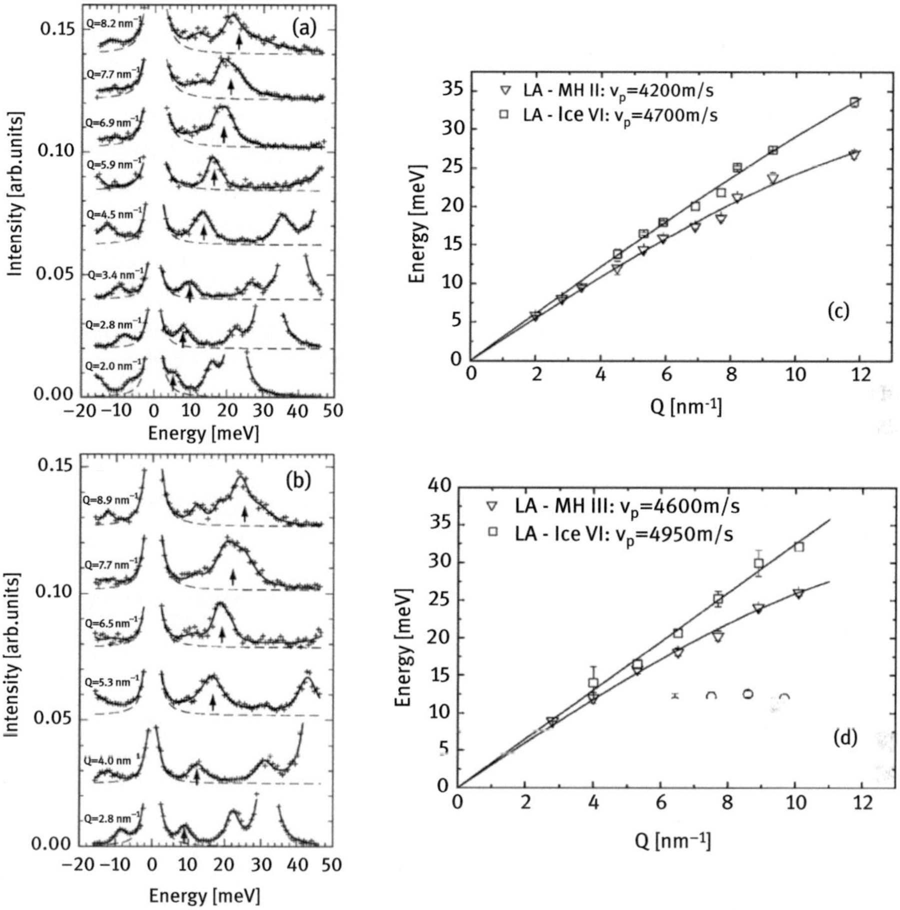

Figure 22: Momentum-resolved IXS spectra of (a) MH-II and (b) MH-III. Plots of the phonon dispersion of (c) MH-II and (d) MH-III.

Although not at high pressure, the versatility of IXS to derive element-specific VDOS independently is perhaps best illustrated in the studies of an $EuFe_4Sb_{12}$ skutterudite. Skutterudites are guest–host compounds and are weakly metallic. It was suggested that the guest–host interactions will be weak, and the rattling of the guests in the cages may result in resonant scattering of the phonons as in the gas hydrate described above. This will lead to the lowering of the thermal conductivity. A material with reasonably high electrical conductivity but low thermal conductivity is ideal for thermoelectric applications. To this end, an $EuFe_4Sb_{12}$ skutterudite, with the Sb forming the host framework and the Fe and Eu the guest atoms, was synthesized. The crystal structure is shown in Figure 23. The rationale is that Fe, Sb, and Eu are Mossbauer nuclei; thus, NRIXS measurements can be used to extract individual element-specific VDOS to shed light on the phonon dynamics. The derived VDOS for Eu, Fe, and Sb is compared in Figure 23. It is clear that the guest Eu and Fe atomic vibrations are localized. As expected, the heavier Eu mode is at 7 meV, and the Fe vibration is split due to the crystallographic symmetry and observed at 29 and 32 meV [66]. Comparison of the ^{151}Eu and ^{57}Fe VDOS clearly shows no mixing of the atomic motions. The ^{121}Sb VDOS of host atoms of $EuFe_4Sb_{12}$ was measured independently [67] and shown in Figure 23. The vibrational profile covers from 0 to 35 meV with peaks observed at 10, 16, and 23 meV. Incidentally, corresponding vibration density was observed from 8 to 20 meV in both the Eu and Fe sites. The overlapping features indicate interactions between the guest and the host lattice. In summary, the NRIXS experiments confirmed the resonant scattering model in which the guest Eu and Fe are localized "rattlers" and interact weakly with the framework. This example illustrated that IXS can provide an in-depth understanding of the atomic dynamics in a complex polyatomic lattice. The use of multielement IXS is a burgeoning technique. There are few examples on the application. A report on the study of the atomic VDOS ^{161}Dy and ^{57}Fe in compressed Dy_3Fe has been reported [68]. The results show gradual mixing of the Dy and Fe atomic vibration with increasing pressure.

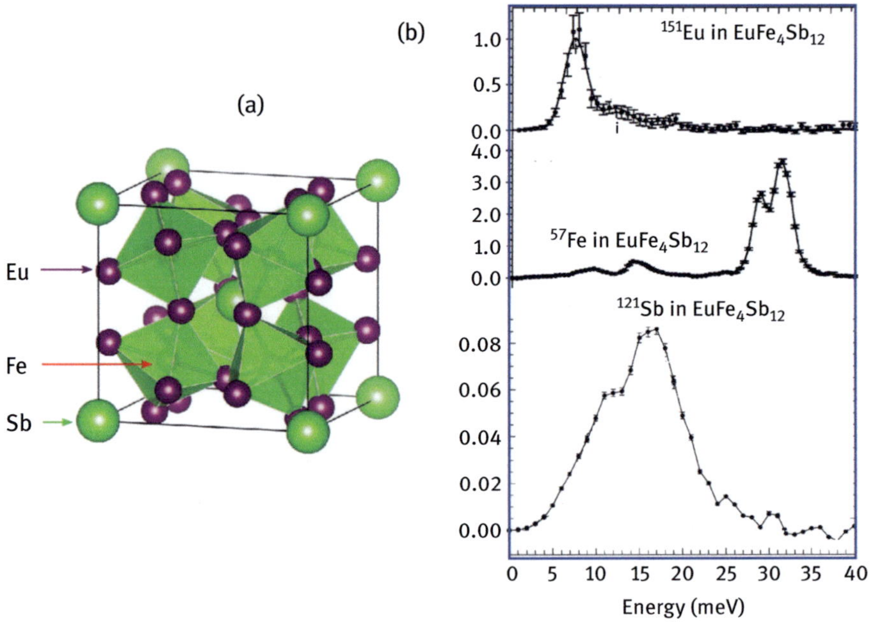

Figure 23: (a) Structure of $EuFe_4Sb_{12}$ skutterudite. (b) Site-specific VDOS from ^{151}Eu, ^{57}Fe, and ^{121}Sb.

3.2.2 Phonon dynamics in liquids

Linking the atomistic description with hydrodynamic limits on material properties is a challenge. The objective is to derive rigorous macroscopic models, such as fluid mechanics, from a macroscopic equation of individual particles' motions. In the length perspective, the task is to establish a connection from nanometer to meter scale, that is, in the long wavelength limit ($\lambda \to \infty$). In the reciprocal space, this is equivalent to investigating the atomic behavior close to zero wave vector, q ($= 2\pi/\lambda$) $\to 0$. In this context, model systems, such as liquid alkali metals, are suitable candidates for detailed studies. These systems exhibit that well-defined collective excitations from the bulk liquid propagate down to the length scale of interatomic distances under ambient pressure. Under normal conditions, these monatomic liquids are well described by the free-electron model. However, this property may deviate from this ideal behavior under high compression. A case of particular interest is liquid cesium. Cesium has a low melting point of 28.4 °C. When the solid is compressed, it undergoes three successive transformations within a very narrow pressure range. It transforms from the bcc structure to fcc at 2.3 GPa, then at 4.2 GPa to a very complex orthorhombic *Cmcm* phase with 84 atoms per unit cell and rapidly back to a simple tetragonal $I4_1/amd$ structure at 4.3 GPa (Figure 24a) [69]. The exact mechanism for these transitions between 4.2 and 4.3 GPa is still unknown. A generally accepted interpretation is attributed to the postulated Cs 6*s* → 5*d* rehybridization (Figure 24a). A diffraction study by Tsuji suggested that the structural changes in liquid Cs about the same pressure range are similar to the solids. This observation raises the possibility of a liquid–liquid phase transition (LLPT) [70]. An ensuing diffraction study by Falconi et al. which measured the diffraction patterns of liquid Cs at 220 °C (494 K) from 0.4 to 4.4 GPa seems to support this conjecture. Assuming the compression is isotropic and the volume contraction is uniform, the specific volumes at each pressure are estimated from scaling the ratio of the first-shape peak of diffraction (q/q_0) to the volume at normal pressure. Selected diffraction patterns and specific volumes plotted as a function of pressure are shown in Figure 24a. A discontinuity in the specific volume is observed at 3.8 GPa. The observation seems to support a first-order LLPT, the first example of an LLPT caused by electron excitation. If the structural transformation in the liquid indeed follows the suggested Cs 6*s* → 5*d* rehybridization, the interatomic interactions between the atoms must also change, perhaps drastically, accompanying the phase transition. To test this suggestion, q-dependent IXS experiments were performed at 493 K in the q-range 1.5–28 nm^{-1} at selected pressures (red dots in Figure 24a) covering the purported LLPT [71, 72]. The spectra are shown in Figure 24b. The IXS spectra can be converted quantitatively to the dynamic structure factor $S(q,\omega)$ that allows detailed analysis using different hydrodynamic models [72]. Analysis of the results reveals a very different picture. No anomaly was found in the sound velocities across the suggested LLPT (Figure 22c). Perhaps, more significant is that a universal behavior on the dispersion of sound velocities was found from combining available data on a large variety of alkali and alkaline liquid metals (Figure 24d). The results suggested that a single model of interatomic potential can describe the dynamical properties of liquid of simple metals. The conclusion is that dynamical studies that are directly related to the interatomic potential do not support an LLPT in liquid cesium. The

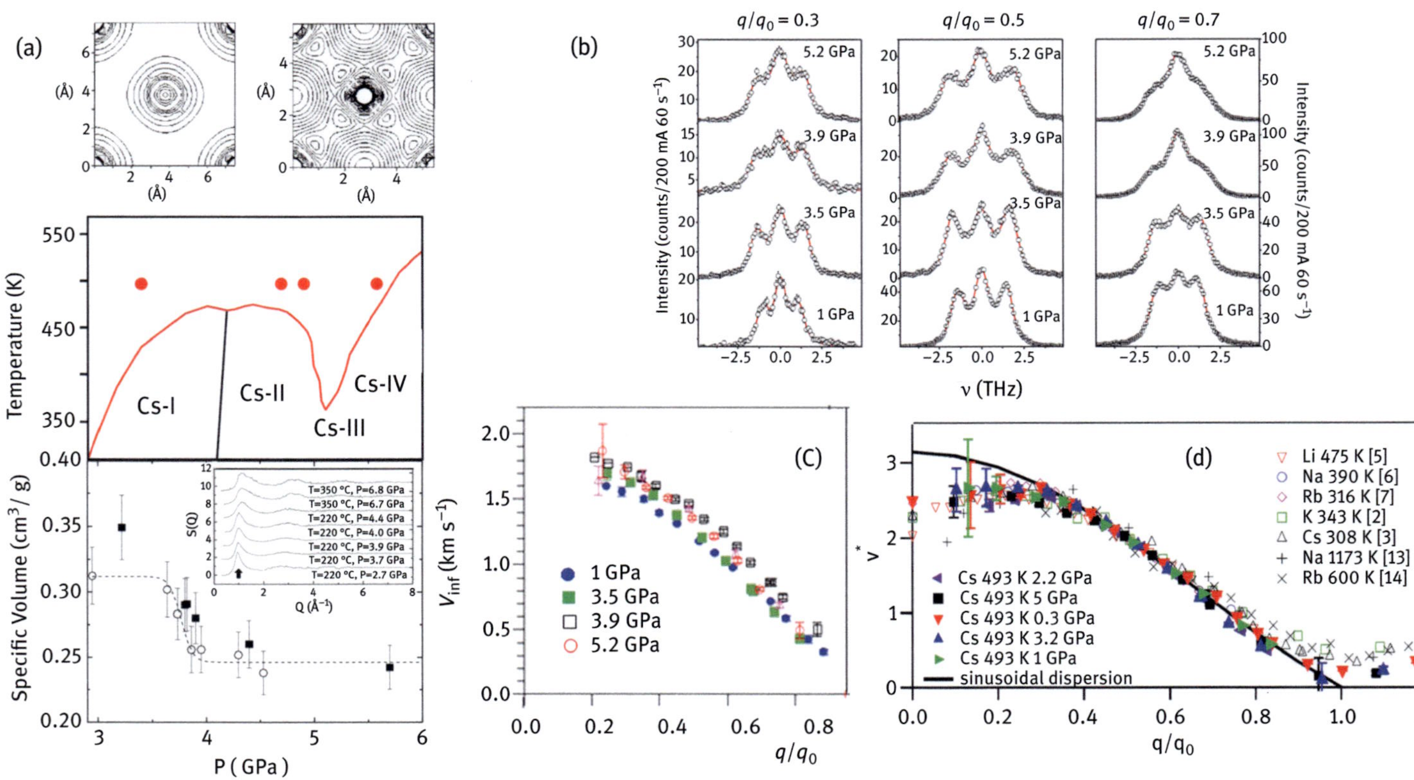

Figure 24: (a) The postulated $6s \rightarrow 5d$ transition in high-pressure Cs and the experimental phase diagram and the volume change associated with the structural transitions. q-Dependent phonon IXS of Cs under pressure. (c) Velocity of sound extracted from the phonon dispersion. (d) Comparison with other liquid alkali metals.

discrepancy with the previous x-ray diffraction studied was later clarified by an *in situ* high-pressure x-ray absorption experiment. The error arises from the assumption of uniform contraction of the liquid such that the volume (density) of the compressed liquid can be estimated from the FSDP of the diffraction pattern.

3.3 Collective dynamics

3.3.1 Momentum-resolved high-energy-resolution inelastic scattering (HERIX) single-crystal IXS and phonon band structure

As shown above, high-resolution IXS can provide valuable information on the atomic dynamics in orientationally disordered materials such as polycrystalline samples and liquids. However, properties dependent on the crystal orientation cannot be obtained. The remedy, of course, is to perform experiments on a single crystal [9]. This can be achieved by measuring energy-resolved scattered radiation $\boldsymbol{k}_f$ from a crystal of known orientation with respect to the incident monochromatized x-ray $\boldsymbol{k}_i$. The momentum transfer q is the difference of the $\boldsymbol{k}_i - \boldsymbol{k}_f$ (Figure 25). From changing the orientation of the crystalline sample and the energy analyzer, the energy loss (ω) as a function of q can be determined. A plot of ω versus $\boldsymbol{q}$ is the phonon band structure of the material. The instrument is known as the triple-axis spectrometer. This method is illustrated in Figure 26 with the study of the phonon dispersion of α-quartz [73]. The calculated phonon dispersions in the Γ→A direction are shown in Figure 26a. One selects a constant wave vector q crustal orientation (second axis), and the energy of the scattered photons from a monochromatized x-ray (first axis) is scanned by tuning the angle of the analyzer (third axis) (Figure 25). The resulting IXS spectrum (Figure 26b) represents a one-dimensional slice of the phonon dispersion at the $\boldsymbol{q}$ along the symmetry line. The full phonon band structure is constructed by changing the values of $\boldsymbol{q}$, to project out the phonon frequencies at different crystal orientations (Figure 25).

The first IXS experiment was performed on two isotopes of solid helium, namely, ^{3}He and ^{4}He [73]. ^{3}He and ^{4}He have very different physical properties. ^{3}He is composed of two protons, one neutron, and two electrons, making it a fermion with an overall spin of ½. On the other hand, ^{4}He has one more neutron; therefore, the net spin is zero, and it is a boson. At low temperature, liquid ^{4}He is a superfluid. The dynamics of liquid and solid ^{4}He and liquid ^{3}He have been studied extensively by INS. In comparison, information on the lattice dynamics of solid ^{3}He has been scarce. This is due to the high absorption cross section of ^{3}He for thermal neutrons and severely reduces the scattered neutron intensity.

IXS can overcome some of the technical difficulties. One of the first applications of this technique was to determine the phonon-dispersive relationship for solid ^{3}He and ^{4}He. Although He is a weak x-ray scatterer, intense and focused synchrotron x-ray compensates for the low scattering intensity. It helps alleviate some of the experimental

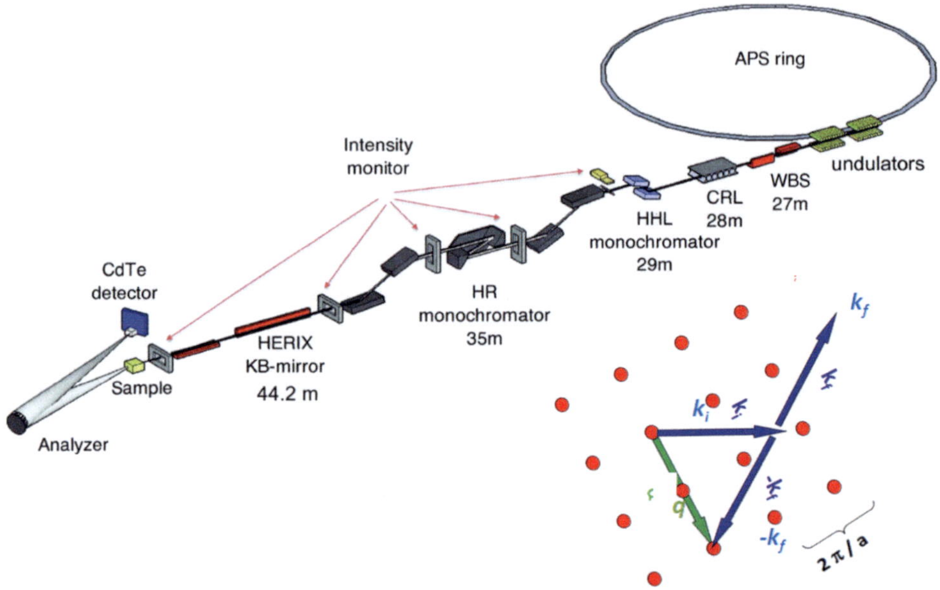

Figure 25: Schematic of the experimental scattering geometry for the measurement of the inelastic scattering from a single crystal. Said, A.H, et.al., High-energy-resolution inelastic X-ray scattering spectrometer at beamline 30-ID of the Advanced Photon Source J. Synchrotron Rad. (2020). 27, 827–835.

challenges associated with INS. For example, a high-energy x-ray can penetrate through a pressurized vessel. Small single crystals of ^{3}He and ^{4}He were grown inside a beryllium pressure cell under external gas pressure of 70 MPa for ^{3}He and 90 MPa for ^{4}He at 6.5 K. The IXS spectra recorded from a ω-q scan along the [001] direction of the hexagonal ^{3}He crystal are shown in Figure 27a. The phonon dispersion curves are constructed from the data. The longitudinal phonon dispersion along the A-Γ-A symmetry line in the [001] direction and along the Γ-M-Γ symmetry line in the [001] direction is shown in Figure 27b and c, respectively. In a similar manner, the dispersion of the longitudinal phonon of ^{4}He has been determined and shown in Figure 27d and e. A useful piece of information obtained from the phonon dispersion study is to understand the anharmonicity of the crystals. Solid helium is a highly anharmonic system due to the very weak He–He interaction. The ideal theoretical ratio of the frequency shift between ^{4}He and ^{3}He is simply the square root of their mass ratio, that is, $\sqrt{4/3} = 1.55$. The experimental values were found to be lower, between 1.11 (5) and 1.15 (5), depending on the dispersion direction.

Solid argon (Ar) was the first IXS measurement of phonon dispersions on a single crystal in a DAC [74]. The counting statistics available at the time allowed the determination of the phonon dispersion of the longitudinal (L) and the degenerated acoustic (A) modes up to 20 GPa. The IXS spectra obtained at 3.1 and 20 GPa at selected q are

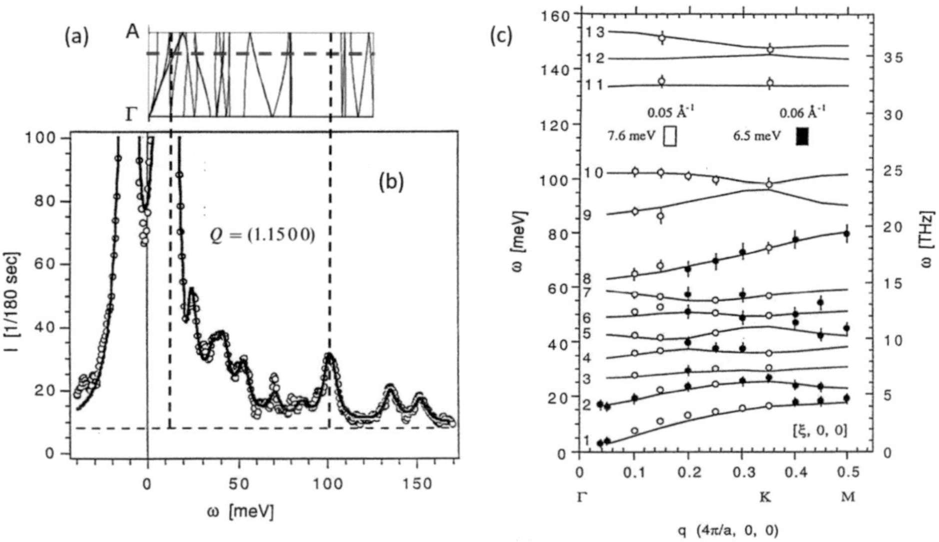

Figure 26: Experimental measurement of the phonon dispersion of α-quartz. (a) Theoretical phonon dispersion branches along the symmetry line Γ→A. IXS spectrum at lattice *Q* vector (1.15 0 0). (c) Comparison of experimental and theoretical phonon dispersion curves.

shown in Figure 28a. The phonon dispersions obtained at 3.1 GPa are shown in Figure 28b. This proof-of-principle experiment demonstrated that the phonon band structure could be determined by IXS on a single crystal under high pressure in a DAC. Since then, the instrumentation of high-resolution IXS has greatly improved. A more intense, better resolved, and highly focused beam is now available. The advancements open an avenue to investigate the atomic dynamics under extreme conditions on samples of micron size. For example, elastic constants of a grain of microstructured fcc δ-Pu-Ga alloy have been determined from high-resolution IXS. Materials relevant as *in situ* pressure calibrants such as MgO [75] and NaCl [76] have also been studied. The most interesting application of high-resolution IXS is the studies on elemental solids under pressure. An example is the use of hcp Co as a surrogate for hcp ε-Fe. Single crystal of the quintessential hcp Fe cannot be obtained from the pressurization of bcc Fe. Therefore, no specific information on the elastic property of high-pressure ε-Fe is available. Co is the closest element to Fe in the periodic table, and it is expected that the elastic properties of hcp Co may be a good representation for ε-Fe [77]. The elastic moduli of a hexagonal crystal are related to the sound velocities to different acoustic phonon branches in different crystal directions. Namely $C_{11} = v(\text{LA})[100]$, $C_{33} = v(\text{LA})[001]$, $C_{44} = v(\text{TA})[110]_{(001)}$, and $C_{66} = v(\text{TA})[110]_{\{100\}}$ C_{13} can be calculated from the longitudinal sound velocity along the [101] direction, which is a function of all five elastic moduli. From IXS measurements of the relevant phonon branches, the full set of elastic constants has been determined. The main finding is that no anomaly was found. The elastic moduli of hcp Co behave in a "regular" manner up to 39 GPa. The same

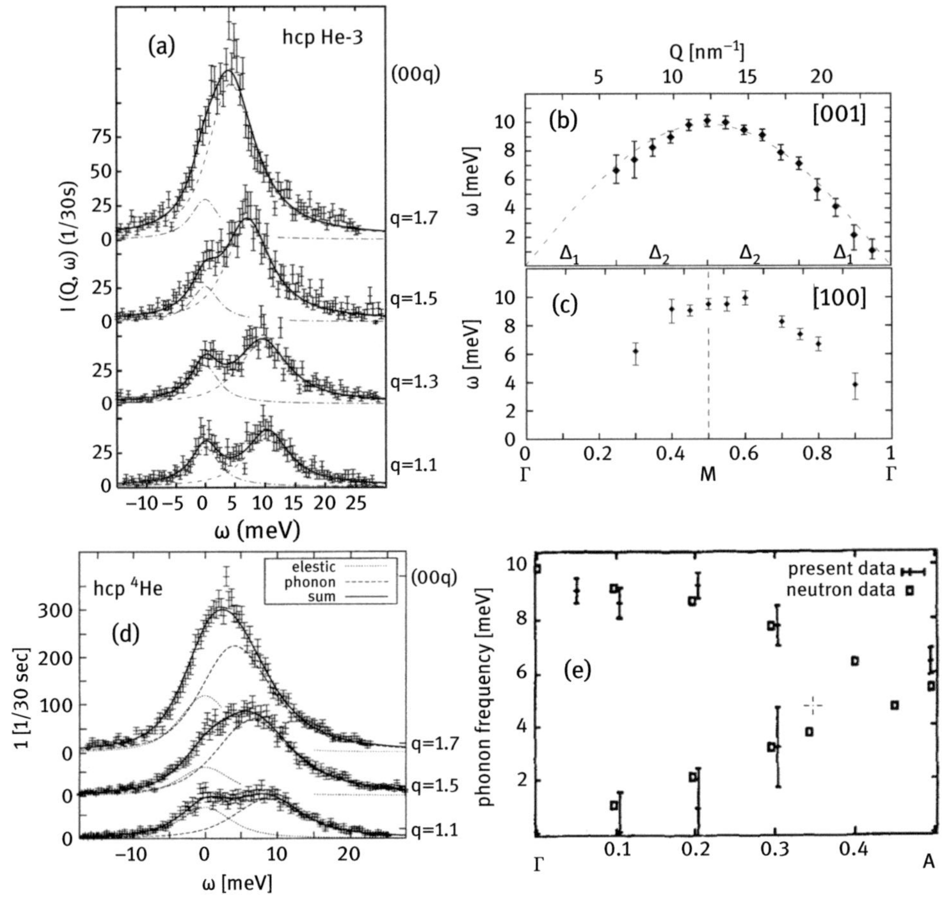

Figure 27: (a) IXS spectra of solid ^{3}He. (b) Phonon dispersion along (b) A→Γ→A and (c) Γ→M→Γ symmetry directions. (d) The IXS spectra for solid ^{4}He. (d) The derived longitudinal phonon dispersion in ^{4}He.

conclusion may apply to ε-Fe. The elastic moduli of platinum (Pt), an important pressure calibrant for diffraction experiments, have also been determined similarly.

Figure 29a–c summarizes the experimental phonon dispersion curves of tantalum (Ta) [78], molybdenum (Mo) [79], and Ce (cerium) [80] at high pressure. In Ta [78], compared to 0 GPa, the LA mode at $q \sim 0.7$ along the [100] direction is softened upon compression (Figure 29a). This is due to the change in the electronic structure, leading to the nesting of the Fermi surface. When the phonon wave vector q matches the nesting vector, $2k_F = q$, strong electron–phonon coupling weakens the vibrational energy, a sign of possible structural phase transition. The phonon softening can be defined quantitatively from calculations of the mode Grüneisen parameters (γ_q). This is shown in the study of Mo [79]. In this case, an anomalous

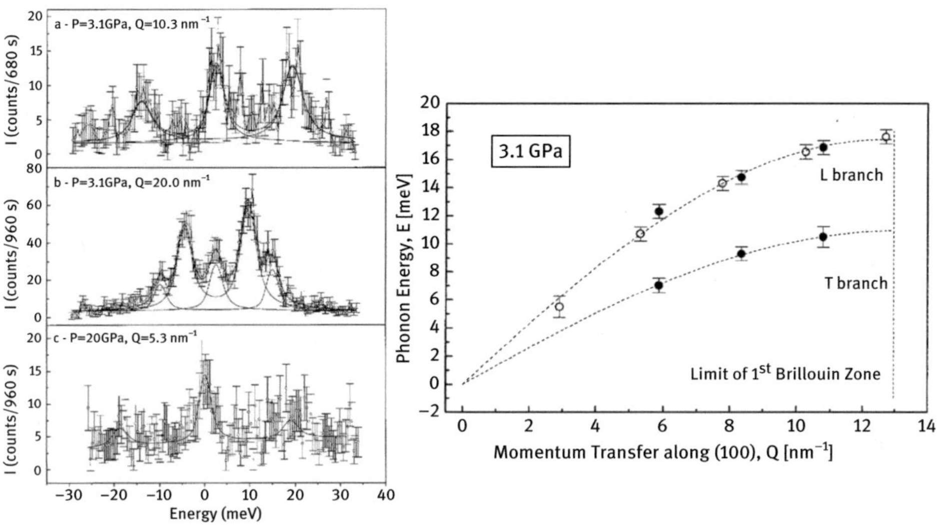

Figure 28: (a) IXS spectra of solid Ar measured at 3.1 GPa at two momentum transfers and at 30 GPa. (b) The longitudinal (L) and transverse (T) phonon dispersion curves of solid Ar at 3.1 GPa.

softening of the phonon mode at the *H*-symmetry point at ambient pressure is well known. In comparison, the lattice frequency at *H* is stiffened (increase) under pressure (Figure 29b) as shown by the increase of the *H*-point Grüneisen parameter with increasing density at the reduced wave vector $q = 0.6$ along the reciprocal *c*-axis. The increase is attributed to reduced electron–phonon coupling, contrary to Mo.

An exemplary application of HERIX is the determination of the phonon dispersion of the incommensurate guest–host complex of rubidium (Rb) [81]. Rb is one of the simple alkali metals exhibiting interesting structural transformations upon compression. The bcc (Rb-I) to fcc (Rb-II) transition is found at 7 GPa. A modulated phase similar to Cs(III) structure is found at 17 GPa. Further compression to 17–20 GPa, a new incommensurate guest–host structure first discovered in Ba was found (Figure 30). The guest Rb atoms form a linear chain inside the Rb lattice channels and belong to a subunit cell incommensurate with the crystal structure of the framework host lattice. The chain length is found to be very sensitive to pressure. As in most guest–host complexes, the guest–host interactions are weak. This expectation is confirmed by x-ray diffraction from the extremely weak modulation reflections that arise from the host–guest interactions. It is of great interest to investigate the atomic dynamics of the Rb in the host lattice and within the chains. Selected IXS spectra of Rb-IV at 17.3 GPa for three momentum transfers are shown in Figure 30a. The dispersion relations of the LA phonons are plotted in Figure 30b. Two distinct vibrational branches were observed. Both can be fitted to parabolic functions. The first branch is assigned to the LA phonon of the Rb host lattice. The second branch with a minimum at $\boldsymbol{q} = (0,\ 0,\ 3.7)$ was observed. The appearance of two independent vibrational

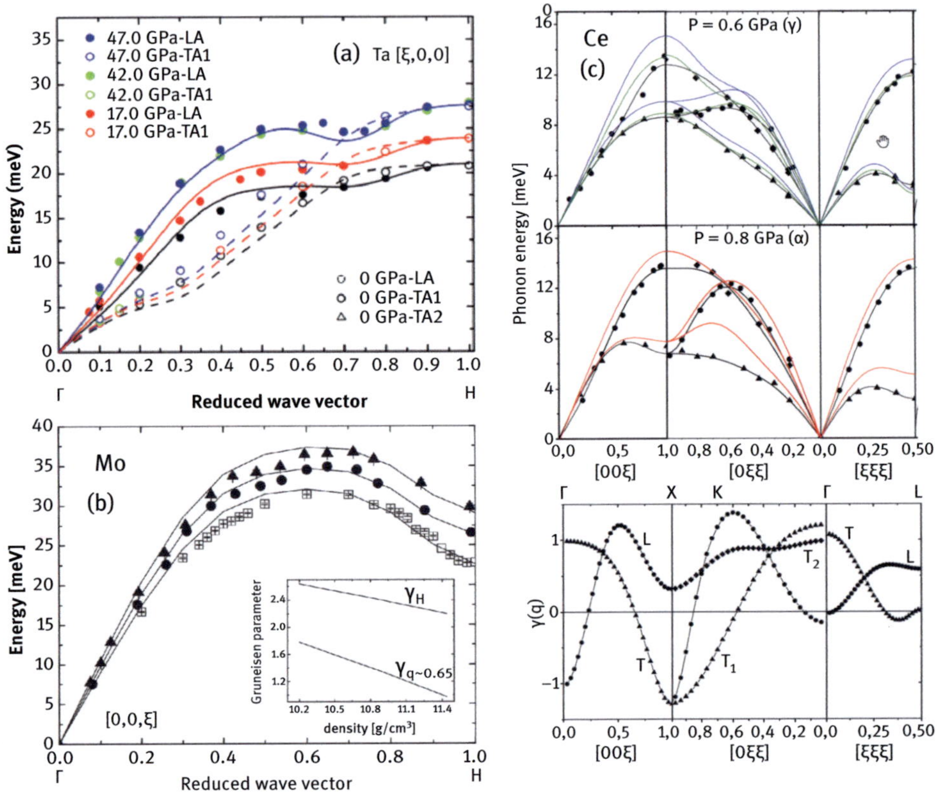

Figure 29: Phonon dispersion curves obtained from IXS experiments on (a) Ta, (b) Mo, and (c) Ce.

components is the signature of an incommensurate host–guest crystal with weak coupling. The shift of the center of the guest phonon branch is due to a rigid translation of the origin of the guest subunit cell with respect to the host. The linear chain of Rb atoms may be the experimental realization of the hypothetical "monatomic linear chain" (*supra infra*) [23] often used in standard textbooks. In this simple model, the velocity of sound is the tangent of the dispersion at $q = 0$, that is, $v = (d\omega/dq)_{q=0}$. The frequency is a sinusoidal function of q, and the derivative gives $v = d\sqrt{K_g/M}$, where d is the length of the primitive cell of the linear chain, M is the mass of Rb, and K_g is the interatomic force constant. Within the pressure range studied, d and K_g can be determined from x-ray diffraction data. Experimental sound velocities for the host and the guest are determined from the respective initial slope of fitted phonon dispersion relations. The dispersion of the Rb guest vibration is compared with the velocities estimated from the simple monatomic model at several pressures in Figure 30c. The agreement is astonishingly good, confirming that the interaction between the host and Rb atoms in linear chains is weak, and the two subsystems vibrate independently.

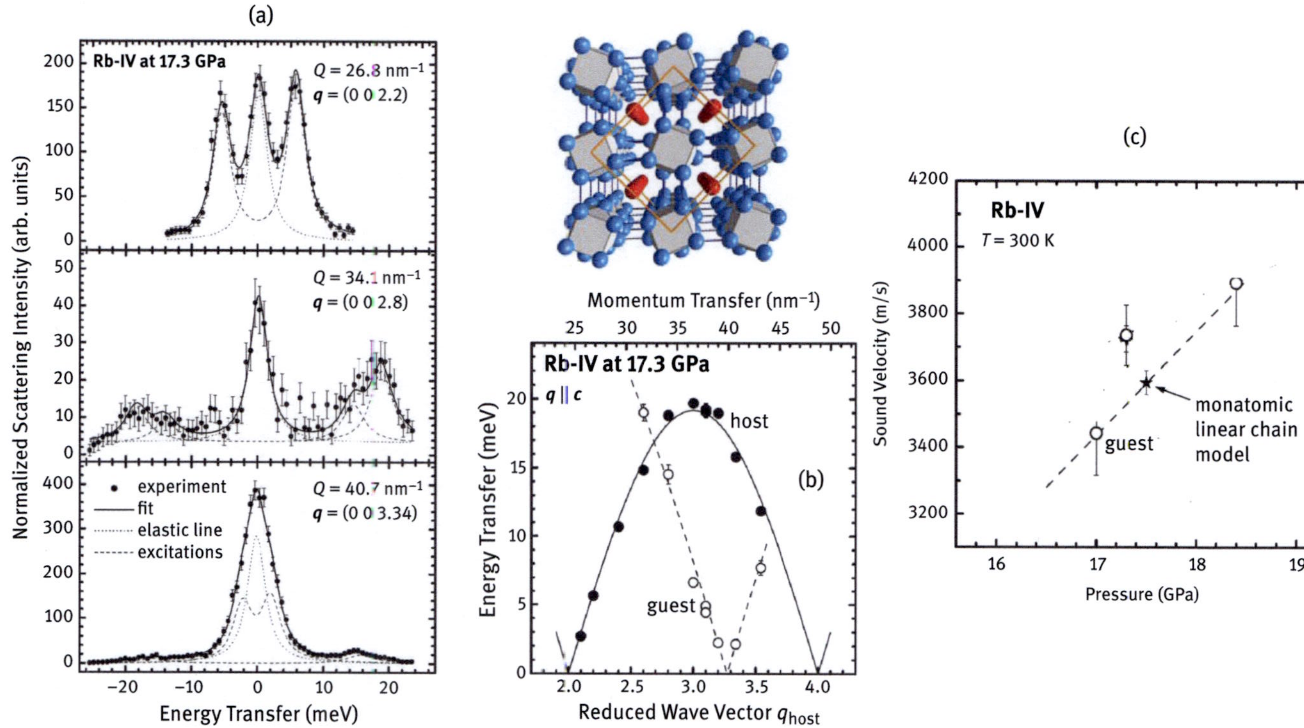

Figure 30: Phonon dispersion in high-pressure Rb-IV. (a) Experimental UXS spectra at different momentum transfers. (b) A plot of the phonon dispersion showing distinct host and guest vibrational branches. (c) Comparison of the guest sound velocity with the one-dimensional monatomic chain mode.

3.4 Electron dynamics

Electronic excitations, such as bound–bound and photoionization, are in the order of several to thousands of eV. Different information on the electronic property of the material can be obtained in different energy regimes (see Figure 1). Low energy excitation, such as bound–bound transition, provides information on the band gap. Higher energy (soft x-ray) leads to ionization of the valence electrons that are responsible for chemical bonding. The core binding energies of atoms are spread over a broad energy range from 90 eV to 89 keV. The large energy separation of the core energy levels between different elements makes x-ray (250 eV to 70 keV) photoabsorption an atom-specific method that offers information on the probe atom's immediate environment. At even higher energy (γ-ray), momentum exchange between photon and electron (Compton scattering) can be used to investigate electron correlation effects.

3.4.1 Plasmon excitation

A phenomenon of low-energy electron dynamics is the collective oscillations of the electron density of weakly bound electrons in metal when coupled to electromagnetic radiations. In a free-electron model, the resonance (plasma) frequency is related to the valence electron concentration and effective mass. The bonding and dynamic (transport) behavior of the valence electrons can be deduced from the plasma frequency. Sodium is the archetypal free-electron metal. When compressed, this behavior changes dramatically. Above 190 GPa, Na is a wide bandgap insulator [82]. So how a free-electron metal change to an insulator? The study of the plasma frequency as a function of pressure can provide insight into the puzzle. As mentioned above, some high-pressure polymorphs of alkali metals exist in exotic host–guest structures. In lighter elements, Li and Na, the transformation sequence is less spectacular. The ambient bcc first transforms to fcc and then to slightly more complicated structures at pressures above 40 GPa for Li and 110 GPa for Na. Li becomes an insulator at 70 GPa and re-entrant to the metallic state at 120 GPa in a large unit cell structure. In comparison, both K and Rb transform to the host–guest structure at relatively low pressures. On the other hand, after the bcc to fcc transition at only a few GPa, Cs exists in a *Cmcm* structure with 84 atoms per unit cell in a very narrow pressure window and then back to a simple tetragonal structure. At present, there is no systematic and satisfactory theoretical explanation of these observations. Investigation of the evolution of the electronic structures with pressure will shed light on this problem. The first reported high-pressure HERIX experiment on the plasmon frequency of Na was up to 8 GPa [83]. Loa et al. extended the plasmon frequency measurement to 43 GPa [84], within the stability field of the bcc structure. Similar experiments were performed on K and Rb, both to 10 GPa. The IXS spectra of Na at $q = 5\ nm^{-1}$ at different pressures are shown in Figure 31a. The q-dependent spectra at 16 GPa are shown in Figure 31b. At a constant q, the plasmon frequencies (ω_p) shift to higher energy. The trend follows very well with the simple nearly free-electron (NFE) relationship, $\omega_p = \sqrt{ne^2/\varepsilon_0\epsilon_s m}$, where n is the electron number density, e is the electronic charge, m is the effective mass, ε_0 and ε_s are the dielectric in free space and in the material, respectively. Na is very compressible, and the electron concentration increases with pressure due to the volume reduction. In the NFE theory, the dispersion of ω_p is quadratic in q, $\omega_p(q) = \hbar\omega_p + (\hbar^2/m)\alpha q^2$, with the dispersion coefficient $\alpha = \frac{3}{5}E_F/\hbar\omega_p$, where E_F is the Fermi energy. A plot of the plasmon frequency versus q^2 for several pressures (Figure 31c) verified this approximation. At the highest pressure reported (43 GPa), Na remains in the bcc structure. Compression simply reduced the volume by 65% and increased the electron number density. The NFE model is qualitatively correct in this pressure regime. In comparison, the plasmon frequency of Rb shows a maximum

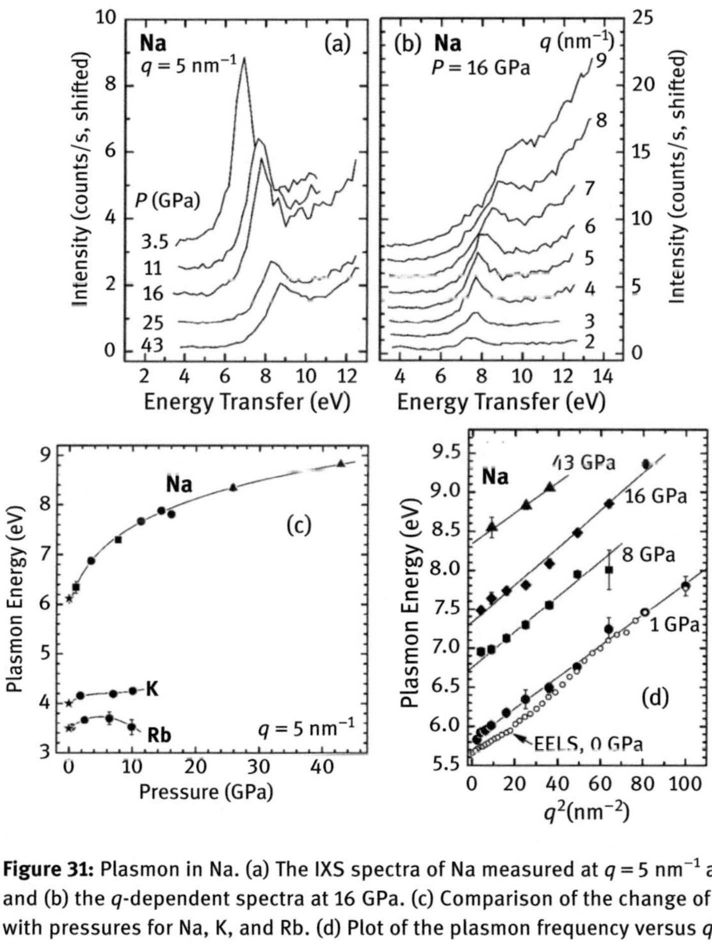

Figure 31: Plasmon in Na. (a) The IXS spectra of Na measured at $q = 5$ nm^{-1} at different pressures and (b) the q-dependent spectra at 16 GPa. (c) Comparison of the change of plasmon frequency with pressures for Na, K, and Rb. (d) Plot of the plasmon frequency versus q^2 at several pressures.

at 4–6 GPa. This pressure corresponds to the onset of bcc to fcc transition. Since the electron concentration increases when the volume is decreased, if the NFE model is still valid, the drop in the plasmon frequency indicates a heavier effective mass. The valence band dispersion in the fcc phase may not be as free electron like as in the bcc phase. Below 10 GPa, K is in the bcc phase. The plasmon frequency only increases slightly and almost linearly with pressure. The plasmon frequency contains rich information on the electronic structure. It is desirable to extend the investigation into Li and Cs in which the sequence of phase transformations and the crystal structures is different from Na, K, and Rb.

3.4.2 Core electron excitation

3.4.2.1 Water and ice

The binding energies of the core electrons are well separated from the valence levels. The excitation spectra provide information on the chemical environment and the electronic structure of the empty states. The sensitivity of the core absorption spectra to the local structural environment is demonstrated in theoretical studies of the oxygen 1*s* (O-*K*) x-ray absorption spectra (XAS) of water clusters [85]. The calculated spectra using the standard density functional at different H-bond environments are compared in Figure 32a and b [69]. The theoretical absorption profiles are found to change substantially with the arrangement of the nearest neighbor water molecules. Note that the pre-edge feature at 535 eV is absent when the H-bond with the two nearest waters is collinear to the O–H bond of the central water (Figure 30a). In contrast, a distinctive peak is predicted if the H-bond deviates from linearity. It was suggested that this feature is a unique fingerprint to characterize the local bonding in liquid water. By matching the calculated spectra with observed XAS (Figure 32c), it was concluded that the local environment of water is not fully coordinated with H-bonds (Figure 32). This proposal contradicts the accepted picture that water is nearly tetrahedrally H-bonded to the nearest neighbors in the liquid state. The XAS technique is subjected to several shortcomings due to the low probing depth and absorption effect of the soft x-ray. IXS at low momentum transfer is an alternative way to measure bulk core-level spectra of a material.

To put this hypothesis to test, IXS spectra of several ices and liquid water forms were measured *in situ* on the same spectrometer [86]. The experiment took advantage of the properties of high-density amorphous ice (HDA) made from the pressurization of crystalline ice Ih at 80 K. It is known upon heating that the recovered HDA ice transforms successively to the low-density form (LDA), then to ice Ic and ice Ih, and eventually melts. Therefore, there is no ambiguity in energy calibration and instrument resolution that may alter the shape of the pre-edge feature. The results of the study are reported in Figure 33. The spectra can be partitioned into three regions: a pre-edge

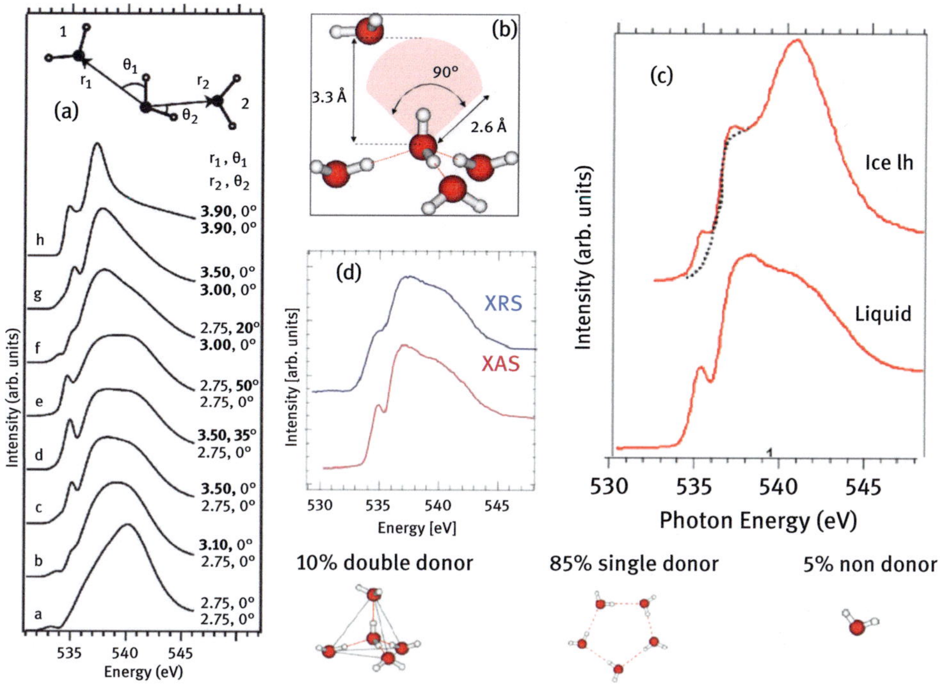

Figure 32: (a) Predicted x-ray absorption spectra (XAS) of water clusters at different local environment. (b) Experimental XAS of liquid water. (c) Comparison of the XAS between ice Ih and water.

region (I) starts from the absorption threshold at 533–536 eV; a near-edge region (II) at 537–539 eV; and a post-edge region (III) from 539 eV and beyond. The intensities of the pre-edge peak in IXS spectra of HDA ice and liquid are very similar. The observation indicates that the local environment of the water in HDA and in the liquid is also similar. It is noteworthy that HDA has been proposed to be a model of quenched water.

Is the pre-edge is a reliable indicator of a "broken" H-bond network? Ice has a rich phase diagram with several dense crystalline forms (Figure 34). Further investigation of the IXS of high-pressure crystalline ices, namely, ice VI, ice VII, and ice VIII is done by compressing water in a DAC at room temperature [90]. The IXS spectra at low and high momentum transfers (q) are shown in Figure 33a and 33b, respectively. The low-q IXS spectrum of the high-pressure dense crystalline ices is very different from ice Ih. In fact, the IXS for both ice VII and ice VIII shows a strong resemblance to HDA and liquid water. Since all the water in the dense crystalline ices is fully four-coordinated, the pre-edge peak cannot be attributed to the disrupted H-bond. In comparison, the intensity distribution of the IXS spectrum for ice IX (Figure 33b), a proton-ordered form of ice III, is quite similar to ice Ih. The main structural difference between ice IX (III) and denser ices is that the water framework of ice III transit from a single H-network to an interpenetrating two independent H-

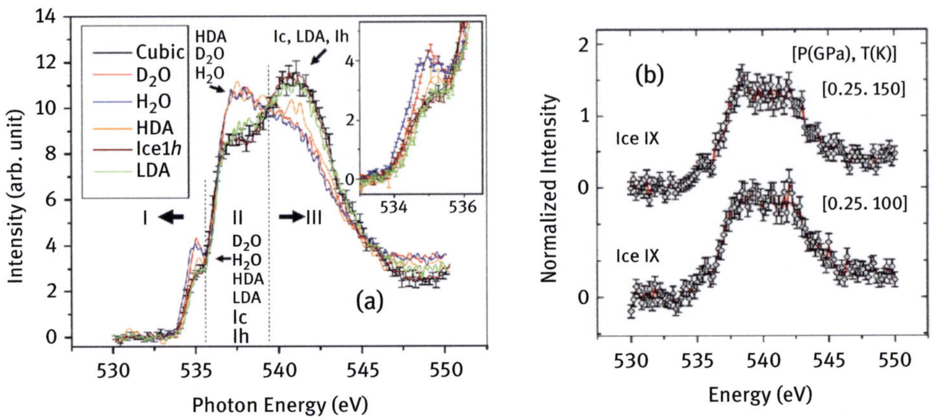

Figure 33: O-*K* IXS spectra of several polymorphs of solid water (a) at ambient pressure and (b) under compression.

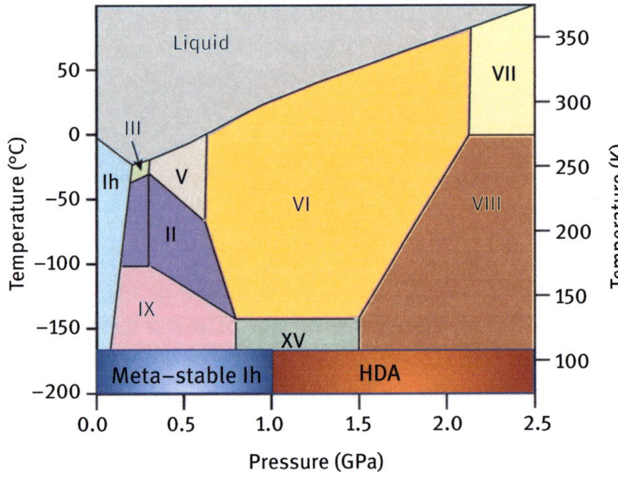

Figure 34: The phase diagram of water (adopted from [87]. Absence of amorphous forms when ice is compressed at low temperature Nature 569, 542–545 (2019)).

bond networks in ice VI and higher pressures. The low and high q IXS of ice VI, VII, and VIII (Figure 35a and b) show no substantial differences. This observation is not surprising because the oxygen atom does not have low-lying orbitals of d-symmetry (l = 2). Therefore, contributions from non-dipolar excitations (e.g., $s \rightarrow d$) are small.

The end member of liquid water is a single uncoordinated molecule. The O-*K* XAS of water in the gas phase is shown in Figure 36a [88]. Low-density water can be produced near the critical point. The IXS for supercritical water at several densities are compared in Figure 36b [89]. The gradual migration of the spectral profiles to water

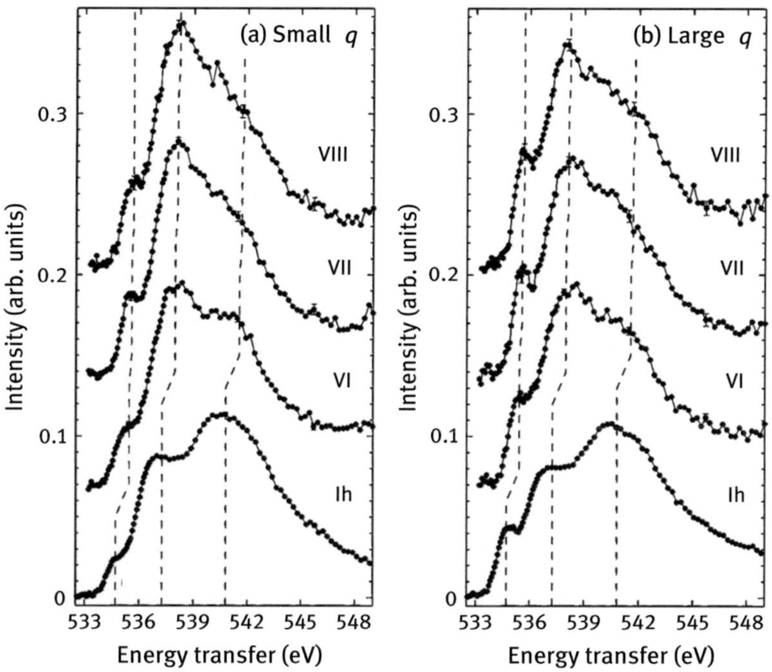

Figure 35: O-*K* IXS of dense ices, ice VI, VII, and VII. Comparison of measured and computed IXS spectra of a-SiO_2.

vapor is evident. The pre-edge is becoming more distinctive. This observation is reproduced very well by the theoretical spectra (Figure 36c) [91]. The instantaneous water structures at two densities are presented in Figure 36d. The sharpening and increasing intensity of the pre-edge peak at a lower density and higher temperature are due to the more localized lowest energy excitation to the antibonding orbital of the more isolated water molecules.

The results and discussion presented earlier show that the pre-edge feature in the O-*K* IXS is not unique to liquid water. It is also observed in other crystalline LDA ices at almost the same excitation energy. The water in the crystalline ices at ambient and high pressures undoubtedly adopts a fully four-coordination local structure. The pre-edge feature is presented in all the cases studied. This shows the peak is not a unique feature for characterizing the nature of the H-bonding network. The excitation cannot be correlated to an electronic transition to the antibonding orbital as in the free molecule. In the condensed phases, the wave functions of excited states are not confined to a single site.

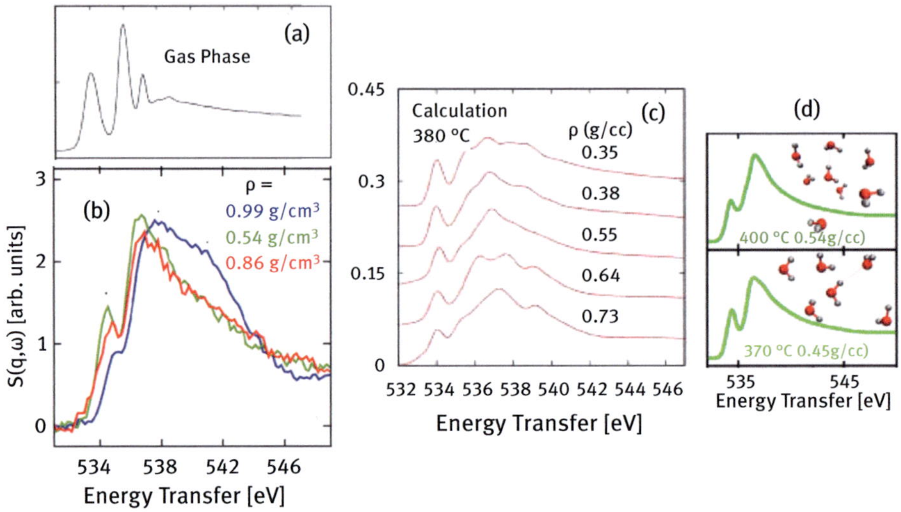

Figure 36: O-*K* IXS of (a) water molecule. (b) Supercritical water at different densities. (c) IXS spectra calculated from density functional theory. (d) The structure of supercritical water at two densities.

3.4.2.2 Oxide glasses

Glasses are often used as proxies to infer structures of the corresponding melts. The conjecture is particularly useful for mineral melts composed of mostly silicates and difficult, if not impossible, to study under extreme pressure and temperature. The validity of the assumption that there is a direct correspondence between the structure of glass and the melt has not been convincingly validated. Boric oxide glass is an essential ingredient for many industrial processes. It has a relatively low melting point, and so it also serves as a convenient model for higher melting oxide glasses such as silica and germania. Boron oxide is a trivalent oxide (B_2O_3) with a trigonal planar local structure in the normal state. Under compression, the oxygen environment around the B can change from three- to four-coordination. Knowledge of the pressure-induced structural change is essential to understand the macroscopic properties and extrapolate to the melt. Boron has three valence electrons and adopts sp^2 hybridization when bonded to three oxygens. Therefore, a vacant *p*-orbital (p_π) is perpendicular to the trigonal plane. When boron oxide glass is compressed, the empty orbital can be utilized to accept the lone pair from neighboring oxygen, and the local structure becomes tetrahedral. Governed by the dipole selection rule, the structural change is easily detected in the B-*K* [92] and O-*K* [93] core electronic spectra. The associated changes in the IXS spectra are clearly shown in Figure 37a and b. In particular, at zero pressure, the dominant feature is the transition from B 1s to the low-energy vacant p_π orbital with a broad band assigned to the excitations to the higher energy antibonding B–O σ^{*-} orbitals. As pressure increases, the planar structure of B starts to distort and becoming more tetrahedral. When the additional bond with the lone pair oxygen is fully established, the intensity of the π transition decreases substantially, and excitations of the core electron to σ^* become the most intense band in the IXS spectra. From 16 to 110 GPa, no dramatic change in the IXS profiles is observed. When the pressure is increased to 119.4 GPa, a band appears at 205 eV. This was attributed to an increase in oxygen coordination greater than 4. Since the first row elements have no low-lying empty orbitals with angular moment (l) higher than the p-orbital, a structural change associated with higher coordination is not expected. This is the underlying reason that diamond is so stable,

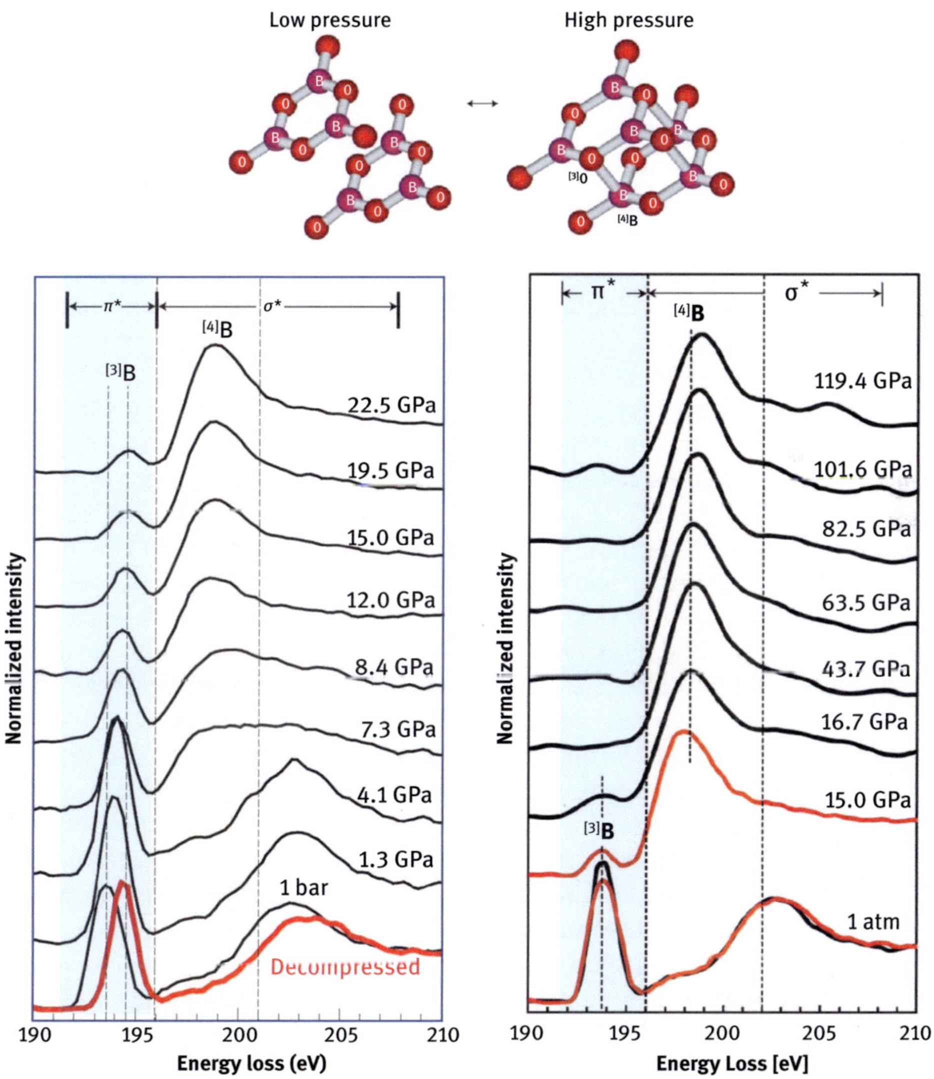

Figure 37: (a) B-*K* and (b) O-K IXS of compressed B_2O_3 glass. The appearance of the high-energy feature in both cases indicates the transformation from 3- to 4-B–O coordination.

and no change in the local coordination was found in the predicted high-pressure ice polymorphs.

3.4.2.3 Silica

Silica is the vitreous form of silicon dioxide (a-SiO_2). It is an important prototype of silicates found commonly in Earth's minerals. It is of utmost importance to understand the structural change of this model system under extreme pressure and temperature to elucidate the properties of the melt. The properties of a-SiO_2 have been studied extensively. There is doubt on the local Si–O structure at very high pressure exceeding 100 GPa. Using low-q O-K IXS as a probe, it was found that the emergence of a new band at 454 eV, indicating a possible structural change with increased Si–O coordination, has occurred above 14 GPa (Figure 38a) [94]. This intensity of this band increases gradually with increasing pressure (51 GPa). However, a study of the Si-L IXS arrives at a different conclusion [95]. The high-q Si-L IXS spectra up to 74 GPa are shown in Figure 38b. No significant change in the spectral profile was observed, notwithstanding the inferior quality of the spectra. This experiment suggested that there is no structural change, and the Si–O coordination remains around 4. Intuitively, one would expect that as the Si–O coordination change from 4 to 6 at high pressure, the electronic structure at the Si site will be affected most and the Si-L IXS should be more susceptible to the structural change. We wish to comment on why a higher q was used to measure the core-level IXS of Si. Beryllium, which is used as the gasket material for the DAC, has a K-shell binding energy of 111.5 eV. The energy is very close to the Si-L edge of 107 eV, and the scattered photon intensity from Si is highly suppressed. Besides, the typical x-ray energy for IXS experiments is 10 keV. At this energy, Compton scattering dominates the low energy part of the spectrum at small q. For practical purposes, it is advantageous to measure the Si-L IXS at higher momentum transfer to avoid this problem. The dipole selection rule ($\Delta l = \pm 1$) is no longer valid at high momentum transfer. For Si-L edge, non-dipole excitations ($\Delta l \neq \pm 1$), such as $p \to s$ and $p \to p$, can contribute to the spectrum. Figure 38c compares the measured and calculated IXS spectra of Si at $q = 1.2$ (low q) and 5.17 a.u. (high q; 1 a.u. = 1.89 $Å^{-1}$ = 0.189 nm^{-1}). At low q, dipole allowed transitions to s- and d-orbitals to dominate the spectrum. The non-dipole $p \to p$ transition becomes important at high q. Furthermore, the spectral features are more clear at high q.

Recently, O-K and Si-L IXS experiments were repeated together with the state-of-the-art theoretical calculations, including electron–hole dynamics correlation of the q-dependent IXS spectra [96]. The O-K IXS measured at low q are presented in Figure 37a. The theoretically calculated spectra using the structures obtained from the first-principles molecular dynamics (MD) [97] calculations are in excellent agreement with the experiment. All the main features observed are reproduced by the calculations. The examination of the MD trajectories, the higher energy band observed at pressures higher than 14 GPa in the previous O-K IXS experiment, is due to an

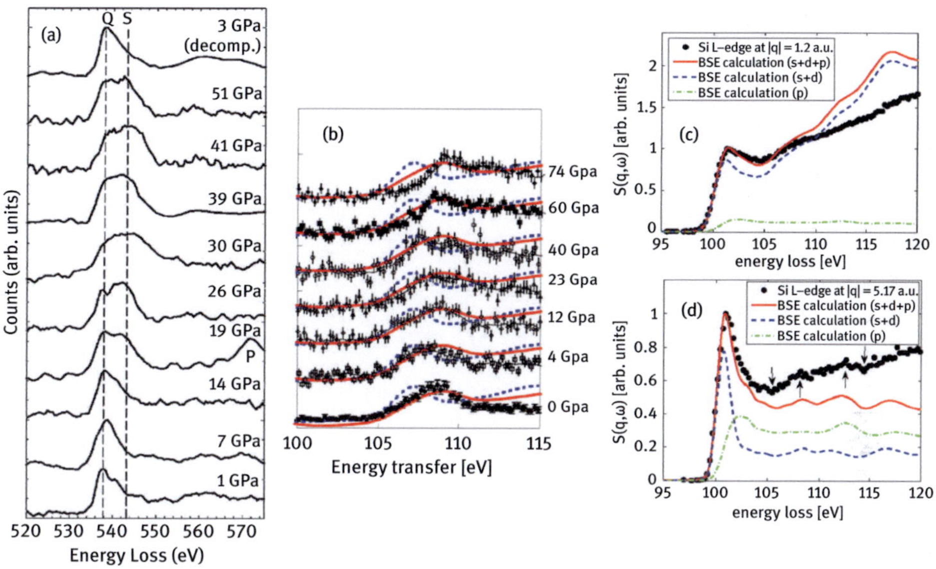

Figure 38: (a) O-*K* and (b) Si-*L* IXS spectra of *a*-SiO_2 under pressure. (c) Si-*L* IXS spectra of pure Si calculated using the Bethe–Salpeter method (BSE) at low ($q = 1.2$ a.u.) and high ($q = 5.17$ a.u.) momentum transfer.

increase in the O–Si coordination. However, it is not a straightforward transition from O bridging two Si atoms to three-coordination, but a mixture of two- and threefold O–Si coordination. Threefold O–Si coordination becomes dominant above 60 GPa. The onset of the O *K*-edge correlated very well with the change in the coordination number predicted by MD calculations (Figure 39b). Comparison of the experimental and theoretical Si *L*-edge IXS provides complementary results. Again, the calculated high *q* Si-*L* IXS spectra are in good agreement with the experiment (Figure 40a). Analysis of MD structures shows that above 20 GPa, the Si–O is a mix of four- and five-coordination, then followed by five- and sixfold coordination. Fully sixfold coordination is established above 60 GPa. Once again, the onset of the Si *L*-edge was found to correlate well with the change in the coordination number (Figure 38b). In view of the favorable comparison between the experimental and theoretical calculations (including electron–electron and electron–hole correlations), the combined IXS and MD study provides a clear understanding of the structural evolution of a-SiO_2 up to 120 GPa. This is by far the most consistent description of the structure of highly compressed a-SiO_2. The results obtained explain the change of compressibility at high pressures in the density data and the variation of the strength of the glass. One may speculate that if the silica melt structure follows that of the glass, then the depolymerized silica melt from silicate minerals at the lower mantle will produce negatively buoyant silicate at the core–mantle boundary during the early Earth's formation.

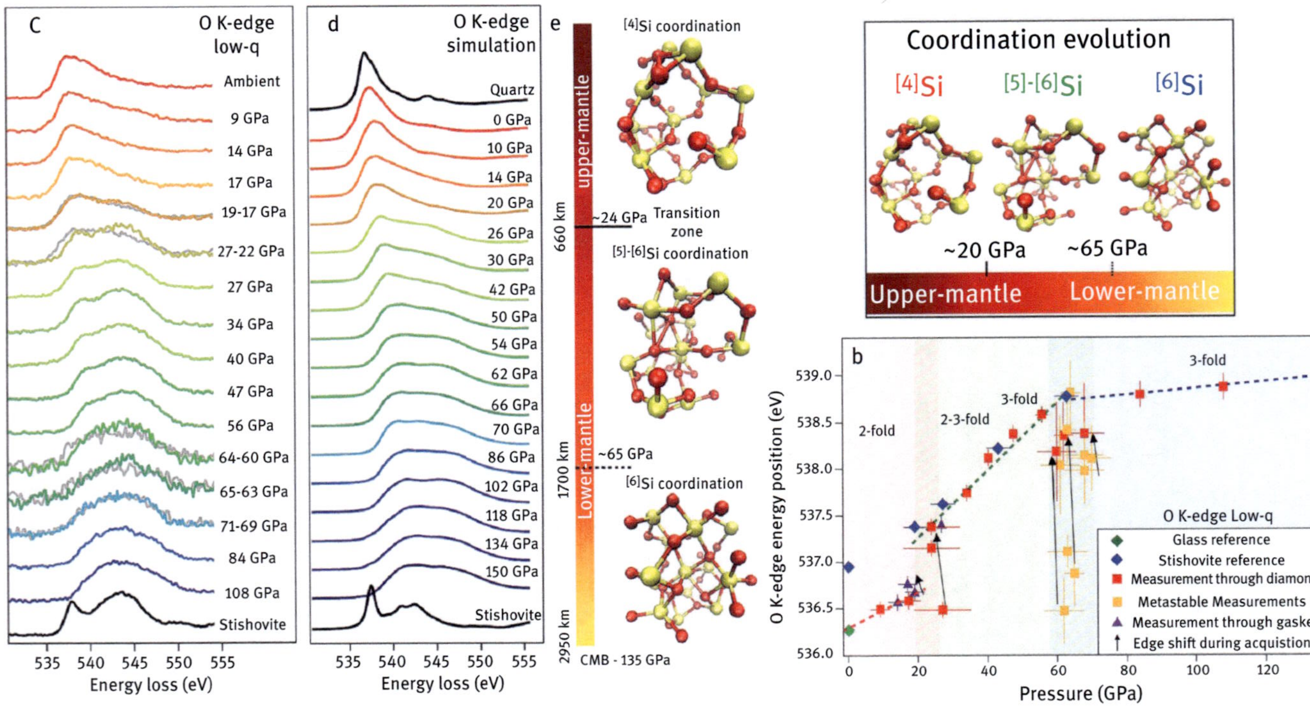

Figure 39: Comparison of measured and calculated O-*K* IXS of *a*-SiO_2 at large momentum transfer. The good agreement between theory and experimental helps to elucidate the structural transformation in the glass associated with increases of Si–O coordination number indicated by the plots of the local structures with pressure.

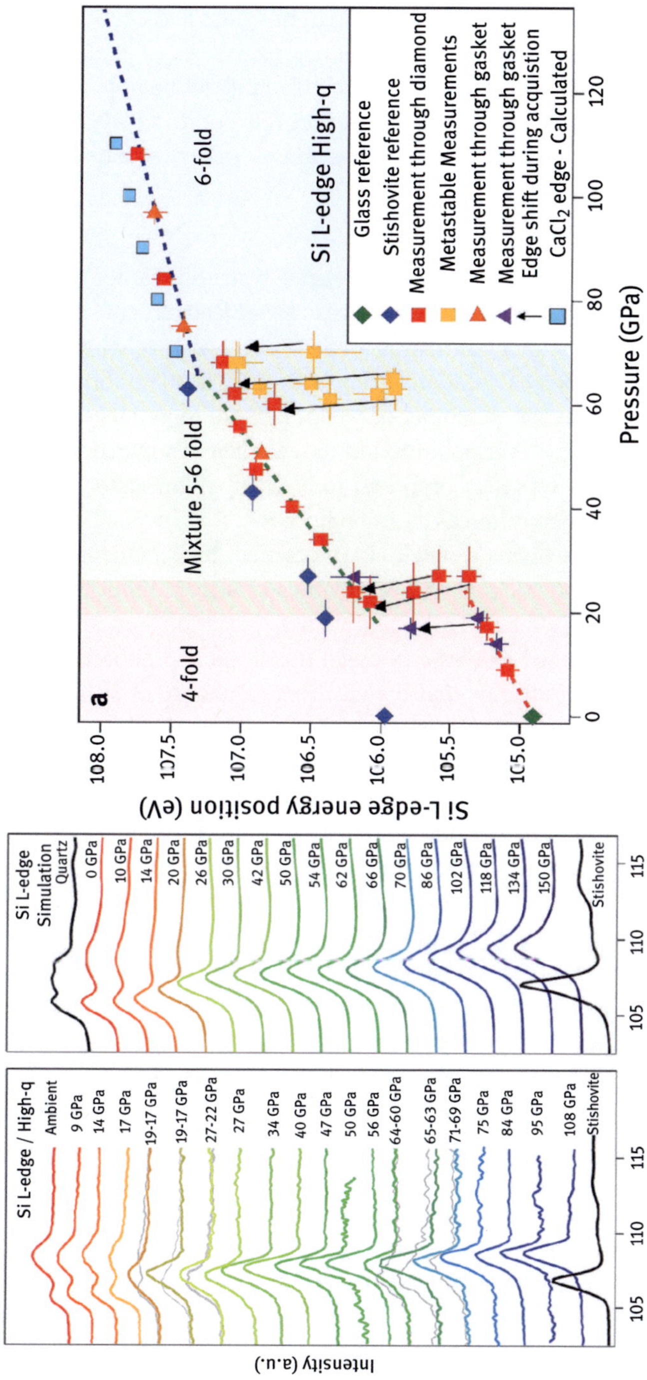

Figure 40: Comparison of measured and calculated Si-*L* IXS of *a*-S O_2 at large momentum transfer and the change of Si–O coordination with pressure.

Reliable knowledge of the evolution of the structure in the glass is essential. Very often, structures of known crystalline phases were used as references to explain the glass structures at similar pressures. On some occasions, this may not be appropriate. For example, studies on a-SiO_2 and a-GeO_2 suggested that five-coordinated Si–O or Ge–O exists in the glasses. No corresponding crystalline phases were ever found.

3.4.2.4 Silicon

Elemental silicon (Si) holds a special place in high-pressure research. Under normal conditions, Si is a small gap semiconductor. The first pressure-induced structural transformation from the fcc (Si-I) to a metallic (Si-II) was predicted at a relatively low pressure of 10 GPa by the first-principles calculations prior to experimental confirmation. The success laid the foundation for the reliability of computational materials science. The metallic phase (Si-II) was the first-ever pressure-induced elemental superconductor [69]. The hydride of silicon, SiH_4, was also proposed to be a superconductor with a high critical temperature (T_c). Superconductivity in compressed SiH_4 was indeed reported later. The latter experiment opens a new field of research in superconductive polyhydrides (superhydrides) that led to recent discoveries of superconductivity in highly compressed SiH_4 [97], H_2S [98] and LaH_{10} [99] with T_c close to room temperature. The ease of the pressure-induced insulator to metal transition and superconductive behavior in the metallic state indicates that the electronic structure of Si is easily manipulated by the application of external pressure. This phenomenon may be rationalized in a simplistic chemical viewpoint [69, 101]. The presence of low-lying empty and spatially extended *d*-orbitals hybridize with valence 3*s* and 3*p* orbitals when compressed, the bonding valence electrons delocalize, turning the insulator into a metal. Moreover, when bonding electrons are promoted to the *d*-orbitals, there are more empty orbitals than electrons can fill. The orbital degeneracy favors electron–phonon coupling and, under certain conditions, leads to superconductivity. The concept of hybridization of valence *s*, *p* with empty *d* orbitals is generally accepted [101]. However, there have not been serious attempts to quantify the change in the electron density topology experimentally. This was accomplished recently by which the electron density of high-pressure polymorphs of Si was extracted from high-resolution powder diffraction data by the maximum entropy method [102]. The results confirm the expectation that *d*-orbital participation increases in the high-pressure phases. Changes in the electron topology are illustrated in Figure 41a. The concurrent change in the electronic structure is characterized by IXS spectra measured at high *q*. Below 10 GPa, Si is an insulator and shows a distinctive low energy band at 101.5 eV (Figure 41b). This band disappears when Si is transformed into the metallic phase. The disappearance is due to the core electron being excited to delocalize empty states in metal as localized antibonding orbitals no longer existed. The calculated IXS spectra (Figure 41c), including core–hole dynamics, agree favorably with experiments.

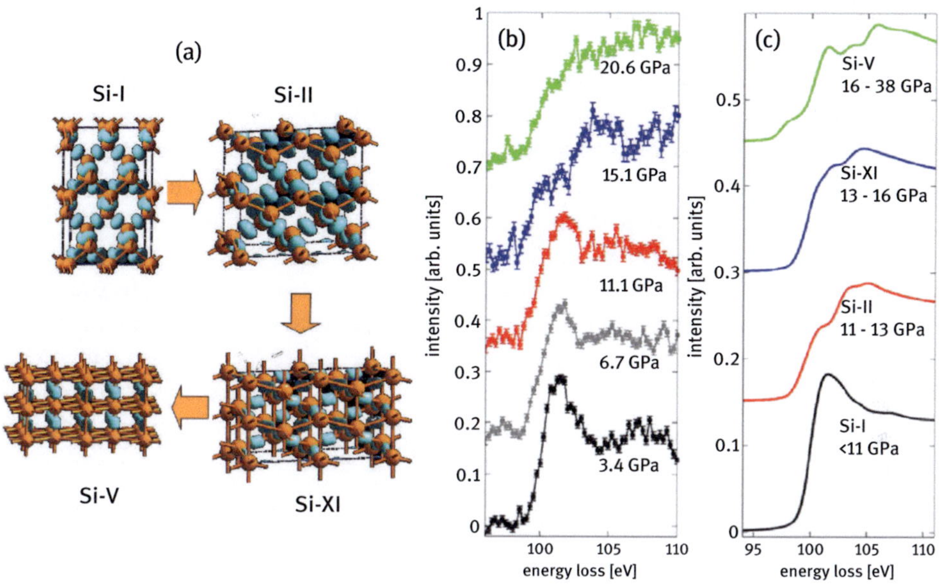

Figure 41: (a) Pressure-induced structural transformations in Si. (b) Observed and (c) computed Si-*L* IXS under pressure.

3.5 Electron spin dynamics

As we already learned, Fe is an important component of the Earth's minerals. In the ground state, Fe atom has an electron configuration [Ar]$3d^6 4s^2$. Fe often exists in the ionic form in minerals, either as Fe^{2+} or Fe^{3+} cations. Since the 4*s* electrons are removed before the 3*d*, in either case, the 3*d* orbitals are not fully occupied. At the presence of a static electric field produced by the surrounding charge distribution, the degeneracy of the five 3*d* orbitals will be lifted. Due to mutual repulsion and depending on the symmetry and the strength of the static electron field [103], the electrons will be distributed in different ways among the five orbitals. This leads to other electronic states with different energies.

In an electric field with the charge distributed in octahedral symmetry, the Fe 3*d* orbitals will split into two degenerated sets, t_{2g} and e_g. Suppose the energy separation between the two sets of orbital is large, and the electron repulsion is small, in this case, electrons prefer to pair up and occupy the lower energy triply degenerate t_{2g} orbitals. An example is Fe^{2+} in a strong electric (crystal) field [103]. The six 3*d* electrons will occupy the three t_{2g} orbitals with spin-pairing. There is no net electron spin and the system is diamagnetic. Conversely, it may be energetically more favorable to distribute the six 3*d* electrons among the t_{2g} and e_g sets in a weak crystal field due to electron repulsion. In this case, not all the electrons are spin-paired. There will be a net spin and the system is paramagnetic or ferromagnetic. Treating the neighboring anions as point charges, Coulomb's law dictates that the electrostatic field experience by the cation is inversely proportional to the distance to the anions. When the separations are large at low pressure, the field will be weak. Compression will shorten the distances and, therefore, the crystal field is dependent on the pressure. At some stage, when the crystal field becomes so large, rearrangement in the distribution of the electrons will occur. This phenomenon is called a spin transition or spin crossover. This phenomenon is common to Fe-bearing minerals and is usually found in the Earth's lower mantle. At low pressure, when the crystal field is weak, Fe atoms in minerals are often found in the high spin (HS) state and the material is magnetic. At high pressure, the transition to low spin (LS) results in a spin-paired electronic state with no magnetic moment and loss of magnetism. This process involves the redistribution of the electrons and affects the electron–electron interactions. The spin crossover is associated with a volume reduction, which significantly changes the compressibility, sound velocities, radiative thermal conductivity, and electrical conductivity. Hence, understanding the spin transition mechanism is highly relevant for understanding the structure and dynamics of the Earth's lower mantle.

The strength of the crystal field is the key factor for a spin transition. The crystal field is characterized by the energy spread of the $3d$ manifold (10 Dq) under the electrostatic field [103]. A typical value of 10 Dq is about 2.5 eV, which falls in the visible region of the electromagnetic spectrum (Figure 1). In principle, optical spectroscopy using ultraviolet radiation can be used to observe the d–d transitions and determine the crystal field since diamond is transparent up to 5 GPa. However, d–d transitions are generally very weak as they are not dipole allowed. Fe $L_{2,3}$-($2p$) edge soft XAS can only be used for samples under ambient pressure as the ionization edge is about 720 eV (Figure 42a). Furthermore, accurate information on the crystal field can only be extracted if the fine details due to electron–hole and spin–spin interactions can be resolved from the spectrum. The Fe-$M_{2,3}$ ($3p$) edge absorption is perhaps more suitable as the $3p$ orbital binding energy is close to $3d$, the $3p$–$3d$ interaction is larger, resulting in a much richer spectrum with many features to analyze.

The sensitivity of the Fe $M_{2,3}$ edge IXS spectroscopy for probing the spin transition at high pressure was explored in a study on FeS as an HS–LS transition is known to occur at 6.5 GPa by x-ray emission experiment [104]. The pressure-dependent measurements of the Fe $M_{2,3}$ edge on FeS at a small momentum transfer for selected pressures between 1.7 and 10.1 GPa [105] are presented in Figure 42b. Differences in the IXS features before and after the spin transition (6.5 GPa) are observed. The spectral intensity of the band at 57 eV is decreased. It is compensated by the increase in the intensity of the broad band centered at 68 eV. The experimental observation agrees with the intuitive expectation. In the LS state, the higher energy e_g orbitals are empty, allowing more excitations into these orbitals and the intensity is enhanced. For a quantitative analysis of the pressure-induced spin transition, difference spectra were calculated by subtracting high-pressure spectra from the reference at 5.3 GPa (Figure 42d). The experimental spectra are compared to the theoretically calculated atomic multiplet

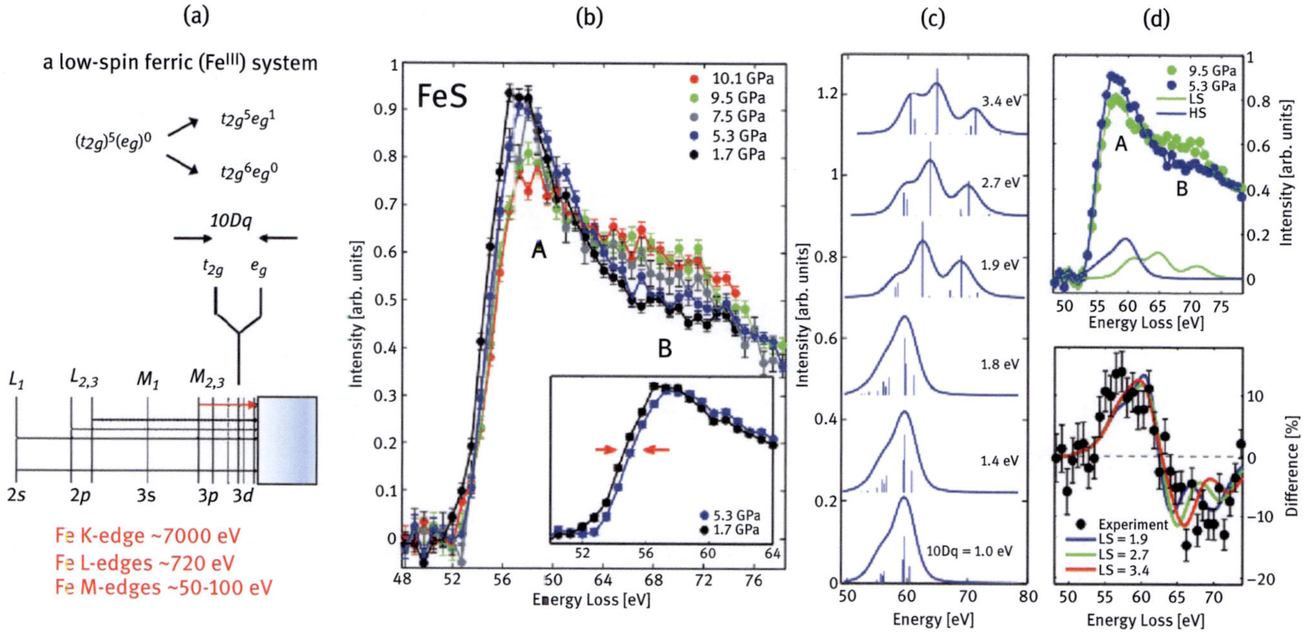

Figure 42: (a) Splitting in the 3*d* manifold of Fe^{2+} in an octahedral crystal field. (b) Fe $M_{2,3}$ edge IXS of FeS under pressure. (c) Calculated Fe $M_{2,3}$ IXS spectra based on a crystal field model. (d) Comparison of calculated and observed IXS spectra.

spectra (Figure 42b and c) at different 10 Dq values to extract the crystal field splitting parameter. From minimizing the spectral difference between experiment and theory, the best fit occurs between 10 Dq values of 1.8 and 1.9 eV. Admittedly, the quality of experimental data is not optimal. The energy resolution of the experiment is only 1.4 eV. The investigation shows that it is feasible to use $M_{2,3}$ edge IXS to extract crystal field parameters under pressure from spectral change analysis. With the improvements in the intensity and energy resolution of the current generation of IXS spectrometers, it is optimistic that Fe $M_{2,3}$ edge IXS will become a suitable tool for *in situ* characterization of the change in crystal field induced by HS to LS transitions at high pressure.

3.6 Compton scattering with hard x-ray

Compton effect is the scattering of photons by electrons [3]. It is an extreme example of IXS of ~ 100 keV x-rays with very high energy transfer in the range of keV and high momentum transfer in the range of 10–100 nm^{-1}. The technique is useful to measure the momentum density of valence electrons which are responsible for forming chemical bonds. The determination of Compton profiles is challenging both technically and in the data reduction. There are few applications of this technique at high pressure. Third-generation synchrotron helps to alleviate some of the potential problems. The use of intense hard x-ray and crystal energy analyzer has greatly improved the momentum resolution.

Compton scattering is an extreme example of IXS of ~100 keV x-rays with very high energy transfer in the range of keV and high momentum transfer in the range of 10–100 nm^{-1}. Two types of Compton scattering measurements have been developed: magnetic Compton scattering and high-resolution Compton scattering. [106]. High-resolution Compton scattering uses a Cauchois-type spectrometer, a bent-crystal analyzer, and a position-sensitive detector [107]. The magnetic Compton scattering, on the other hand, can provide a bulk-sensitive alternative to angle-resolved photoemission spectroscopy for determination of Fermi surface measurements. It measures the momentum distribution function $n(k)$, using back-scattering technique with a high-energy-resolution energy-dispersive detector. This approach requires circularly polarized incident radiation, which can be produced by an elliptical multipole wiggler. This method is applied to image the lithium distribution in Li-ion batteries during *in-operando* conditions [108].

Magnetic Compton scattering under high pressure can elucidate the nature of element-specific localized 3*d* and 4*f* spin magnetic moments and together with conduction electron components. One such study was conducted in Lave-phase $ErCo_2$ [109], where Er 4*f* and Co 3*d* spin magnetic moment contributions were detailed under pressure change. As shown in Figure 43, while the Co 3*d* spin moments show a remarkable reduction with increasing pressure, Er 4*f*-spin moments show weak

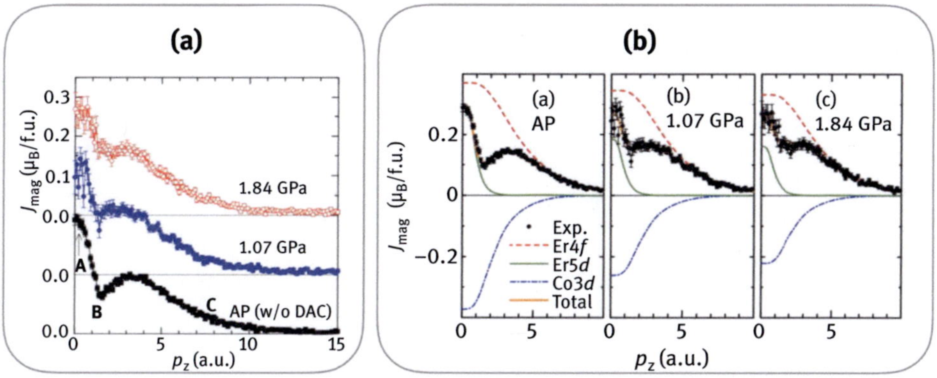

Figure 43: (a) Magnetic Compton profile (MCP) of ErCo2, taken under pressure (AP: ambient pressure), and (b) breakdown of MCP in terms of Er and Co outer orbitals.

variation with pressure. The main challenge in this field remains the ability to focus very high-energy x-rays to a few micrometer spot size.

The first high-pressure experiment was performed on Li embedded between two anvils. The highest pressure achieved was 1.8 GPa [110]. To further increase the signal-to-noise ratio, a Pairs–Edinburgh (PE) press with a larger sample volume was used to study metallic Na in a boron gasket. The Compton profile was measured up to 4.2 GPa. This is limited by the pressure that can be generated by the PE cell. A significant improvement was made later in a study on Si using the DAC [111]. Using a condensed refractive lens made from Al, monochromatized x-ray at 81 keV was focused to a 7 µm (vertical) × 50 µm (horizontal) spot on the Si sample in a DAC through a 250 µm thick Be gasket. The scattered photons were measured with a Ge-solid-state detector. Several spectra were accumulated on both the Si sample and the pure Be gasket. Measurements were made at 9 and 18 GPa. Since no pressure-transmitting medium was used, under the nonhydrostatic condition, Si exists as metastable rhombohedral Si-XII at 9 GPa and the simple hexagonal Si-V at 19 GPa. The Compton profiles at the two pressures are compared in Figure 44a. The strong line is due to quasi-elastic scattering. The results exemplify the difficulty of the measurement of accurate Compton profile. The core electron contribution dominates the Compton profile at high momentum. At first glance, there is no significant difference between the two spectra. As the signal-to -noise ratio is known and the counting statistic was sufficient, a proper correction to the raw data was made. After data reduction and normalization, the Compton profile of Si-XII is found to be slightly different from Si-V at low momentum, 0.9 a.u. The difference between the two pressures is further amplified by subtracting the Compton profile of Si-V from Si-XII shown in Figure 44b. Within the limits of statistical accuracy, the difference is negative between 0 and 1 a.u. with a dip between 0.7 and 1.3 a. u. The observed trend agrees with the theoretical calculated Compton profiles, which predicted a dip at 0.9 a.u. The experiment on Si shows that it is practical to measure

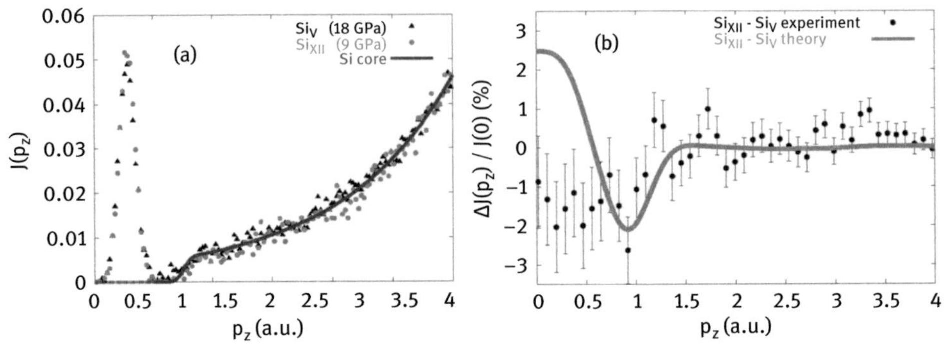

Figure 44: (a) Hard x-ray Compton scattering of compressed silicon at 9 and 18 GPa. (b) Comparison of the experimental and theoretical difference in the Compton profile between Si-XII and Si-V.

Compton profile in a DAC using a tightly focused and intense hard x-ray. With optimized gasket materials and design, Compton profile can be used to probe the electronic structure of materials under high pressure.

3.7 Synchrotron Mossbauer spectroscopy

3.7.1 Time domain: nuclear forward scattering

Nuclear forward scattering [111, 112] measures the coherent decay of the emitted radiation after nuclear excitations. The collective and coherent part of the radiative decay is peaked in the forward direction. When photons impinge on an ensemble of identical nuclei, they are resonantly absorbed, and there are many possible scattering paths in all directions. If the nuclear levels are hyperfine split, various transitions between ground and excited states are also possible. The energy difference between the allowed nuclear hyperfine transitions leads to quantum beats in the time spectrum. The hyperfine splitting (magnetic or quadrupole) of the excited and ground nuclear states and the energy shifts for different chemical environments at different nuclear sites (isomer shift) can be derived from the beating pattern.

3.7.2 Energy domain: pure nuclear Bragg diffraction

Energy-domain synchrotron Mossbauer method is similar to the conventional technique in which the nuclear excitation is prompted by a radiation source with an emission spectrum close to the nuclear energy level [16, 17]. At the synchrotron, the energy spectrum is produced by nuclear Bragg reflection from a suitable crystal

[113, 114]. The Bragg diffraction from a crystal is determined by a geometric factor that depends on the atomic arrangement in the crystal and the nature of the scattering process. In the case of scattering from atoms, which are also Mossbauer nuclei, there are two contributing factors: the coherent scattering by the electrons and the resonant scattering from the nuclei. The latter is dependent on the phase of the wave scattered from the nucleus and the orientation of the nuclear spin with respect to the wave vector. Due to magnetic hyperfine interaction, the spin orientation is related to the direction of the magnetic moment of the atom.

An 14.4 keV x-ray with a width in the order of neV has been achieved [110] from pure nuclear Bragg diffraction from the electronic forbidden (333) reflection of a single crystal of $FeBO_3$ near the Néel temperature of 75.9 °C in an external magnetic field. Interaction with the internal electric field gradient splits the ^{57}Fe nuclear level into several magnetic sublevels. Near the magnetic transition (from antiferromagnetic to paramagnetic), the magnetic dipole and electric quadrupole hyperfine interactions are strongly mixed. The magnetic hyperfine interaction becomes weaker and eventually disappears, and the quadrupole transitions lose intensity due to destructive interference producing a single-line x-ray beam close to the natural linewidth.

Both methods have their own merits and disadvantages. The advantage of the time-domain method is that the energy resolution is not limited by the emitter's lifetime (width). Instead, the energy resolution is determined by the length of the time employed to measure the decay processes. Therefore, the energy resolution of the calculated Mossbauer spectrum is not restricted by the natural linewidth of the emitter. However, the duration is limited by the time structure of the synchrotron radiation bunch. Furthermore, the quantum beat pattern can be very complicated for a sample with nuclei at different chemical distinct sites The extraction of Mossbauer parameters from the deconvolution of the beating pattern may be difficult. In comparison, the energy-resolved method is not dependent on the synchrotron bunch mode. It is possible to study Mossbauer nuclei that would otherwise require extreme timing conditions of the synchrotron. The Mossbauer spectra of an Fe_3S_4 sample [115] measured using laboratory ^{60}Co source (Figure 45b) and synchrotron x-ray in energy domain (Figure 45a) and time domain are compared in Figure 42. It can be seen that details on the magnetic hyperfine structure of multiple inequivalent Fe sites (Figure 45c) are revealed in the simulated pattern from the parameters extracted fitting of the beating patterns (Figure 45d). The time-domain method has also been used to study the high-pressure spin transition in hematite (Fe_2O_3). Fe_2O_3 is an antiferromagnetic insulator under ambient conditions. A structural transformation from the corundum to Rh_2O_3(II) accompanied by magnetic collapse has been reported to occur between 55 and 75 GPa. Figure 46 compares the ^{57}Fe Mossbauer spectra of Fe_2O_3 derived from time-domain nuclear forward scattering and the energy domain using an $FeBO_3$ source measured at high pressure. The quantum beat patterns of the time spectra (Figure 46a) are rather complicated below 63 GPa. The pattern becomes much simpler above 73 GPa. The corresponding simulated Mossbauer spectra are shown in Figure 46b. It is clear from the hyperfine patterns that Fe_2O_3 is magnetic at low pressure. It becomes nonmagnetic (NM) above 73 GPa. The calculated spectra also reveal that the NM component starts to appear at 43 GPa and become more dominant with increasing pressure until 83 GPa when Fe_2O_3 has fully transformed to the NM Rh_2O_3(II) phase. The energy-domain Mossbauer spectra (Figure 46c) show a similar trend. However, no impurity due to the NM phase was observed in the spectrum recorded at 43 GPa. The apparent absence of the NM phase may be attributed to the slightly lower resolution of the energy-domain spectrum.

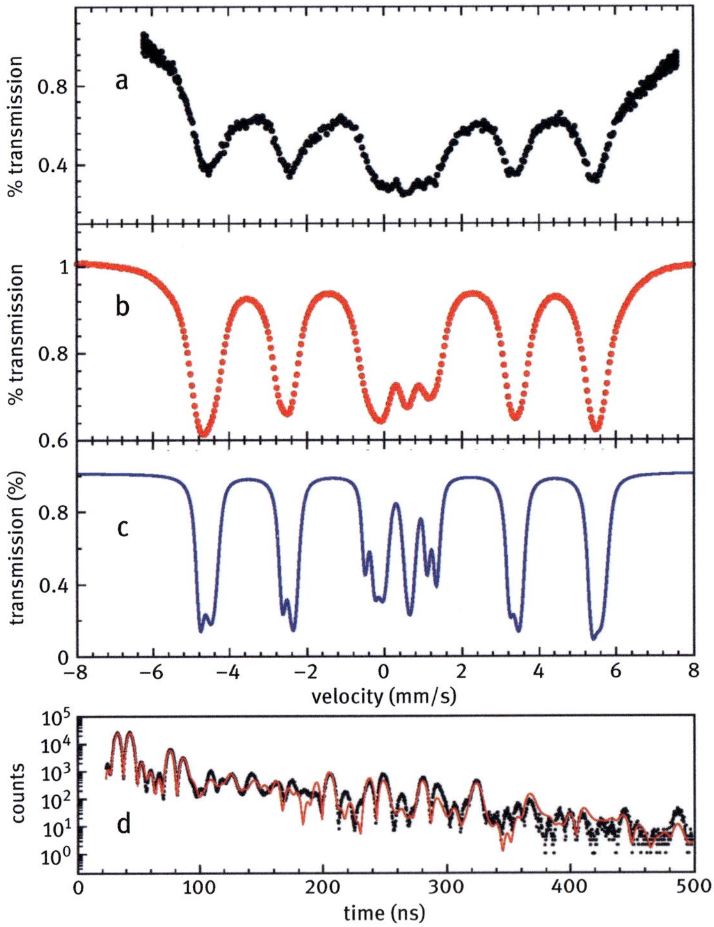

Figure 45: Mossbauer spectrum of Fe_3S_4 measured using (a) synchrotron x-ray in the energy domain; (b) laboratory ^{60}Co source in energy domain; (c) simulated pattern from the fitting of the (d) beating pattern in the time domain.

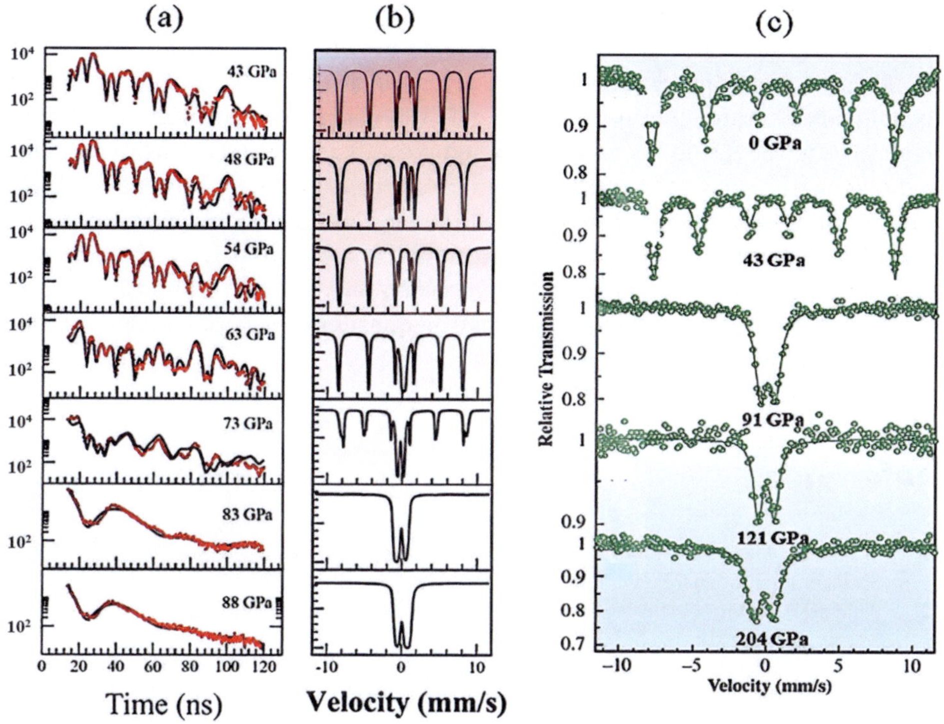

Figure 46: (a) Time-domain beating pattern of compressed Fe_2O_3 and (b) the simulated Mossbauer spectra compared to (c) the Mossbauer spectra measured in the energy domain.

4 Summary

This chapter provided an overview of the application of IXS to high-pressure research. The selected examples highlight the versatility of the method with applications from the study of very low frequency vibrations in biological systems to high-energy electron–phonon scattering (Compton effect) of condensed matter. We illustrated the single particles (VDOSs) and collective dynamics (phonons and hydrodynamics) as well as electron dynamics (excitations) can be investigated based on a general theoretical principle [3] but with a specialized instrument designed for the specific purpose [116]. Synchrotron radiation has been an indispensable tool in high-pressure research. Starting from the 1970s, elastic scattering of x-ray (diffraction) had offered a window to interrogate the internal structure of highly compressed materials. The advent of third-generation synchrotron in the late 1990s and early 2000s greatly improved the diffraction data quality and propelled the realization of inelastic scattering techniques. High-pressure research has benefited substantially in the ability to acquire both static and dynamic information, leading to the rapid growth of this field.

Inelastic scattering by synchrotron x-ray is still a rapidly developing technique. Exploitation of the unique characteristic of fourth-generation synchrotron and x-ray free-electron laser (XFEL) is expected to significantly increase the sensitivity, time, and energy resolution for experiments under extreme conditions. Recently, a high-resolution IXS setup with a resolution of ~45 meV has been incorporated into the High Energy Density scientific instrument at the European XFEL [117]. The availability of tunable intense radiation with femtosecond pulse lengths will push the frontier to query the most challenging scientific questions of matter under extreme static and dynamic compression [118].

References

[1] Cramer SP. X-Ray Spectroscopy with Synchrotron Radiation: Fundamentals and Applications. Springer International Publishing; Springer, 2020.

[2] Willmott P. An Introduction to Synchrotron Radiation: Techniques and Applications. 2nd Edition, Wiley, 2019.

[3] Schulke W. Electron Dynamics by Inelastic X-ray Scattering. Oxford University Press, 2007.

[4] Sinha SK. Theory of inelastic x-ray scattering from condensed matter. Journal of Physics Condensed Matter 2011;13:7511.

[5] Baron AQ. Introduction to high-resolution inelastic X-ray scattering In: Jaeschke E., editor et al, Synchrotron Light Sources & Free Electron Lasers. Springer International Publishing, 2016, 1643–1757.

[6] Wang S-X, Zhu L-F. Non-resonant inelastic X-ray scattering spectroscopy: A momentum probe to detect the electronic structures of atoms and molecules. Matter and Radiation at Extremes 2020;5:054201.

[7] Toellner TS, Hu MY, Sturhahn W, Quast K, Alp EE. Inelastic nuclear resonant scattering with sub-meV energy resolution. Applied Physics Letters 1997;71:2112.

[8] Burkel E. Inelastic Scattering of X-Rays with Very High Energy Resolution By. Berlin Heidelberg, Springer, 2013.

[9] Burkel E. Phonon spectroscopy by inelastic x-ray scattering. Instrumentation & Science 2000;63:171–212.

[10] Pauling L. The Nature of the Chemical Bond and the Structure of Molecules and Crystals: An Introduction to Mode. 3rd Edition. Cornell University Press, 1960.

[11] Wilson EB, Decius JC, Cross P. Molecular Vibrations: The Theory of Infrared and Raman Vibrational Spectra (Dover Books on Chemistry). Revised ed. Edition, 1980.

[12] Huotari S, Sahle CJ, Henriquet C, Al-Zein A, Martel K, Simonelli L, Verbeni R, Gonzalez H, Lagier M-C, Ponchut C. A large-solid-angle X-ray Raman scattering spectrometer at ID20 of the European synchrotron radiation facility. Journal of Synchrotron Radiation 2017;24.

[13] Moretti Sala M, Krisch M, Monaco G, Nardone M, Sette F, Verbeni R. Molecular vibrational spectroscopy by inelastic x-ray scattering: Experimental determination of the absolute vibrational cross section in liquid nitrogen. Physical Review B 2001;64:212102.

[14] Alp EE, Sturhahn W, Sinn H, Toellner T, Hu M, Sutter J, Alata A. Inelastic scattering of synchrotron radiation from electrons and nuclei for lattice dynamics studies. AIP Conference Proceedings 2000;506:479.
[15] Sturhahn W, Toellner TS, Alp EE, Zhang X, Ando M, Yoda Y, Kikuta S, Seto M, Kimball CW, Dabrowski B. Phonon density of states measured by inelastic nuclear resonant scattering. Physical Review Letters 1995;74(19):3832–3835.
[16] Fultz B. In: Kaufmann E Editor. Mössbauer Spectrometry, in Characterization of Materials. New York, John Wiley, 2011.
[17] Bancroft GM. Mossbauer Spectroscopy: An Introduction for Inorganic Chemists and Geochemists. New York, John Wiley and Sons, 1973.
[18] Gerdau E, Ruffer R, Winkler H, Tolksdorf W, Klages CP, Hannon JP. Nuclear Bragg diffraction of synchrotron radiation in yttrium iron garnet. Physical Review Letters 1985;54:835.
[19] Struhahn W. Nuclear resonant spectroscopy. Journal of Physics Condensed Matter 2004;16: S497.
[20] Klug DD, Zgierski M, Tse JS, Liu Z, Kincaid JR, Czarnecki K, Hemley RJ. Doming modes and dynamics of model heme compounds. Proceedings of the National Academy of Sciences 2002;99:12526.
[21] Leu BM, Silvernail NJ, Zgierski MZ, Wyllie GRA, Ellison MK, Scheidt WR, Zhao J, Sturhahn W, Alp EE, Sage JT. Quantitative dynamics of iron in carbonyl porphyrins. Biophysical Journal 2007;92:3764–3783.
[22] Smith MC, Xiao Y, Wang H, George SJ, Coucouvanis D, Koutmos M, Sturhahn W, Alp EE, Zhao J, Cramer SP. Normal-mode analysis of FeCl4 and Fe2S2Cl42 via vibrational Mossbauer, resonance Raman, and FT-IR spectroscopies. Inorganic Chemistry 2005;44:5562–5570.
[23] Dove MT. Lattice Dynamics: An Atomic View of Materials, Oxford Master Series in Physics. Cambridge, England, Oxford University Press, 2003.
[24] Srivastava G. The Physics of Phonons. Boca Raton, Florida, CRC Press, 1990.
[25] Chen Y-L, Yang D-P. Mössbauer Effect in Lattice Dynamics: Experimental Techniques and Application. Weinheim, Germany, Wiley-VCH Verlag GmbH & Co, 2007.
[26] Schiller M, Bizzarro M, Siebert J. Iron isotope evidence for very rapid accretion and differentiation of the proto-Earth. Science Advances 2020;6(7):eaay7604. DOI: 10.1126/sciadv.aay7604.
[27] Cormier V, Bergman M, Olsen P, van der Pluijm B, Marshak S. Earth's Core, Geophysics of a Planet's Deepest Interior. Netherlands, Elsevier, 2021.
[28] Dziewonski AM, Anderson DL. Preliminary reference earth model. Physics of the Earth and Planetary Interiors 1981;25:297–356.
[29] McDonough WF, Sun S. The composition of the earth. Chemical Geology 1995;120:223–253.
[30] https://en.wikipedia.org/wiki/Kola_Superdeep_Borehole.
[31] Fedotenko T, Dubrovinsky L, Aprilis G, Koemets E, Snigirev A, Snigireva I, Barannikov A, Ershov P, Cova F, Hanfland M, Dubrovinskaia N. Laser heating setup for diamond anvil cells for *in situ* synchrotron and in house high and ultra-high pressure studies. Review of Scientific Instruments 2019;90:104501.
[32] Lübbers R, Grünsteudel HF, Chumakov AI, Wortmann G. Density of phonon states in iron at high pressure. Science 2000;287:1250–1253.
[33] Mao HK, Xu J, Struzhkin VV, Shu J, Hemley RJ, Sturhahn W, Hu MY, Alp EE, Vocadlo L, Alfè D, Price GD, Gillan MJ, Schwoerer-Böhning M, Häusermann D, Eng P, Shen G, Giefers H, Lübbers R, Wortmann G. Phonon density of states of iron up to 153 Gigapascals. Science 2001;292:914–916.
[34] Sturhahn W. CONUSS and PHOENIX: evaluation of nuclear resonant scattering data. Hyperfine Interactions 2000;125:149–172.

[35] Dauphas N, Hu M, Baker EM, Hu J, Tissot FLH, Alp EE, Roskosz M, Zhao J, Bi W, Liu J, Lin J-F, Nie NX, Heard A. SciPhon: A data analysis software for nuclear resonant inelastic X-ray scattering with applications to Fe, Kr, Sn, Eu and Dy. Journal of Synchrotron Radiation 2018;25(5):1581–1599.
[36] Giles PM, Longenbach MH, Marder AR. High-pressure α⇄ε martensitic transformation in iron. Journal of Applied Physics 1971;42:4290.
[37] Birch F. The alpha-gamma transformation of iron at high pressures, and the problem of the earth's magnetism, Am. Journal of Science 1940;238:192–211.
[38] Jephcoat A, Olson P. Is the inner core of the Earth pure iron? Nature 1987;325:332–335.
[39] Murphy CA, Jackson JM, Sturhahn W, Chen B. Melting and thermal pressure of hcp-Fe from the phonon density of state. Physics of the Earth and Planetary Interiors 2011;188:114–120.
[40] Birch F, Roy GJ. Astronomical Society 1961;4:295. Journal of Geophysical Research 1961;66:2199.
[41] Chung DH. Birch's law: Why is it so good? Science 1972;177:261–263.
[42] Lin A-F, Sturhahn W, Zhao J, Shen G, Mao H-K, Hemley RJ. Sound velocities of hot dense iron: Birch's law revisited. Science 2005;308:1892–1894.
[43] Alp E, Mooney TM, Toellner T, Sturhahn W, Witthoff E, Rohlsberger R, Gerdau E, Homma H, Kentjana M. Time resolved nuclear resonant scattering from Sn nuclei using synchrotron radiation. Physical Review Letters 1993;70:3351.
[44] Barla A, Rüffer R, Chumakov AI, Metge J, Plessel J, Abd-Elmeguid MM. Direct determination of the phonon density of states in β–Sn. Physical Review B 2000;61:R14881(R).
[45] Salamat A, Briggs R, Bouvier P, McMillan P. High-pressure structural transformations of Sn up to 138 GPa: Angle-dispersive synchrotron x-ray diffraction study. Physical Review B 2013;88:104104.
[46] Giefers H, Tanis EA, Rudin SP, Greeff C, Ke X, Chen C, Nicol MF, Pravica M, Pravica W, Zhao J, Alatas A, Lerche M, Sturhahn W, Alp E. Phonon density of states of metallic Sn at high pressure. Physical Review Letters 2007;98:245502.
[47] Giefers H, Koval S, Wortmann G, Sturhahn W, Alp EE, Hu MY. Phonon density of states of Sn in textured SnO under high pressure: Comparison of nuclear inelastic x-ray scattering spectra to a shell model. Physical Review B 2006;74:094303.
[48] Tse JS, Klug DD. Pressure amorphized ices – An atomistic perspective. Physical Chemistry Chemical Physics 2012;14:8255–8263.
[49] Haines J, Leger J, Gorelli F, Hanfland M. Crystalline Post-Quartz Phase in Silica at High Pressure. Physical Review Letters 2001;87:155503.
[50] Hamaya N, Sato K, Usui-Watanabe K, Fuchizaki K, Fujii Y, Ohishi Y. Amorphization and Molecular Dissociation of SnI4 at High Pressure. Physical Review Letters 1997;79:4597.
[51] Pasternak MP, Taylor RD, Kruger M, Jeanloz R, Itie J-P, Polian A. Pressure induced amorphization of GeI4 molecular crystals. Physical Review Letters 1994;72:2733.
[52] Liu H, Tse JS, Hu MY, Bi W, Zhao J, Alp EE, Pasternak M, Taylor RD, Lashley JC. Mechanisms for pressure-induced crystal-crystal transition, amorphization, and devitrification of SnI4. Journal of Chemical Physics 2015;143:164508.
[53] Houben K, Jochum JK, Couet S, Menéndez E, Picot T, Hu MY, Zhao JY, Alp EE, Vantomme A, Temst K, Van Bael MJ. The influence of phonon softening on the superconducting critical temperature of Sn-nanostructures. Scientific Reports 2020;10:5729.
[54] Roskosz M, Amet Q, Fitoussi C, Dauphas N, Bourdon B, Tissandier L, Hu MY, Said A, Alatas A, Alp EE. Redox and structural controls on tin isotopic fractionations among magmas. Geochimica et Cosmochimica Acta 2020;268:42–55.
[55] Fiquet G, Badro J, Guyot F, Requardt H, Krisch M. Sound velocities in Iron to 110 Gigapascals. Science 2001;291:468–471.

[56] Bosak A, Krisch M. Phonon density of states probed by inelastic x-ray scattering. Physical Review B 2005;72:224305.
[57] Ohtani E, Shibazaki Y, Sakai T, Mibe K, Fukui H, Kamada S, Sakamaki T, Tsutsui YS, Baron AQR. Sound velocity of hexagonal close-packed iron up to core pressures. Geophysical Research Letters 2013;40:5089–5094.
[58] Sakamaki T, Ohtani E, Fukui H, Kamada S, Takahashi S, Sakairi T, Takahata A, Sakai T, Tsutsui S, Ishikawa D, Shiraishi R, Seto Y, Tsuchiya T, Baron AQR. Constraints on Earth's inner core composition inferred from measurements of the sound velocity of hcp-iron in extreme conditions. Science Advances 2016;2(2016):e1500802.
[59] Ripmeester JA, Alavi S., editors. Clathrate Hydrates: Molecular Science and Characterization. Weinheim, Germany, Wiley-VCH Verlag GmbH, 2022.
[60] Tse JS, White MA. Origin of glassy crystalline behavior in the thermal properties of clathrate hydrates: A thermal conductivity study of tetrahydrofuran hydrate. Journal of Physical Chemistry 1998;92(17):5006–5011.
[61] English NJ, Tse JS. Mechanisms for thermal conduction in methane hydrate. Physical Review Letters 2009;103:015901.
[62] Tse JS, Shpakov VP, Belosludov VR, Handa YP, Trouw F, Press W. Coupling of localized guest vibrations with the lattice modes in clathrate hydrates. Europhysics Letters 2001;54:354–360.
[63] Baumert J, Gutt C, Shapkov VP, Tse JS, Krisch M, Muller M, Requardt H, Klug DD, Jansen S, Press W. Lattice dynamics of methane and xenon hydrate, observation of symmetry-avoided crossing by experiment and theory. Physical Review B 2003;68:174301.
[64] Loveday JS, Nelmes RJ, Guthrie M, Belmonte SA, Allan DR, Klug DD, Tse JS, Handa YP. Stable methane hydrate above 2 GPa and the source of Titan's atmospheric methane. Nature 2001;410:661–663.
[65] Baumert J, Gutt C, Krisch M, Requardt H, Muller M, Tse JS, Klug DD, Press W. Elastic properties of methane hydrate at high pressure. Physical Review B 2005;72:054302.
[66] Long GJ, Hermann RP, Grandjean F, Alp EE, Sturhahn W, Johnson CE, Brown DE, Leupold O, Rüffer R. Strongly decoupled europium and iron vibrational modes in filled skutterudites. Physical Review B 2005;71:140302-1–140302-4.
[67] Wille H-C, Hermann RP, Sergueev I, Leupold O, van der Linden P, Sales BC, Grandjean F, Long GJ, Rüffer R, Shvyd'ko YV. Antimony vibrations in skutterudites probed by ^{121}Sb nuclear inelastic scattering. Physical Review B 2007;76:14030.
[68] Tanis EA. Partial Phonon Density of States of 57-iron and 161-dysprosium in $DyFe_3$ by Nuclear Resonant Inelastic X-ray Scattering under High. Pressure, M.Sc. dissertation, Dept. Phys. Las Vegas, University of Nevada, 2010.
[69] Tse JS, Boldyreva EV. Electron density of crystalline solids at high pressure. In: Gatti C, Macchi P., editors. Modern Charge Density Analysis. Heidelberg, Germany, Springer-Verlag, 2011, 573–623.
[70] Falconi S, Ackland GJ. Ab initio simulations in liquid caesium at high pressure and temperature. Physical Review B 2006;73:184204.
[71] Giordano VM, Monaco G. High frequency dynamics in liquid Cs at high pressure. Journal of Chemical Physics 2009;131:014501.
[72] Giordano VM, Monaco G. Through a liquid-liquid transition: The case of Cs. Advances in Chemical Physics 2013;152:101–112.
[73] Burkel E, Seyfert C, Halcoussis C, Sinn H, Simmons RO. Phonon dispersion curves in α-SiO_2 and hcp ^{3}He and ^{4}He determined by inelastic X-ray scattering. Physica B 1999;264:412–415.
[74] Occelli F, Krisch M, Loubeyre P, Sette F, Le Toullec R, Masciovecchio C, Rueff J-P. Phonon dispersion curves in an argon single crystal at high pressure by inelastic x-ray scattering. Physical Review B 2001;63:224306.

[75] Ghose S, Krisch M, Oganov AR, Beraud A, Bosak A, Gulve R, Seelaboyina R, Yang H, Saxena S,K. Lattice dynamics of MgO at high pressure: Theory and experiment. Physical Review Letters 2006;96:035507.
[76] Fukui H, Yoneda A, Kamada S, Uchiyama H, Tsutsui S, Hirao N, Baron AQR. Elasticity of single-crystal NaCl under high-pressure: Simultaneous measurement of x-ray inelastic scattering and diffraction. High Pressure Research 2002. DOI: 10.1080/08957959.2020.1806260
[77] Antonangeli D, Krisch M, Fiquet G, Farber DL, Aracne CM, Badro J, Occelli F, Requardt H. Elasticity of cobalt at high pressure studied by inelastic X-Ray scattering. Physical Review Letters 2004;93:215505.
[78] Zhang Y, Yang C, Alatas A, Said AH, Salke NP, Hong J, Lin J-F. Pressure effect on Kohn anomaly and electronic topological transition in single-crystal tantalum. Physical Review B 2019;100:075145.
[79] Farber DL, Krisch M, Antonangeli D, Beraud A, Badro J, Occelli F, Orlikowski D. Lattice dynamics of molybdenum at high pressure. Physical Review Letters 2006;96:115502.
[80] Krisch M, Farber DL, Xu R, Antonangeli D, Aracne CM, Beraud A, Chiang T-C, Zarestky J, Kimh DY, Isaev EI, Ahuja R, Johansson B. Phonons of the anomalous element cerium. Proceedings of the National Academy of Sciences 2011;108:9342–9345.
[81] Loa LF, Lundegaard MI, McMahon SR, Evans AB, Krisch M. Lattice dynamics of incommensurate composite Rb-IV and a realization of the monatomic linear chain model. Physical Review Letters 2007;99:035501.
[82] Ma Y, Eremets M, Oganov AR et al. Transparent dense sodium. 2009;458(2009):182.
[83] Mao H-K, Kao CC, Hemley RJ. Inelastic x-ray scattering at ultrahigh pressures. Journal of Physics Condensed Matter 2001;137:847–7858.
[84] Loa I, Syassen K, Monaco G, Vanko´ G, Krisch M, Hanfland M. Plasmons in sodium under pressure: increasing departure from nearly free-electron behavior. Physical Review Letters 2011;107:086402.
[85] Wernet P, Nordlund D, Bergmann U, Cavalleri M, Odelius M, Ogasawara H, Naslund LA, Hirsch TK, Ojamae L, Glatzel P, Pettersson LGM, Nilsson A. The structure of the first coordination shell in liquid water. Science 2004;304:995–999.
[86] Tse JS, Shaw DM, Klug DD, Patchkovskii S, Vankó G, Monaco G, Krisch M. X-ray Raman spectroscopic study of water in the condensed phases. Physical Review Letters 2008;100:095502.
[87] Tulk, C. A. et al. Absence of amorphous forms when ice is compressed at low temperature. Nature 569, 542–545 (2019).
[88] Schirmer ABT, Randall KJ, Feldhaus J, Bradshaw AM, Ma Y, Chen CT, Sette F. K-shell excitation of the water, ammonia, and methane molecules using high-resolution photoabsorption spectroscopy. Physical Review A: Atomic, Molecular, and Optical Physics 1993;47:1136–47.
[89] Sahle CJ, Sternemann C, Schmidt C, Lehtola S, Jahn S, Simonelli L, Huotari S, Hakala M, Pylkkänen T, Nyrow A, Mende K, Tolan M, Hämäläinen K, Wilke M. Microscopic structure of water at elevated pressures and temperatures. Proceedings of the National Academy of Sciences 2013;110:6301–6306.
[90] Pylkkanen T, Giordano VM, Chervin J, Sakko A, Hakala M, Soininen JA, Hamalainen K, Monaco G, Huotari S. Role of non-hydrogen-bonded molecules in the oxygen K-edge spectrum of ice. Journal of Physical Chemistry B 2010;114:3804–08.
[91] Shaw D, Tse JS. unpublished results.
[92] Lee SK, Eng P, Mao H, Meng Y, Newville M, Hu MY, Shu J. Probing of bonding changes in B_2O_3 glasses at high pressure with inelastic X-ray scattering. Nature Materials 2005;4:851–854.

[93] Lee SK, Kim Y, Chow P, Xiao Y, Ji C, Shen G. Amorphous boron oxide at megabar pressures via inelastic X-ray scattering. Proceedings of the National Academy of Sciences 2018;115:5855–5860.
[94] Lin JF, Fukui H, Prendergast D, Okuchi T, Cai YQ, Hiraoka N, Yoo C,S, Trave A, Eng P, Hu MY, P. Chow P. Electronic bonding transition in compressed SiO2 glass. Physical Review B 2007;75:012201.
[95] Fukui H, Kanzaki M, Hiraoka N, Cai Y. Coordination environment of silicon in silica glass up to 74 GPa: An x-ray Raman scattering study at the silicon L edge. Physical Review B 2008;78:012203.
[96] Petitgirard S, Sahle CJ, Weis C, Gilmore K, Spiekermann G, Tse JS, Wilke M, Cavallari C, Cerantola V, Sternemann C. Magma properties at deep Earth's conditions from electronic structure of silica. Geochemical Perspectives Letters 2019;9:32–37.
[97] Wu M, Liang Y, Jiang J-Z, Tse JS. Structure and properties of dense silica glass. Nature Scientific Reports 2012;2:398–402.
[98] Eremets IA, Trojan IA, Medvedev SA, Tse JS, Yao Y. Superconductivity in hydrogen dominant materials: Silane. Science 2008;319:1506.
[99] Drozdov AP, Eremets MI, Troyan IA, Ksenofontov V, Shylin SI. Conventional superconductivity at 203 kelvin at high pressures in the sulfur hydride system. Nature 2015;3(525):73–76.
[100] Drozdov AP, Kong PP, Minkov VS, Besedin SP, Kuzovnikov MA, Mozaffari S, Balicas L, Balakirev FF, Graf DE, Prakapenka VB, Greenberg E, Knyazev DA, Tkac Z M, Eremets MI. Superconductivity at 250 K in lanthanum hydride under high pressures. Nature 2019;569:528–531.
[101] Tse JS, Chemical A. Perspective on high pressure crystal structures and properties. Nature Science Review 2020;7:149–169.
[102] Tse JS, Hanfland M, Flacau R, Desgreniers S, Li Z, Mende K, Gilmore K, Nyrow A, Moretti-Sala M, Sternemann C. Pressure-induced changes on the electronic structure and electron topology in the direct FCC → SH transformation of silicon. Journal of Physical Chemistry C 2014;118:1161–1167.
[103] Jørgensen: CK. Modern Aspects of Ligand Field Theory. North-Holland Publishing Co, 1971.
[104] Rueff J-P, Kao -C-C, Struzhkin VV, Badro J, Shu J, Hemley RJ, Mao H-K. Pressure-Induced High-Spin to Low-Spin Transition in FeS Evidenced by X-Ray Emission Spectroscopy. Physical Review Letters 1999;82:3284.
[105] Nyrow A, Tse JS, Hiraoka N, Desgreniers S, Buning T, Mende K, Tolan M, Wilke M, Sternemann C. Pressure induced spin transition revealed by iron M2,3 -edge spectroscopy. Applied Physics Letters 2014;104:262408.
[106] Yamase H, Sakurai Y, Fujita M, Wakimoto S, Yamada K. Fermi surface in La-based cuprate superconductors from Compton scattering imaging. Nature Communications 2021;12:2223.
[107] Hiraoka N et al. A new X-ray spectrometer for high-resolution Compton profile measurements at SPring-8. Journal of Synchrotron Radiation 2001;8:26.
[108] Barbiellini KB, Orikasa Y, Kaprzyk S, Itou M, Yamamoto K, Wang YJ, Hafiz H, Uchimoto Y, Bansil A, Sakurai Y, Sakurai H. Non-destructive measurement of *in-operando* lithium concentration in batteries via x-ray Compton scattering. Journal of Applied Physics 2016;119:025103.
[109] Watanabe S, Ishimatsu N, Maruyama N, Kobayashi H, Itou H, Kawamura M, Sakurai N. Instability of Co spin moment in $ErCo_2$ probed by magnetic Compton scattering under high pressure. Journal of the Physical Society of Japan 2011;80:097305.
[110] Hämäläinen K, Huotari S, Laukkanen J, Soininen A, Manninen S, Kao C-C, Buslaps T, Mezouar M. Free electron gas under high pressure. Physical Review B 2000;62:R735.

[111] Tse JS, Klug DD, Jiang DT, Sternemann C, Volmer M, Huotari S, Hiraoka N, Honkimaki V, Hamanlainen K. Compton scattering of elemental silicon at high pressure. Applied Physics Letters 2005;87:191905.

[112] Alp EE, Sturhahn W, Toellner T. Synchrotron Mossbauer spectroscopy of powder samples. Nuclear Instruments and Methods in Physics Research Section B: Beam Interactions with Materials and Atoms 1995;97:526–529.

[113] Mitsui T. Synchrotron Mossbauer spectroscopy measurement. In: Kono Y, Sanloup C., editors. Magmas under Pressure. Chap. 7. Elsevier, 2018, 179–210.

[114] Arthur J, Brown GS, Brown DE, Ruby SL. Resonance energy shifts during nuclear Bragg diffraction of x rays. Physical Review Letters 1989;63:Published 9 October 1989.

[115] Alp EE, Bi W, Tse JS. unpublished.

[116] Bi W, Zhao J, Lin J-F, Jia Q, Hu MY, Jin C, Ferry R, Yang W, Struzhkin VV, Alp EE. Nuclear resonant inelastic X-ray scattering at high pressure and low temperature. Journal of Synchrotron Radiation 2015;22:760–765.

[117] Wollenweber L, Preston TR, Descamps A, Cerantola V, Comley A, Eggert JH, Fletcher LB, Geloni G, Gericke DO, Glenzer SH, Göde, Hastings J, Humphries OS, Jenei A, Karnbach O, Konopkova Z, Loetzsch R, Marx-Glowna, McBride, McGonegle, Monaco G, Ofori-Okai K, Palmer CAJ, Plückthun C, Redmer R, Strohm C, Thorpe I, Tschentscher T, Uschmann I, Wark JS, White TG, Appel K, Gregori G, Zastrau U. High-resolution inelastic x-ray scattering at the high energy density scientific instrument at the European X-ray free-electron laser. Review of Scientific Instruments 2021;92:013101.

[118] Cerantola V, Dorothea Rosa A, Konôpková Z, Torchio R, Brambrink E, Rack A, Zastrau U, Pascarelli S. New frontiers in extreme conditions science at synchrotrons and free electron lasers. Journal of Physics Condensed Matter 2021;33:274003.

Israel Roger Montoya Matos, Alvaro Carrillo Herrera,
Silvio Buchner, Naira Maria Balzaretti

Lanthanide-doped glasses under high pressure

According to Zanotto and Mauro [1], "glass is a nonequilibrium, noncrystalline condensed state of matter that exhibits a glass transition. The structure of glasses is like that of their parent supercooled liquids. Their fate, in the limit of infinite time, is to crystallize." Glasses are usually formed by quenching of a molten mixture in such a way that the atoms do not have time to reach the equilibrium spatial configuration, forming an amorphous solid with short-range order. The glass transition temperature, T_g, consists in the temperature range between the rigid solid-state structure and a more flexible, viscous structure. Good candidates to make glasses must have high viscosity in the molten state, where the atomic rearrangement is constrained. Materials that are not able to form glasses are, sometimes, used as network modifiers. In this case, they induce the formation of dangling bonds in the primary three-dimensional glass network, usually decreasing density. The physical properties of glasses can be tuned by adjusting their chemical composition.

Silica (SiO_2), the major constituent of sand, has been the raw material of glasses for thousands of years. It is formed by SiO_4 tetrahedra sharing four oxygen atoms (bridging oxygen atoms) in a random spatial configuration. To overcome the high melting point of silica (1,723 °C), alkali metal oxides such as Na_2O, K_2O, and Li_2O are used as network modifiers. For instance, the addition of 33 mol% of Na_2O decreases the melting temperature to 875 °C [2, 3]. The presence of alkali metal positive ions as modifiers induces the formation of nonbridging oxygen (NBO) atoms, so the SiO_4 tetrahedra would share less than four oxygen atoms, affecting the local structure and properties of the glass [4–8].

Germania (GeO_2) is also a good glass former, like SiO_2, consisting of GeO_4 tetrahedra sharing four oxygen atoms in a random spatial configuration. It was first synthesized by Dennis and Laubengayer in 1926 [9]. The incorporation of glass modifiers induces an atypical behavior, known as *germanate anomaly*: thermophysical properties present maxima or minima depending on the amount of the modifier. In the atomic level, the anomaly is related to modifications in the coordination number of Ge atoms or to the formation of small rings of GeO_4 [10–13].

Israel Roger Montoya Matos, Department of Civil Engineering, University of Lima, Lima, Peru
Alvaro Carrillo Herrera, Department of Physics, Universidad de Antioquia, Medellin, Colombia
Silvio Buchner, Naira Maria Balzaretti, Department of Physics, Universidade Federal do Rio Grande do Sul, Porto Alegre, Brazil

https://doi.org/10.1515/9783110668612-007

PbO is an interesting modifier for germania glasses, forming the so-called *heavy metal oxide* $PbO{-}GeO_2$ glass. The high electronic polarizability and ionic mass of Pb ion are responsible for increasing density, refractive index, and electro-optical properties of the glass. The local structure and glass properties are strongly dependent on the PbO concentration [14].

The combination of the mechanical and optical properties of glasses with the luminescent properties of lanthanide ions has been explored for potential technological applications, for instance, in telecommunications as optical fibers and amplifiers in the near-infrared (NIR) spectral range, in solar cells, and also in medical devices. Rare-earth (RE) lanthanides have incomplete 4f shells with electronic configuration given by $1s^2 2s^2 2p^6 3s^2 3p^6 3d^{10} 4s^2 4p^6 4d^{10} 4f^n 5s^2 5p^6 6s^2$. Electronic transitions between unfilled states within 4f shell can take place, resulting in very rich and unique absorption and emission spectra, covering from UV to NIR range. Since the $4f^n$ shell is closer to the nucleus than the 5s, 5p, and 6s shells, the optical transitions have a narrow spectral width of about 10 cm^{-1} [15]. Dieke and Crosswhite presented a comprehensive study about the spectra of doubly and triply ionized RE based on the emission spectra of the free ions and theoretical calculations [16].

The energy-level diagram for these electronic transitions is slightly disturbed by the Stark effect from the local electric field when RE ions are dopants in glass or crystalline hosts, increasing the complexity of the absorption and emission spectra. Judd–Ofelt's theory enables calculating the radiative transition rate and the fluorescence branching ratio between two states of the $4f^n$ shell when the RE ion is embedded in different hosts. There are many articles in the literature dealing with the luminescence of RE dopants in different hosts, such as borate, silicate, phosphate, germanate, tellurite, and fluorophosphate glasses. The effect of the incorporation of metallic nanoparticles, usually of gold or silver, has also been investigated. These nanoparticles act as nanoantennas whose electric field interacts with the RE ions, also affecting the emission spectra [17, 18]. In this context, the area of research devoted to luminescent properties of RE ions in different hosts is quite vast. For the same ion, it is possible to investigate the luminescence in a multitude of different hosts, glass or crystalline, in the presence or absence of metallic nanoparticles.

The luminescence of RE ions can also be used as a *high-pressure calibrant*, since the modification of the interatomic distances and bond angles would affect the ligand field and, therefore, disturb the energy levels of the $4f^n$ electrons. In fact, RE-doped yttrium–aluminum–garnets (YAG), Eu:YAG and Sm:YAG, are alternatives to ruby as high-pressure calibrants since the decade of 1980 [19, 20]. The electronic transitions of Eu and Sm ions are nearly temperature independent, while the wavelengths of the fluorescence peaks shift almost linearly with hydrostatic pressure.

In 1956, Anderson [21] investigated the effect of pressure on glass structures, considering the anelastic response in terms of the random network hypothesis. He considered that the glass has a short-range structure and a long-range disorder. The volume change induced by high pressure would be mostly related to the long-range disorder, even when the short-range structure remains practically the same.

In 2001, Bray [22] published a comprehensive study about high-pressure probes of electronic structure and luminescence properties of transition metals and lanthanide systems. The study was based on using hydrostatic pressure to influence the bonding environment of luminescent centers, creating new opportunities for understanding the behavior of transition metals and lanthanide ions. He pointed out the advantages of using high-pressure studies as complementary to the conventional chemical composition approach: first, the possibility of continuous variation of the structure compared to discrete modifications induced by changing chemical composition; second, the ability to study a given composition over a wide range of structure and bonding conditions; and third, high pressure can reach structures and electronic states that are not achievable by other routes.

Kanzaki et al. [23] investigated the structure of high-pressure silicate glasses up to 14 GPa using ^{29}Si NMR analyses for $Na_2O–SiO_2$ and $K_2Si_4O_9$ systems. The increase of pressure induced reduction of Si–O–Si angles at lower pressures and formation of octahedral SiO_6 through conversion of NBO into bridging oxygens at higher pressures. They suggest that the local structure of high-pressure glasses in alkali silicate systems is like those of alkali germanate systems. Fuss et al. [24] and Buchner et al. [25] investigated the crystallization of $Li_2O–SiO_2$ system at high pressure, also observing irreversible modifications in the interatomic distances and bond angles.

In this chapter, RE ions were used as local probes to evaluate the effect of densification of silica and germania glasses. At high pressures (7.7 GPa), irreversible modifications in the interatomic distances due to the degrees of structural freedom of the glass compared to the crystalline matrices should be expected. These changes should affect the local field in the vicinity of the RE ions and, therefore, their absorption and emission spectra.

1 Silicate glasses

A set of *alkali metal silicate glasses* was prepared using conventional melt and quenching route. The compositions are presented in Table 1. Analytical reagent-grades Li_2CO_3, Na_2CO_3, K_2CO_3, SiO_2, Sm_2O_3, Nd_2O_3, and Pr_2O_3 from Sigma-Aldrich (high purity) were used for sample preparation after drying at 200 °C during 2 h. The mixtures were melted in a Pt crucible up to 1,550 °C during 2 h in air and then poured into a preheated mold heated at 350 °C for 4 h to release the thermal stress induced by quenching. Then the samples were cooled down slowly to room temperature.

For high-pressure processing, the samples were shaped as cylinders with 3 mm in diameter and 2.3 mm height, placed inside hBN cups used as soft pressure-transmitting medium, and processed at 7.7 GPa during 15 min at room temperature.

Table 1: Composition of the alkali metal silicate glasses produced by melt and quenching route, where M = Nd, Sm, and Pr.

Glass label	Composition
LS-M	33 Li_2O + 66 SiO_2 + 1 M_2O_3
NS-M	33 Na_2O + 66 SiO_2 + 1 M_2O_3
KS-M	33 K_2O + 66 SiO_2 + 1 M_2O_3

The high-pressure apparatus was of toroidal type and a 400-ton hydraulic press was used to increase pressure. Specific information about the processing and pressure calibration is given in ref. [26].

Table 2 shows the refractive index and density of Pr-, Nd-, and Sm-doped alkali silicate glasses before and after high-pressure densification at room temperature.

Table 2: Density and refractive index of the RE-doped alkali silicate glasses before and after high-pressure densification during 15 min at 7.7 GPa and room temperature.

Sample	Density (±0.0001 g/cm^3)		Refractive index (at 585 nm)	
	Pristine	7.7 GPa	Pristine	7.7 GPa
LS-Pr	2.8255	2.8721	1.476	1.773
LS-Nd	2.4951	2.5312	1.611	1.713
LS-Sm	2.4582	2.4816	1.493	1.629
NS-Pr	2.6224	2.6632	1.464	1.668
NS-Nd	2.4809	2.5205	1.586	1.685
NS-Sm	2.5396	2.5635	1.449	1.585
KS-Pr	2.5253	2.5674	1.428	1.469
KS-Nd	2.4694	2.4934	1.445	1.546
KS-Sm	2.5598	2.5742	1.431	1.503

The results for samples containing potassium are less affected by the RE ion than samples containing lithium and sodium, probably due to its larger ionic size. As should be expected, both density and refractive index of all samples increased after densification at 7.7 GPa, revealing irreversible modifications in the local structure of the glasses induced by high pressure. It is important to point out that the measurements of density and refractive index were done after processing the glasses at 7.7 GPa at room temperature during 15 min (not *in situ*). These irreversible structural changes are the interesting and intriguing feature of amorphous materials compared to crystalline structures that present elastic behavior under hydrostatic pressure inside the regime of stability of a given crystalline phase. The irreversible structural changes in the glass hosts would affect the local electric field in the surrounding of the dopant RE ions and this effect should impact the luminescent properties of these ions.

Previous works [27, 28] described the effect of high pressure in Raman and Fourier-transform infrared spectra, as well as the absorbance and emission spectra in the visible and NIR range of Nd- and Sm-doped alkali silicate glasses listed in Table 1. Matson et al. [29] presented a comprehensive investigation about the effect on Raman spectrum of silica glasses due to the addition of different amounts of alkali oxides as glass modifiers. The Raman spectra for the samples listed in Table 1 are consistent with the spectra shown in [28] for the same glass compositions. The signal-to-noise signal in the spectrum for sodium silicate glass is larger, suggesting that this glass structure is more homogeneous than lithium and potassium silicate glasses [27, 28]. The Raman spectra also indicate that doping with RE ions did not affect the glass structure but densification at 7.7 GPa at room temperature seems to induce irreversible modifications in the local structure of the glasses, especially in the one containing potassium in the region close to 1,000 cm^{-1}, assigned to Si-NBO stretching in SiO_4 tetrahedral units [28].

For Nd-doped alkali silicate glasses [27], it was observed that, after densification, the effect of the alkali ion in the absorption spectra was less relevant probably due to an increase in the symmetry of the local field due to the close packing of the SiO_4 tetrahedra, with a small effect of the alkali ion itself. The Stark effect was more pronounced for Nd-doped potassium silicate both in the absorption and emission spectra probably due to its larger ionic size compared to lithium and sodium ions. The relative intensity of the emission bands depended both on the alkali ion and on the effect of high-pressure densification. For Sm-doped glasses [28], the densification induced irreversible changes in optical absorbance but the effect on the emission bands, related to inner energy levels, was very small. After densification, the radiative lifetime decreased for LS:Sm glass, increased for NS:Sm glass, and remained practically the same for KS:Sm glass, which pointed out that alkali ions play a relevant role in the luminescence of the RE-doped alkali silicate glasses.

After densification at high pressure, a broad background was observed for Nd [27], Sm [28], and Pr ions, maybe related to active defects induced in the alkali silicate glass host. A clear blueshift was observed only for KS-Sm sample for absorption bands in the NIR range (above 1,200 nm), indicating that potassium plays a distinct role compared to sodium and lithium, consistent with the results from Raman spectroscopy. For Pr-doped alkali silicates, the effect of densification in the visible range was practically the same for Li, Na, and K silicates. The only exception was the appearance of a shoulder in the larger wavelength side of the $^3H_4 \rightarrow {}^3P_0$ transition, which is more intense for KS-Pr sample. In the NIR range, the effect of densification depended on the alkali ion: for LS-Pr, the effect was a small blueshift and the appearance of a large plateau from 1,000 to 1,400 nm. For NS-Pr, the blueshift was also observed while the background increased for smaller wavelengths. For KS-Pr, the shoulder in the $^3H_4 \rightarrow {}^3F_2$ transition increased considerably. The background was like the one observed for LS-Pr sample.

It is interesting to point out that the energy levels corresponding to the absorption bands in the NIR range are related to the external shells of the 4f orbitals which should be more affected by the effect induced by high pressure due to the decrease in the interatomic distances.

For Sm-doped alkali silicate glasses [28], the effect of densification on the luminescence spectra was relatively small compared to the effect on the absorption spectra, independent of the alkali ion. A small effect was observed in the splitting of the bands, due to the Stark effect, probably related to slight modifications in the local field surrounding Sm ions. For Pr-doped alkali silicate glasses, the effect of densification on luminescence was also small for Li and Na silicate glasses. However, the intensity of $^3P_0 \rightarrow {}^3F_2$ transition decreased considerably for K silicate glass after densification, and a broad emission background appeared in the range from 500 to 600 nm. Considering the energy levels for Pr^{3+} ions, maybe the increase in the population levels $^3P_1$ and $^3P_2$ in Pr-doped KS glass is responsible for the broad background observed in the emission spectrum from 500 to 600 nm, competing with the emission at 649 nm.

Table 3 presents the *Judd–Ofelt parameters* [15, 30] obtained from the oscillator strengths calculated from the absorption spectra of the alkali silicate glasses doped with Nd, Sm, and Pr ions, before and after high-pressure densification. Small values of Ω_2 indicate that the surroundings of the RE ion are symmetric (ionic nature), while large values are related to covalent nature. As can be seen, the local electric field for Sm ions is more symmetric and almost independent of the alkali ion for pristine glasses. The effect of densification at 7.7 GPa, on the other hand, is different for each alkali ion, probably due to the different ionic sizes: for lithium, the symmetry decreased; for sodium, the symmetry increased; and for potassium, the symmetry remained practically the same. For Nd ions, the symmetry is smaller in the pristine glasses, especially for KS:Nd glass. After densification, it increased for all alkali ions, remarkably for LS:Nd glass. The Ω_2 values for Pr-doped glasses are significantly larger than the values for Sm- and Nd-doped glasses, indicating a larger asymmetry in the local electric field, which depend on the alkali ion, being larger for potassium. The effect of densification was not so evident, except for NS:Pr, where the symmetry increased. Parameters Ω_4 and Ω_6 are usually related to the rigidity of the glass. It is interesting to note that the values for potassium silicate glasses are smaller than for lithium and sodium silicate glasses, corroborating the idea that potassium plays a distinct role in the silicate glass, as pointed out by vibrational spectroscopy and luminescent properties of RE-doped glasses.

The comprehensive study considering Li, Na, and K alkali silicate glasses doped with Nd, Sm, and Pr ions, before and after processing at 7.7 GPa during 15 min at room temperature, revealed that density and refractive index are strongly affected by the alkali ion and by densification at high pressure. Absorption spectra of RE ions also depended on the alkali ion and densification while the emission spectra did not. The effect of densification was more pronounced for potassium silicate

Table 3: Judd–Ofelt parameters for rare-earth-doped alkali silicate glasses before and after densification at 7.7 GPa and room temperature during 15 min (±0.05 × 10^{-20} cm^2) [27, 28].

	Ω_2		Ω_4		Ω_6	
	Pristine	7.7 GPa	Pristine	7.7 GPa	Pristine	7.7 GPa
LS:Pr	12.7	12.2	4.97	9.74	14.4	20.8
NS:Pr	20.6	15.4	5.12	4.54	14.1	9.54
KS:Pr	23.9	23.7	2.10	2.55	7.21	4.91
LS:Nd	3.71	0.92	3.68	0.84	4.14	1.56
NS:Nd	3.84	3.04	3.06	2.16	2.72	3.29
KS:Nd	5.46	3.15	2.81	1.05	1.85	2.46
LS:Sm	1.28	2.24	3.14	4.70	4.53	3.22
NS:Sm	1.27	0.65	2.69	1.45	3.15	1.63
KS:Sm	1.04	1.13	1.12	1.29	1.61	1.22
GP:Sm	3.2	0.7	7.49	5.68	6.17	4.8

The last row of the table shows the values for Sm-doped GeO_2–PbO glass.

glass, as revealed by Raman spectroscopy and by the luminescence properties of the RE ions, corroborated by the Judd–Ofelt analysis.

2 Germania glasses

A series of experimental investigations was performed on *heavy metal oxide glasses* of composition 99 (59GeO_2–41PbO) – 1M_2O_3 glasses, where M correspond to RE elements (Sm and Pr), considering the effect of high-pressure densification at 7.7 GPa and the incorporation of Au nanoparticles on the luminescence properties [31–33]. It is interesting to point out that, while the coordination number of SiO_2 remains stable (4) up to 7.7 GPa, for GeO_2 it starts to change from 4 to 6 above 4 GPa. Therefore, the densification of germania glasses at 7.7 GPa should induce significant structural changes in the surroundings of RE ions.

Glass samples were prepared by melting the powder mixture at 1,250 °C during 1 h followed by quenching at a brass mold preheated at 330 °C. High-pressure densification at 7.7 GPa was performed using the same experimental procedure for alkali silicate glasses described in the previous section.

For Sm-doped GeO_2–PbO glass (GP:Sm), a novel NIR band was observed in the range 1,250–1,350 nm under excitation at 488 nm, related to $^4G_{5/2}$ to $^6F_{11/2}$ transition [31]. After densification at 7.7 GPa at room temperature, the glass transition temperature of GP:Sm increased from 411 to 435 °C while the melting temperature decreased from 761 to 759 °C. Irreversible changes were also observed in the refractive index, Raman spectrum, and extended x-ray absorption fine structure, probably due to transformation of GeO_4 to GeO_6 structural configuration induced by pressure.

The analysis of the absorption and emission spectra of GP:Sm glass indicated that, after densification, the contribution of cross-relaxation processes between Sm and Sm ions increased, inducing quenching in the luminescence related to $^6H_{5/2} \rightarrow {}^4G_{5/2}$ transition [32]. The Judd–Ofelt parameters are presented in the last row of Table 3. Comparing the values for Sm-doped alkali silicate glasses and Sm-doped GeO_2–PbO glass, it is possible to see that the symmetry of the local field is strongly influenced by the glass host, being larger for silicate glasses.

A similar investigation was done for Pr-doped GeO_2–PbO glass containing gold nanoparticles [33]. About 0.1 mol% of Pr_2O_3 and 0.1 mol% of Au_2O_3 were incorporated in the glass of composition $59GeO_2$–41PbO. The melting and quenching conditions, as well as the high-pressure apparatus, were similar to those used to prepare GP:Sm glass. To nucleate and grow nanoparticles of gold, the glass was heated at 400 °C during 6 h. Transmission electron microscopy revealed gold nanoparticles with average size of (3 ± 1) nm. For Pr-doped GeO_2–PbO (GP:Pr) glass, novel absorption bands at 342 and 412 nm were observed. After densification at 7.7 GPa, these bands were not observed, and the intensity of the luminescence bands in the visible range was quenched considerably. The densification of GP:Pr samples containing gold nanoparticles induced quenching in the luminescence of Pr ions in the visible range under excitation at 488 nm and a strong amplification of the emission in the NIR range under excitation at 445 nm. The synergistic effect between metallic nanoparticles (nanoantenna effect) and the luminescence of RE ions in a glass host depend not only on the distance among the metallic particles and the ions but also on the size of the nanoparticles. According to Eichelbaun and Rademann [34], nanoparticles larger than 5 nm would have strong plasmon absorption that dominates the mechanism responsible for enhancement or quenching of the luminescence of RE ions in the surroundings, closer than 10 nm. Smaller nanoparticles do not have surface plasmon resonance, which is the case for the present study, where the nanoparticles have, in average, 3 nm in diameter. In this case, the effect of densification under high pressure would decrease the average distance among nanoparticles and Pr ions, increasing the efficiency of the energy transfer processes in the NIR range under excitation at 445 nm. The same effect would be responsible for the strong quenching after densification in the visible range under excitation at 488 nm.

3 Concluding remarks

The structure of glasses depends on the glass former and glass modifiers. In principle, there are a very large number of possible combinations among them, each one resulting in glasses having their own mechanical, thermal, structural, and optical properties. In principle, the glass structure has several degrees of freedom due to the lack of crystalline constraints. In fact, irreversible structural modifications were detected for alkali silicate glasses and GeO_2–PbO glasses after densification at 7.7 GPa. Density and refractive index, for instance, changed significantly. RE ions embedded in a glass host as dopants are probes of the local field since the electronic transitions in the 4f orbitals are affected by this field. The results presented in this chapter showed that the absorption and emission spectra of Sm, Nd, and Pr ions in alkali silicate glasses depend on the alkali modifier (Li, Na, and K) and on the densification at 7.7 GPa during 15 min at room temperature. Irreversible changes were induced by pressure, affecting density, refractive index, and luminescence properties. The same behavior was observed for Sm- and Pr-doped GeO_2–PbO glasses. In this case, the incorporation of gold nanoparticles helped to enhance the emission of Pr ions in the NIR range due to the antenna effect. The densification at high pressure probably decreased the distance among the nanoparticles and Pr ions, affecting the luminescence properties. The main conclusion is that, in fact, RE ions are interesting probes to evaluate local modifications in glass structures after densification under high pressure. These modifications, in turn, can result in amplification of emission bands in the NIR range, useful, for instance, for telecommunication or solar cell applications.

References

[1] Zanotto ED, Mauro JC. The glassy state of matter: Its definition and ultimate fate. Journal of Non-Crystalline Solids 2017;471:490–495.
[2] Yamane M, Asahara Y. Glasses for Photonics. Cambridge, Cambridge University Press, 2000.
[3] Rawson H. Properties and Applications of Glass. Amsterdam, Elsevier Scientific, Publishing Company, 1980.
[4] Greaves GN. EXAFS and the structure of glasses. Journal of Non-Crystalline Solids 1985;71:203–217.
[5] La Course WC. A defect model for the mixed alkali effect. Journal of Non-Crystalline Solids 1987;95-96:905–912.
[6] Greaves GN, Smith W, Giulotto E, Pantos E. Local structure, microstructure and glass properties. Journal of Non-Crystalline Solids 1997;222:13–24.
[7] Ali F, Chadwick AV, Greaves GN, Jermy MC, Ngai KL, Smith ME. Examination of the mixed-alkali effect in (Li,Na) disilicate glasses by nuclear magnetic resonance and conductivity measurements. Solid State Nuclear Magnetic Resonance 1995;5:133–143.
[8] Sitarz M, Mozgawa W, Handke M. Rings in the structure of silicate glasses. Journal of Molecular Structure 1999;511:281–285.
[9] Laubengayer AW, Tabern DL. Germanium. XIV. Journal of Physical Chemistry 1926;30:1047–1048.
[10] Hannon AC, Di Martino D, Santos LF, Almeida RM. A model for the Ge-O coordination in germanate glasses. Journal of Non-Crystalline Solids 2007;353:1688–1694.

[11] Evstropiev KS, Ivanov AO. In: Matson FR., editors. Advances in Glass Technology, Part 2, Tech. Papers 6th Intern. Congr. Glass, New York, Plenum Press, 1963, 79.

[12] Henderson GS, Fleet ME. The structure of glasses along the $Na_2O.GeO_2$ join. Journal of Non-Crystalline Solids 1991;134:259–269.

[13] Henderson GS, Wang HM. Germanium coordination and the germanate anomaly. European Journal of Mineralogy 2002;14:733–744.

[14] Červinka L, Bergerová J, Sigaev VN, Rocca F. Structure of $(GeO_2)1–x(PbO)x$ glasses by X-ray scattering. Journal of Non-Crystalline Solids 2001;293-295:502–509.

[15] Ofelt JS. Intensities of crystal spectra of rare earth ions. Journal of Chemical Physics 1962;37:511–520.

[16] Dieke GH, Crosswhite HM. The spectra of the doubly and triply ionized rare earths. Applied Optics 1963;3:675–686.

[17] Rivera VAG, Ledemi Y, Messaddeq Y, Marega Jr E. Plasmonic emission enhancement from Er^{3+} doped tellurite glass via negative-nanobowtie. In: Photonic and Phononic Properties of Engineered Nanostructures IV, 2014, 899421–1.

[18] De Araujo CB, Kassab LRP, Tolentino Dominguez C, Ribeiro SJL, Gomes ASL, Reyna AS. Photoluminescence and nonlinear optical phenomena in plasmonic random media – A review of recent works. Journal of Luminescence 2016;169:492–496.

[19] Arashi H, Ishigame M. Diamond anvil pressure cell and pressure sensor for high-temperature use. Japanese Journal of Applied Physics 1982;21:1647–1649.

[20] Hess NJ, Exarhos GJ. Temperature and pressure dependence of laser induced fluorescence in Sm: YAG– A new pressure calibrant. High Pressure Research 1989;2:57–64.

[21] Anderson OL. Effect of pressure on glass structure. Journal of Applied Physics 1956;27:943–949.

[22] Bray KL. High pressure probes of electronic structure and luminescence properties of transition metal and lanthanide systems. Topics in Current Chemistry 2001;213:1–94.

[23] Kanzaki M, Xue X, Stebbins JF. Phase relations in Na_2O-SiO_2 and $K_2Si_4O_9$ systems up to 14 GPa and ^{29}Si NMR study of the new high-pressure phases: Implications to the structure of high-pressure silicate glasses. Physics of the Earth and Planetary Interiors 1998;107:9–21.

[24] Fuss T, Mogus-Milankovic A, Ray CS, Lesher CE, Youngman R, Day DE. Ex situ XRD, TEM, IR, Raman and NMR spectroscopy of crystallization of lithium disilicate glass at high pressure. Journal of Non-Crystalline Solids 2006;352:4101–4111.

[25] Buchner S, Balzaretti NM. Pressure and temperature stability range of crystalline lithium metasilicate in a binary $Li_2O.2SiO_2$ glass. Journal of Physics and Chemistry of Solids 2013;74:1179–1183.

[26] Eremets MI. High Pressure Experimental Methods. New York (NY), Oxford University Press, 1996.

[27] Matos IRM, Balzaretti NM. High pressure effect in the near-infrared emission of Nd^{3+}-doped alkali silicate glasses. High Press. Res. (published online: https://doi.org/10.1080/08957959.2021.2012570, 2021).

[28] Matos IRM, Balzaretti NM. High pressure effect on structural and spectroscopic properties of Sm^{3+} doped alkali silicate glasses. High Pressure Research 2017;37:296–311.

[29] Matson DW, Sharma SK, Philpotts JA. The structure of high-silica alkali-silicate glasses: A Raman spectroscopic investigation. Journal of Non-Crystalline Solids 1983;58:323–352.

[30] Judd BR. Optical absorption intensities of rare earth ions. Physical Review 1962;127:750–761.

[31] Herrera A, Becerra A, Balzaretti NM. Novel NIR emission $^4G_{5/2}$ to $^6F_{11/2}$ and efficient multichannel emissions on Sm^{3+} doped GeO_2-PbO glass. Journal of Luminescence 2017;188:193–198.

[32] Herrera A, Azevedo GM, Balzaretti NM. High-pressure effect in Vis-NIR emission of Sm^{3+}-doped GeO_2-PbO glasses. Journal of Physical Chemistry 2017;121:28745–28483.
[33] Herrera A, Balzaretti NM. Effect of high-pressure in the luminescence of Pr^{3+}-doped GeO_2-PbO glass containing Au nanoparticles. Journal of Physical Chemistry 2018;122:27829–27835.
[34] Eichelbaum M, Rademann K. Plasmonic enhancement or energy transfer? On the luminescence of gold-, silver-, and lanthanide-doped silicate glasses and its potential for light- emitting devices. Advanced Functional Materials 2009;19:2045–2052.

Christian Reber, Chahinez Dab, Pierre-François Blanc

Metal–ligand bonding in six-coordinate d-block complexes probed by luminescence spectroscopy at variable pressure

1 Introduction

Luminescence spectroscopy is a key technique to advance our understanding of the electronic structure of many compounds and materials, with important applications ranging from molecular photochemistry to materials science and optical technology. Spectroscopic studies using variable pressure can provide additional insight on specific effects, even in the modest pressure range up to approximately 30 kbar discussed here [1]. Many luminescence processes involve only the transition from the lowest energy excited state to the ground state, leading to relatively simple spectra without overlapping bands, therefore, allowing the observation of weak, forbidden transitions.

Several reviews of luminescence properties of transition metal compounds and doped materials at high pressure have been published [2–7]. In contrast to these general reviews, we focus here on variable-pressure luminescence spectra of two specific categories of molecular complexes: first, spin-flip luminescence bands, recently reviewed in depth for six-coordinate d-block complexes [8], and second, luminescence spectra of metal–oxo complexes [9]. The first type of transition has been studied in great detail for many solids doped with chromium(III) ions [3], with ruby being the prototype and its spin-flip luminescence as the prevalent standard for pressure calibration [10–13], as discussed in the chapter "Diamond anvil cell molecular spectroscopy of solid materials" in this volume. Isoelectronic molecular chromium(III) complexes show a much wider variety of variable-pressure luminescence properties, most importantly stronger shifts of the band maxima to longer wavelength than reported for the doped solids [3]. The variable-pressure luminescence spectra of metal–oxo complexes of rhenium(V) and molybdenum(IV) with the d^2 configuration represent the second category summarized in this chapter. Their luminescence

Acknowledgments: We thank all group members and colleagues who have contributed to these results over the years. Financial support from the Natural Sciences and Engineering Research Council (NSERC) of Canada and the Fonds de recherche du Québec – Nature et technologies is gratefully acknowledged. Djeva Production SA, Industrie de pierres scientifiques, Monthey, Switzerland, generously donated the synthetic ruby used to calibrate pressure in our laboratory.

Christian Reber, Département de chimie, Université de Montréal, Canada,
e-mail: christian.reber@umontreal.ca
Chahinez Dab, Pierre-François Blanc, Département de chimie, Université de Montréal, Canada

https://doi.org/10.1515/9783110668612-008

maxima can shift to either longer or shorter wavelength, and this wide variation provides insight on well-established metal–ligand bonding concepts.

Both families of complexes discussed here are six-coordinate with only the t_2 (*O* point group) molecular orbitals populated in the ground and emitting states. In chromium(III) complexes, these orbitals are approximately degenerate, as illustrated in Figure 1. Both the ground and emitting states arise from the t_2^3 electron configuration. Intraconfigurational electronic transitions between the $^4A_2$ and ^2E states do not cause structural changes, and luminescence spectra with narrow overall bandshapes are observed. In contrast, the t_2 levels are no longer degenerate in the metal–oxo compounds with D_4 point group symmetry, as schematically illustrated for *trans*-dioxo rhenium(V) complexes in Figure 1. The energy difference between these levels has been denoted as Δ_π, in formal analogy to the ligand-field splitting Δ_o, also shown in Figure 1. The luminescence transition in the metal–oxo complexes involves changes in electron populations along both the short metal–oxo bonds, oriented along the molecular *z*-axis in Figure 1, and along the longer bonds to ancillary ligands in the *x,y*-plane. As a consequence, spectra with broad bands are observed.

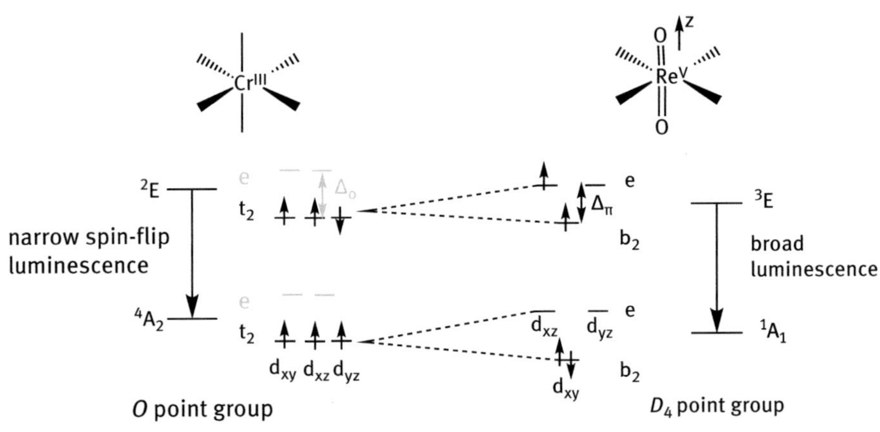

Figure 1: Schematic structures of complexes compared in this chapter with ground and emitting state assignments and relevant molecular orbital energy levels in *O* and D_4 point group symmetry.

The luminescence spectra of both categories of molecular systems compared in the following formally involve only electronic states arising from the t_2 molecular orbitals, but the observed spectroscopic phenomena vary widely, providing an interesting comparison to illustrate the detailed insight that can be gained from variable-pressure luminescence spectroscopy.

2 Variable-pressure spin-flip luminescence of molecular chromium(III) complexes

The red R-line luminescence of ruby is among the most studied spectroscopic transitions, used to probe many different physical processes and relevant as the first source of laser light [14]. The pressure-induced shift of the R lines, seminal spin-flip luminescence bands, has been studied over a wide range of pressures [10–13]. Such shifts for ruby and numerous other doped oxides are always to lower wavenumber (longer wavelength, redshift) by approximately −0.1 to −1 cm^{-1} $kbar^{-1}$ [3]. This is schematically illustrated in Figure 2 and straightforward to understand, as both the $^4A_2$ ground state and the ^2E emitting state arise from the same t_2^3 electron configuration, and their energies are therefore almost parallel in Figure 2, similar to a conventional Tanabe–Sugano diagram with the ligand field strength Δ_o on the horizontal axis. All spin-flip transitions are represented by parallel ground and emitting states in such diagrams [8]. No population change of orbitals with varying bonding characteristics occurs in these transitions, leading to very small changes of the transition energy caused by decreases of the Racah parameters *B* and *C*, a consequence of the bond compressions induced by pressure.

In contrast, significant shifts to higher wavenumber (shorter wavelength, blueshift) are observed for the interconfigurational transition involving the $^4T_2$ excited state, observable in absorption spectra and for some systems also as luminescence bands. These shifts range from +10 to +30 cm^{-1} $kbar^{-1}$ [3], absolute values higher by an order of magnitude than for the spin-flip transitions of ruby and other doped solids. Figure 2 schematically shows this stronger blueshift for the $^4T_2$ excited state, again qualitatively similar to a Tanabe–Sugano diagram for the d^3 configuration. Pressure-induced metal–ligand bond length decreases lead to a strong destabilization of the σ antibonding (σ*) e orbitals and therefore to a higher ligand field splitting Δ_o. This is the most important factor defining the observed shifts of band maxima for interconfigurational transitions involving a population change of the e levels. Reported values are similar in magnitude for numerous doped solids and molecular compounds [3, 15]. The effect of pressure on the t_2 levels is smaller and cannot be unambiguously identified from ${}_4T_2 \leftrightarrow {}_4A_2$ bands of chromium(III) complexes with *O* point group symmetry, as other factors, such as compressibilities, contribute to the total observed shift.

Only a handful of molecular chromium(III) complexes have been studied by variable-pressure luminescence spectroscopy, due to their low quantum yields under ambient conditions, both in solution and for solid samples. Recent developments in synthetic chemistry have led to optimized ligand designs in order to overcome this limitation [16–18]. Representative luminescence spectra of such a complex in solution and in the solid state are shown in Figure 3 [19]. This $[Cr(ddpd)_2]^{3+}$) complex has been named molecular ruby and is six-coordinate with two expanded terpyridine-type *N,N′*-dimethyl-*N,N′*-dipyridine-2-yl-pyridine-2,6-diamine (ddpd) ligands, resulting

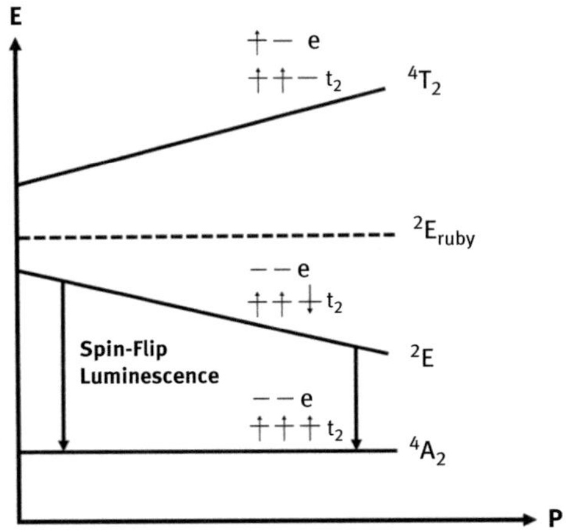

Figure 2: Schematic pressure-induced energy variations for luminescent states of chromium(III) complexes. The ^2E initial state for spin-flip luminescence is denoted by solid and dotted lines for molecular chromium(III) complexes and ruby, respectively. The $^4T_2$ emitting state of some chromium(III) compounds is shown for comparison. The $^4A_2$ ground-state energy is defined by a horizontal line. $t_2^n e^m$ (*O* point group) electron configurations are given for all electronic states shown.

in a nearly perfect octahedral coordination geometry. The strong-field ddpd ligands cause a very high energy difference between the ^2E and $^4T_2$ excited states, an important factor contributing to the intense luminescence observed at ambient conditions.

The luminescence of molecular ruby occurs at longer wavelength than the ruby R-lines, and spectra measured at variable pressure are similar in solution and in the solid state, indicative of an absence of intermolecular effects, for example, due to hydrogen bonding [20]. Transitions from both the ^2E and the thermally populated $^2T_1$ states are observed. The luminescence maxima in Figure 3 show a strong redshift with increasing pressure, on the order of $-15\ cm^{-1}\ kbar^{-1}$, a trend also illustrated in Figure 2. The absolute value of this shift is therefore comparable in magnitude, but different in sign, to the effect of pressure on interconfigurational transitions, such as those involving the $^4T_2$ state in Figure 2. This is a surprising observation and shows that variable-pressure luminescence spectroscopy can reveal properties beyond the expectations based on qualitative transition assignments and electron configurations of the ground and emitting states. Variable-pressure luminescence spectroscopy shows that the electronic structure of the emitting state is distinctly different from the ground state, even for spin-flip transitions as illustrated by the spectra in Figure 3.

We rationalize this strong shift as a combination of several effects: the first contribution is the decrease of the Racah parameters; the second contribution is structural distortions, leading to a split ^2E state and a lower energy emitting state; and finally vibronic effects involving the extended electronic system of the tridentate ligands can also contribute to the sizeable redshift. All three effects are beyond a

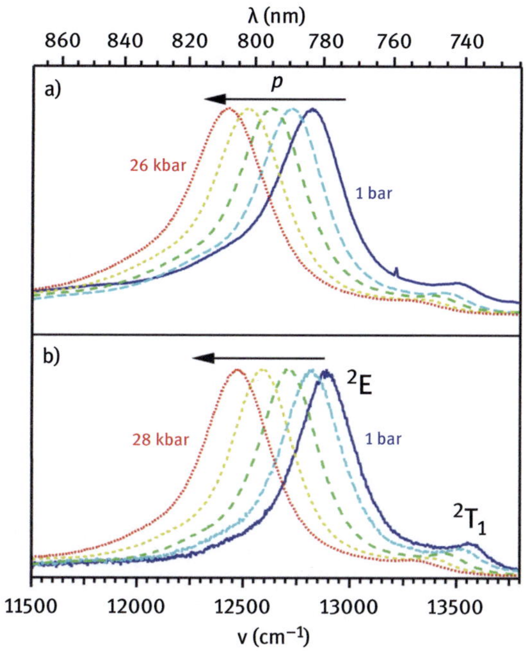

Figure 3: Variable-pressure luminescence spectra of molecular ruby, $[Cr(ddpd)_2][BF_4]_3$, at room temperature in the solid state (a) and in aqueous solution (b) (Copyright Wiley-VCH Verlag GmbH & Co. KGaA, Weinheim © 2018 [19] – reprinted with permission).

traditional Tanabe–Sugano diagram, and the strong redshift measured for molecular ruby in Figure 3 cannot be rationalized in terms of a pressure-induced increase of the ligand field strength Δ_o. The second and third effects are negligible in doped ionic solids and probably small for molecular complexes with simple ligands, leading to lower absolute values of the pressure-induced shifts. The spectra in Figure 3 reveal therefore a unique aspect of variable-pressure luminescence from modern molecular chromium(III) complexes. Table 1 shows that pressure-induced redshifts of spin-flip luminescence maxima for molecular complexes are generally stronger in absolute value than for ruby, with molecular ruby in Figure 3 displaying the strongest redshift reported to date for chromium(III) complexes.

Even stronger pressure-induced redshifts might be observed for chromium(III) complexes with longer wavelength near-infrared luminescence. Spectra of two representative complexes have recently been reported [21, 22]. It is for such molecular systems that variable-pressure luminescence spectroscopy can provide useful complementary information and insight.

Table 1: Pressure-induced shifts $\Delta E/\Delta P$ of spin-flip luminescence maxima E_{max} for chromium(III) compounds.

Type	Compound	E_{max} (cm^{-1})	$\Delta E/\Delta P$ (cm^{-1} kbar^{-1})	References
Doped solids	Al_2O_3:Cr^{3+} (Ruby)	14,404	−0.75 to −0.77	[10, 11, 13]
Molecular complexes	$[Cr(NH_3)_6](NO_3)_3$	15,225	−4	[23]
	{rhodo}$Cl_5 \cdot H_2O$‡	~14,585 ~14,680 ~14,165[14 K]†	−5.6 ± 0.6 −4.8 ± 0.4 −4.2 ± 0.4	[24, 25]
	{triol}$(ClO_4)_3$‡	~14,165 ~14,475 ~14,595[14 K]†	−8.1 ± 0.5 −5.4 ± 0.3 −4.1 ± 0.2	[24, 25]
	$[Cr(urea)_6](ClO_4)_3$	14,191 [1.3 K]	−5.4 ± 0.5	[26]
	$[Cr(urea)_6]I_3$	–	−3.1 ± 0.5	[26]
	$[CrF_6](NH_4)_3$	15,583†	−7.8 [>88 kbar]	[27]
	$[NaCr(ox)_3][Ru(bpy)_3]$	~14,380	−2.5 ± 1.2	[28]
	$[Rh(bpy)_3][NaCr(ox)_3]ClO_4$	~14,380	−2.6	[28]
	$[Cr(bpy)_3][NaRh(ox)_3][ClO_4]$	~13,714	−6.1	[29]
	$[Cr(bpy)_3][NaAl(ox)_3][ClO_4]$	~13,685	−7.9	[29]
	$[Cr(H_2tpda)_2][ClO_4]_3$	12,761	−13.0	[19]
	$[Cr(ddpd)_2][PF_6]_3$	12,872	−13.0	[19]
	$[Cr(ddpd)_2][BF_4]_3$	12,876	−14.5	[19]

{rhodo}, $[(NH_3)_5Cr(\mu\text{-}OH)Cr(NH_3)_5]^{5+}$; {triol}, tris-(μ-hydroxo)-bis-[(1,4,7-trimethyl-1, 4,7-triazacyclononane) chromium(III)]$^{3+}$; ox, oxalate; bpy, 2,2′-bipyridine; H_2tpda, 2,6-bis (2-pyridylamino)pyridine; ddpd, *N,N'*-dimethyl-*N,N'*-dipyridine-2-yl-pyridine-2,6-diamine.

† Not observed under ambient conditions; position has been extrapolated from measurements at higher pressure.

‡ Bimetallic complexes with exchange coupling, resulting in multiple electronic origins.

3 Variable-pressure luminescence of metal–oxo complexes with the d^2 configuration

The second family of d-block metal complexes discussed in this chapter is also six-coordinate, but with distinctly different bonds along one axis due to short metal–oxo multiple bonds [9]. The oxo complexes of both rhenium(V) and molybdenum(IV) compared here have the d^2 electron configuration, leading to singlet and triplet electronic states [9]. The difference between the short, multiple metal–oxo bonds along the molecular *z*-axis is defined in Figure 1, and the longer, weaker bonds to ancillary ligands in the *x*,*y*-equatorial plane lift the degeneracy of the t_2 orbitals (*O* point group), and the effect is sufficient in size to lead to a singlet ground state with the electrons paired in the lower energy b_2 (d_{xy}) orbital (D_4 point group) [30–33]. The lowest-energy excited state is a triplet, with one electron occupying the higher energy e orbitals with π^* metal–oxo character. The energy difference between these states leads to luminescence in the orange-red wavelength range. The bonding characteristics of the b_2 orbital can be either π^* for donor ligands or π for acceptor ligands. The resulting pressure-induced energy variations are distinctly different from the Tanabe-Sugano diagram for the d_2 configuration in O point group symmetry.

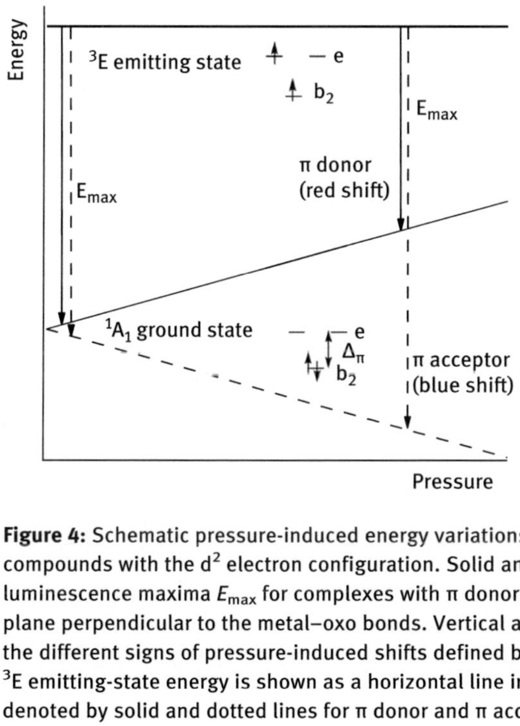

Figure 4: Schematic pressure-induced energy variations for luminescence transitions of metal–oxo compounds with the d^2 electron configuration. Solid and dotted vertical arrows correspond to luminescence maxima E_{max} for complexes with π donor and π acceptor ligands in the equatorial plane perpendicular to the metal–oxo bonds. Vertical arrows at low and high pressure illustrate the different signs of pressure-induced shifts defined by metal–ligand bonding characteristics. The ^3E emitting-state energy is shown as a horizontal line in this representation. The $^1A_1$ ground state is denoted by solid and dotted lines for π donor and π acceptor ligands, respectively. $b_2{}^n e^m$ (D_4 point group) electron configurations are given for all electronic states shown.

Figure 4 schematically illustrates the pressure-induced energy difference between the ground and emitting states, with the energy of the ^3E emitting state held constant, analogous to the $^4A_2$ ground state in Figure 2, allowing to visualize the qualitative effects of pressure on the luminescence band maximum for metal–oxo complexes. The diagram is based on the qualitative assumption that the metal–oxo multiple bonds are less compressed than the metal–ligand single bonds in the equatorial plane, and it is therefore the pressure effect on the b_2 orbitals that defines the shift of the luminescence maxima. The effect of pressure on the ground state energy depends on the type of π bonding with the ligands in the equatorial plane: for π donor ligands, the b_2 level is antibonding (π^*), and destabilization occurs as bond lengths shorten; in contrast, stabilization is expected for π acceptor ligands, where the b_2 level is π bonding. This difference leads to pressure-induced shifts of luminescence maxima with different signs, as shown in Figure 4, likely one of the clearest experimental distinctions between π donor and π acceptor ligands. Figures 5 and 6 show experimental spectra documenting this difference.

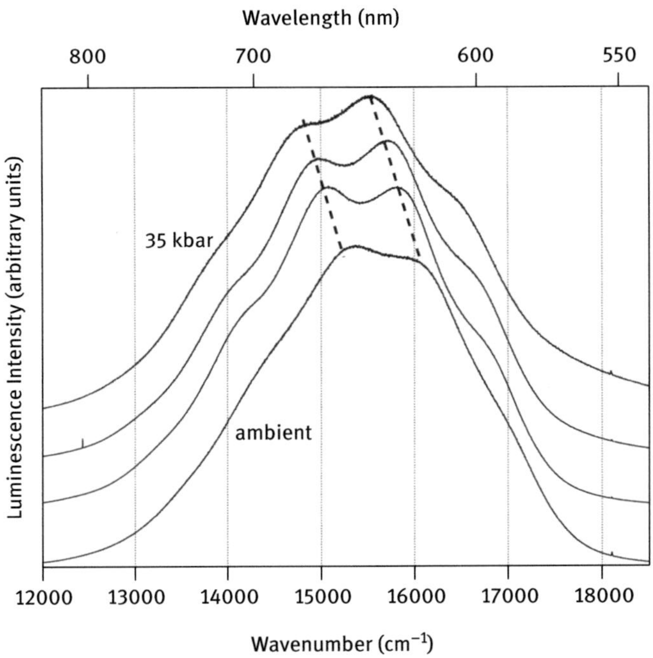

Figure 5: Variable-pressure luminescence spectra of *trans*-$[ReO_2(tmen)_2]Cl$ (tmen: tetramethylethylenediamine). The pressure-induced redshifts of the two most intense vibronic maxima are represented by the dotted lines, as schematically illustrated by the solid arrows in Figure 4.

The luminescence spectra for these metal–oxo complexes are broad bands, due to vibronic progressions involving metal–ligand vibrational modes [30, 31, 33]. The strongest vibronic progression involves metal–oxo stretching modes with frequencies on the order of 900 cm^{-1}, leading to the resolved maxima in Figure 5 [34]. The spectra in Figure 5 clearly show a pressure-induced redshift, easily seen from the dotted lines connecting the two most intense vibronic peaks. Such redshifts are observed for many similar complexes with a variety of nitrogen donor ligands [35, 36], as summarized in Table 2.

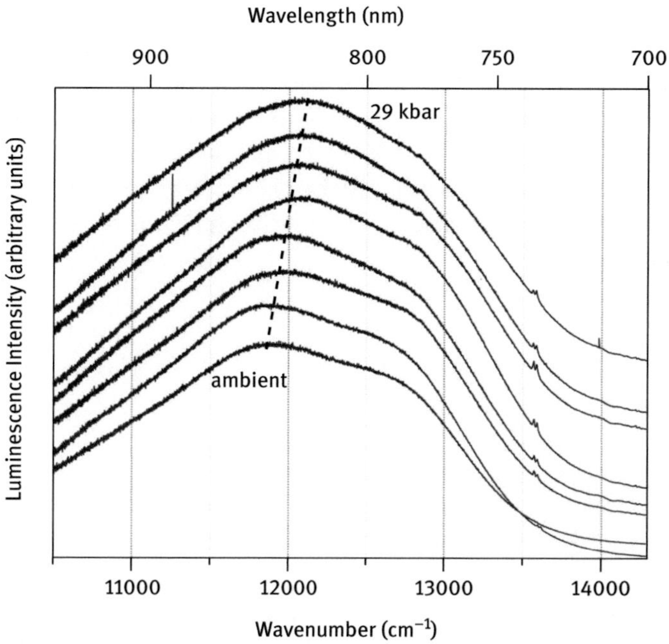

Figure 6: Variable-pressure luminescence spectra of [MoOCl(CN-*t*-Bu)$_4$][BPh$_4$]. The dotted line is a guide for the eye to show the blueshift of the luminescence maximum with increasing pressure, as schematically illustrated by the dotted arrows in Figure 4.

The variable-pressure luminescence spectra of an isoelectronic molybdenum(IV) oxo complex with π acceptor ligands are shown in Figure 6 [37]. The luminescence maxima show a clear blueshift between ambient pressure and 29 kbar, as indicated by the dotted line. The observed shifts in Figures 5 and 6 therefore correspond to the qualitative expectations from Figure 4. The values in Table 2 document that it is indeed possible to change the sign of the pressure-induced shift of the luminescence maximum through the type of ligands coordinated in the equatorial plane, illustrating the detailed insight that can be gained from luminescence spectra measured at variable pressure.

Table 2: Pressure-induced shifts $\Delta E/\Delta P$ of luminescence maxima E_{max} for metal–oxo compounds of rhenium(V) and molybdenum(IV).

Compound	E_{max} (cm^{-1})	$\Delta E/\Delta P$ (cm^{-1} kbar^{-1})	References
trans-[ReO$_2$(tmen)$_2$]Cl	15,590	−4.6	[34]
trans-[ReO$_2$(pyridine)$_4$]I	15,360	−15.7	[36]
trans-[ReO$_2$(en)$_2$]Cl	13,780	−6.8	[35]
[MoOF(pyridine)$_4$]BPh$_4$	13,000	−7.5	[37]
[MoOCl(CN-*t*-Bu)$_4$]BPh$_4$	11,950	+12.0	[37]

tmen, tetramethylethylenediamine; en, ethylenediamine.

4 Conclusion

The comparison of two categories of luminophores with related electron configurations illustrates the insight that can be gained from variable-pressure luminescence spectroscopy. Shifts of luminescence spectra reveal detailed information on metal–ligand bonding and emitting-state properties. The recent discoveries of many new luminescent molecular compounds make such experiments even more useful in the future.

References

[1] Grochala W, Hoffmann R, Feng J, Ashcroft NW. The chemical imagination at work in very tight places. Angewandte Chemie International Edition 2007;46:3620–3642.

[2] Kenney III JW. Pressure effects on emissive materials. In: Roundhill DM, Fackler JP Jr., editors. Optoelectronic Properties of Inorganic Compounds. Modern Inorganic Chemistry. New York and London, Plenum Press, 1999, 231–268.

[3] Bray KL. High pressure probes of electronic structure and luminescence properties of transition metal and lanthanide systems. Topics in Current Chemistry 2001;213:1–94

[4] Grey JK, Butler IS. Effects of high external pressures on the electronic spectra of coordination compounds. Coordination Chemistry Reviews 2001;219–221:713–759

[5] Reber C, Grey JK, Lanthier E, Frantzen KA. Pressure-induced change of d-d luminescence energies, vibronic structure and band intensities in transition metal complexes. Comments on Inorganic Chemistry 2005;26:233–254

[6] Valiente R, Renero-Lecuna C, Rodríguez F, González J. Role of high pressure for understanding luminescent phenomena. Journal of Luminescence 2016;169:410–414

[7] Rodríguez F, Valiente R, Lavín V, Rodríguez-Mendoza UR. Optical spectroscopy. In: Recio JM, Menendez JM, Otero de la Roza A, editors. An Introduction to High-Pressure Science and Technology. Boca Raton, CRC Press, 2015, 219–264.

[8] Kitzmann WR, Moll J, Heinze K. Spin-flip luminescence. Photochemical and Photobiological Sciences 2022. Published online March 5.

[9] Miskowski VM, Gray HB, Hopkins MD. Electronic structure of metal-oxo complexes. In: Che C-M, Yam V-W-W, editors. Adv in Trans Met Coord Chem. Greenwich, CT, JAI Press, 1996, Vol. 1, 159–186.

[10] Piermarini GJ, Block S, Barnett JD, Forman RA. Calibration of the pressure dependence of the R_1 ruby fluorescence line to 195 kbar. Journal of Applied Physics 1975;46:2774–2780

[11] Grasset O. Calibration of the R ruby fluorescence lines in the pressure range [0-1GPa] and the temperature range [250–300 K]. High Pressure Research 2001;21:139–157

[12] Chijioke AD, Nellis WJ, Soldatov A, Silvera IF. The ruby pressure standard to 150 GPa. Journal of Applied Physics 2005;98:114905

[13] Syassen K. Ruby under pressure. High Pressure Research 2008;28:75–126

[14] Imbusch GF, Yen WM. Ruby – solid state spectroscopy's serendipitous servant. In: Yen WM, Levenson MD, editors. Lasers, Spectroscopy and New Ideas. Springer Series in Optical Sciences. Berlin, Springer, 1987, Vol. 54, 248–267.

[15] Poirier S, Tailleur E, Lynn H, Reber C. Characterization of Pd. . .H–C interactions in bis-dimethyldithiocarbamate palladium(II) and its deuterated analog by luminescence spectroscopy at variable pressure. Dalton Transaction 2016;45:10883–10886. Table S1

(Supporting Information) gives pressure-induced shifts of interconfigurational transitions in d-block complexes of several metals with different coordination geometries.
[16] Otto S, Grabolle M, Förster C, Kreitner C, Resch-Genger U, Heinze K. $[Cr(ddpd)_2]^{3+}$: A molecular, water-soluble, highly NIR-emissive ruby analogue. Angewandte Chemie International Edition 2015;54:11572–11576
[17] Otto S, Dorn M, Förster C, Bauer M, Seitz M, Heinze K. Understanding and exploiting long-lived near-infrared emission of a molecular ruby. Coordination Chemistry Reviews 2018;359:102–111
[18] Wang C, Otto S, Dorn M, Kreidt E, Lebon J, Sršan L, Di Martino-Fumo P, Gerhards M, Resch-Genger U, Seitz M, Heinze K. Deuterated molecular ruby with record luminescence quantum yield. Angewandte Chemie International Edition 2018;57:1112–1116
[19] Otto S, Harris JP, Heinze K, Reber C. Molecular ruby under pressure. Angewandte Chemie International Edition 2018;57:11069–11073
[20] Pannwitz A, Poirier S, Bélanger-Desmarais N, Prescimone A, Wenger OS, Reber C. Controlling second coordination sphere effects in luminescent ruthenium complexes by means of external pressure. Chemistry – A European Journal 2018;24:7830–7833
[21] Sinha N, Jiménez J-R, Pfund B, Prescimone A, Piguet C, Wenger OS, Near-Infrared-II Emissive A. Chromium(III) complex. Angewandte Chemie International Edition 2021;60:23722–23728
[22] Stein L, Boden P, Naumann R, Förster C, Niedner-Schatteburg G, Heinze K. The overlooked NIR luminescence of $Cr(ppy)_3$. Chemical Communication 2022; 58: 3701–3704
[23] Lee SH, Waltz WL, Demmer DR, Walters RT. Effects of applied pressure on the emission lifetimes and spectroscopic features of some chromium(III) ammonia and amine complex ions in aqueous media. Inorganic Chemistry 1985;24:1531–1538
[24] Riesen H, Güdel HU. Effect of high pressure on the exchange interactions in binuclear chromium(III) complexes. The Journal of Chemical Physics 1987;87:3166–3172
[25] Riesen H, Güdel HU. Pressure tuning of exchange interactions in dinuclear chromium(III) complexes. Inorganic Chemistry 1987;26:2347–2348
[26] Yersin H, Huber P, Gietl G, Trümbach D. Pressure-induced tuning of fluorescence to phosphorescence in $[Cr(urea\text{-}h_4)_6]$ $(Clo_4)_3$ and $[Cr(urea\text{-}d_4)_6]$ $(Clo_4)_3$. Chemical Physics Letters 1992;199:1–9
[27] Kenney III JW, Clymire JW, Agnew SF. Change from fluorescence to phosphorescence and tuning the pseudo-Jahn-Teller effect with pressure for $(Nh_4)_3[crf_6]$. Journal of the American Chemical Society 1995;117:1645–1646
[28] Milos M, Penhouet T, Pal P, Hauser A. Structural study of mixed crystals of $[Zn_{1-x}ru_x(bpy)_3]$ $[NaCr(ox)_3]$ probed by high-resolution absorption spectroscopy and high-pressure experiments. Inorganic Chemistry 2010;49:3402–3408
[29] Milos M, Pal P, Hauser A. Effect of external pressure on the excitation energy transfer from $[Cr(ox)_3]^{3-}$ to $[Cr(bpy)_3]^{3+}$ in $[Rh_{1-x}crx(bpy)_3][nam_{1-y}cry(ox)_3]clo_4$. ChemPhysChem 2010;11:3161–3166
[30] Winkler JR, Gray HB. Emission spectroscopic properties of dioxorhenium(V) complexes in crystals and solutions. Journal of the American Chemical Society 1983;105:1373–1374
[31] Winkler JR, Gray HB. Electronic absorption and emission spectra of dioxorhenium(V) complexes. Characterization of the luminescent ^3E_g state. In: Inorg Chem, 1985, Vol. 24, 346–355.
[32] Savoie C, Reber C, Bélanger S, Beauchamp AL. Spectroscopic observation of coupled metal-ligand vibrational modes in the luminescence spectrum of *trans*-$[ReO_2(1\text{-methylimidazole})_4]I$. Inorganic Chemistry 1995;34:3851–3852
[33] Savoie C, Reber C. Coupled electronic states in *trans*-dioxo complexes of rhenium(V) and osmium(VI) probed by near-infrared and visible luminescence spectroscopy. Journal of the American Chemical Society 2000;122:844–852

[34] Grey JK, Triest M, Butler IS, Reber C. Effect of pressure on the vibronic luminescence spectrum of a *trans*-dioxo rhenium(V) complex. The Journal of Physical Chemistry A 2001;105:6269–6272
[35] Grey JK, Butler IS, Reber C. Effect of pressure on coupled electronic ground and excited states determined from luminescence spectra of *trans*-dioxorhenium(V) complexes. Journal of the American Chemical Society 2002;124:11699–11708
[36] Grey JK, Butler IS, Reber C. Temperature- and pressure-dependent luminescence spectroscopy of the *trans*-$[ReO_2(pyridine)_4]^+$ complex – analysis of vibronic structure, luminescence energies and bonding characteristics. Canadian Journal of Chemistry 2004;82:1083–1091
[37] Lanthier E, Bendix J, Reber C. Pressure-dependent luminescence spectroscopy of molybdenum(IV) oxo complexes. Dalton Transaction 2010;39:3695–3705

Index

https://doi.org/10.1515/9783110668612-009

Bei Fragen zur Produktsicherheit wenden Sie sich bitte an:
If you have any questions regarding product safety,
please contact:

Walter de Gruyter GmbH
Genthiner Straße 13
10785 Berlin
productsafety@degruyterbrill.com